Springer Series in Optical Sciences   Volume 54

Springer Series in Optical Sciences

Editorial Board: D.L. MacAdam A.L. Schawlow K. Shimoda A.E. Siegman T. Tamir

Volumes 1–41 are listed on the back inside cover

Lasers, Spectroscopy and New Ideas

A Tribute to Arthur L. Schawlow

Editors:
W. M. Yen and M. D. Levenson

With 161 Figures

Springer-Verlag Berlin Heidelberg GmbH

Professor William M. Yen

Department of Physics and Astronomy, The University of Georgia,
Athens, GA 30602, USA

Dr. Marc D. Levenson

K69/803 (E), IBM Almaden Research Center, 650 Harry Rd.,
San Jose, CA 95120, USA

ISBN 978-3-662-13608-9   ISBN 978-3-540-47872-0 (eBook)
DOI 10.1007/978-3-540-47872-0

Library of Congress Cataloging-in-Publication Data. Lasers, spectroscopy, and new ideas. (Springer series in optical sciences ; v. 54) Includes index. 1. Lasers. 2. Spectrum analysis. 3. Schawlow, Arthur L., 1921-. I. Levenson, Marc D. II. Yen, W. M. (William M.) III. Schawlow, Arthur L., 1921-. IV. Series. QC688.L37 1987 535.5'8 87-23332

© Springer-Verlag Berlin Heidelberg 1987

Originally published by Springer-Verlag Berlin Heidelberg New York Tokyo in 1987

Softcover reprint of the hardcover 1st edition 1987

2153/3150-543210

To Do Successful Research,
You Don't Need to Know Everything.
You Just Need to Know of
One Thing That Isn't Known.
A.L. Schawlow

Arthur L. Schawlow

Foreword

This volume originated in a happy event honoring Arthur Schawlow on his 65th birthday. As a research physicist, Schawlow has been a major influence on the present nature of physics and of high technology. He has also had a role, through the American Physical Society and other organizations, in shaping policy for the world of physicists. Important as these professional activities have been, the contributions to this volume were not prepared just for these reasons, but more for Art Schawlow the friend, colleague, and teacher. I am one who has had the privilege of knowing and collaborating with Art, probably over a longer period of time than others participating in this volume, and in a number of different enterprises; his friendship and stimulating scientific abilities are a very significant part of my own life. It is hence a pleasure to take part in this volume celebrating his contributions to science and to scientists.

Schawlow's career has been geographically centered at the University of Toronto, Columbia University, the Bell Telephone Laboratories, and Stanford University. But, as is illustrated by the papers of this volume, its effects and his personal influence have diffused widely. In his own work, Art Schawlow is noted for thoughtful imagination, keen physical intuition, and what might be thought an interest in gadgets – not just any gadgets, but beautiful and innovative mechanisms or new techniques in which he characteristically recognizes important potentials. One can say that he has always been a spectroscopist – from a thesis at Toronto in optical spectroscopy, to his work and book on microwave spectroscopy, the first publication on the laser, and most recently his part in the inventive development of laser spectroscopy to remarkable refinement, precision, and power. From this brief list it is obvious that his work has also had great breadth and touched many fields.

Art has also touched many people, and always with consideration, friendship, a delight in scientific discovery, and an infectious sense of humor. I believe the inspiration and motivation for this volume spring largely from these latter warm personal qualities. The contributions it contains will illustrate some of the many fields and individuals indebted to Arthur Schawlow, and it is hoped that here and there the authors will have captured at least some approximation of his almost inimitable humor.

Berkeley, California   March, 1987   *Charles H. Townes*

Preface

It occurred to some of us as the occasion of Art Schawlow's sixty-fifth birthday approached in 1986 that we needed to make an appropriate gesture to honor this man, not only for his well-recognized scientific contributions, but also for the personal legacy which he is leaving to, and the influence he has exerted on, everyone who has had the privilege of coming into contact with him through the years. After some false starts, it was decided that a collection of articles and reminiscences would serve as an appropriate vehicle for such a tribute, and it is thus that this venture came into existence.

To quantify or enumerate Art's contributions to the scientific literature is a relatively easy task, and his articles and reviews, many of which are classics, are clear, concise and numerous. He has co-authored papers with almost one hundred different people and he has worked with nearly seventy-five collaborators with varying functions during his period at Stanford. The range of topics to which his contributions are addressed is impressively wide-ranging and spans subjects as diverse as Doppler-free atomic spectroscopy and the properties of xenon flash lamp discharges. Needless to say, this volume of scientific work has had undeniable influence and impact in a number of areas of scientific and technical importance, which we need not belabor here. The nature of Art's influence is sampled in the articles presented here and is evinced by the accomplishments of the many researchers he has trained.

It is much more difficult to provide an adequate measure of Art's other contributions, especially those concerned with the fostering of scientific ideas and scientific talent and attitudes. Indeed, it is because of the very positive influence he exerted on many of us with respect to our professional growth that we decided to organize this celebratory volume. For those of us who have had the privilege of falling under his tutelage, it is generally agreed that he attempted to teach us (sometimes successfully, many times not) that very simple concepts are normally sufficient to explain even the most complex observations. This principle has served us all well in our subsequent careers. In addition to developing and encouraging new scientific ideas and approaches, Art has always provided a personal touch in his interactions; in these he reveals without fail his patience, his intrinsic kindness, his humanism, and his humor. This touch was most welcome as it nurtured self-confidence in the many raw and inexperienced graduate students and postdocs that joined his effort at Stanford, including the two editors of this volume.

It was the humanistic side of his influence that led us to choose the general tone of this collection of writings. The authors who graciously agreed to participate in this effort represent a sample of the many scientific areas in which Art has left a legacy or made an impact. We suggested to all the contributors that they write their articles in such a way as to include not only some description of some phase of their present area of scientific endeavors but also to include impressions as to how their personal attitudes and development were affected by interactions with Art. Some of the contributions describe work in which the authors are currently engaged, while others are archival, as they are concerned with the evolution of areas in which Art has made seminal contributions. By and large, we are pleased by the results of the effort, and we believe that in this collection a number of the contributions will remain relevant well into the future, especially those which were designed to be historical. We have incorporated, between parts, anecdotes and other items which address only the humanistic side and are exemplary of the joy and humor which normally prevail in any association with Art. Indeed, we would also have liked to provide recordings of his jazz clarinet playing dating from his graduate school days, but unfortunately he would not allow their release for circulation.

The volume is organized as follows: The contributed articles are divided into four areas. The first three parts include material devoted to areas in which Art has had an undeniable role, either in establishing a field of endeavor or in exercising exceptional leadership. These are, in sequence, lasers and laser spectroscopy, spectroscopy of atomic and molecular systems, and spectroscopy in the condensed phases. Each of these parts contain four to six papers from authors who have made recognizable contributions in each of the respective areas and who, following their contact with Art, have gone on to distinguished careers of their own. The fourth part consists of three contributions which are illustrative of areas where Art has had an indirect influence, in these cases by training a cadre of scientists who have advanced other frontiers by utilizing those attitudes which are so characteristic of "The Boss". The picture we have succeeded in presenting in this sampling does not totally summarize all the accomplishments of Art Schawlow. Many of us are cognizant of the fact that Art made a pioneering attempt at laser isotope separation in the early 1960s, that he played a principal role in interpreting the spectra of magnetically ordered materials, and that methods to induce cooling in atoms with lasers were suggested by him in the early 1970s. Regardless of the shortcomings of this collection, for which we, the editors, assume full responsibility, we believe that each of the contributions has its own worth; in some instances the articles are important reviews in their own right, albeit softened somewhat from the usual austere scientific format because of the nature of this enterprise. The advantage in return is that the majority of the contributions are eminently readable and will be understood by a wide range of readers not directly involved in the specific areas of scientific endeavor.

It is always difficult to take time out from the many pursuits which normally engage our time to participate in extracurricular ventures. It is indeed gratifying that so many people readily agreed to contribute to this volume and, for the most part, produced their manuscripts on time. The editors would thus like to take this opportunity to express their thanks to all who participated in this worthwhile cause and also to Dr. Helmut Lotsch and Springer-Verlag for their cooperation, which made this volume possible. Ms. Nancy Bachman of the University of Georgia is thankfully acknowledged for her assistance in sundry editorial tasks. And, of course, Mrs. Fred-a Jurian is acknowledged to be the true "boss of bosses" of the operations at Stanford, and she bears direct responsibility for many of us having survived the vicissitudes of our youth, perhaps at times to her regret. Indeed, this volume is also a tribute to her wisdom, concern and kindness.

San Jose, California   *W.M. Yen*
April 1987   *M.D. Levenson*

Contents

Part III   Solid State Spectroscopy

Part IV   Miscellaneous Ideas

Part I

Lasers and Laser Spectroscopic Techniques

Sign on entrance to
Schawlow's Stanford laboratory

From (Incr)edible Lasers to New Spectroscopy

T.W. Hänsch

Sektion Physik, Universität München, and Max-Planck-Institut
für Quantenoptik, D-8046 Garching, Fed. Rep. of Germany

Arthur L. Schawlow was already a very famous man when I first met him at a summer school in Scotland in 1969, a few months after I had received my doctorate from the University of Heidelberg in Germany. Immediately captivated by his personality, his quick and sharp mind, and his warm humor, I wrote to him, asking if he would take me on as a postdoc for a year or two. Fortunately, he agreed, and I arrived at his laboratory in May 1970 with a NATO Fellowship. Little did I know that I would join the faculty of the Physics Department at Stanford University two years later, and that I would be able to enjoy a close association and friendship with Art Schawlow for the next 16 years.

The early years of this period were most exhilarating, since we found ourselves at the heart of a revolution in laser spectroscopy. Many accounts have been written of the research at Stanford during this time [1-3]. Here I hope to add some personal impressions and anecdotes which capture a little bit of the human side of this great scientist.

The Incredible Laser

After arriving at Stanford, I was fascinated by a special "magic" atmosphere in Art Schawlow's laboratory. Walking down the hallway on the second floor of the Varian Physics building, a futuristic poster on one of the doors caught my eye. It showed an enormous laser gun blasting at some attacking rockets in the sky. The caption in bold letters read: "The Incredible Laser". In smaller letters below, someone had written "for credible lasers, see inside".

There could be no doubt that to Art Schawlow science is great fun. Despite his extremely busy schedule, he would often find the time to treat visitors to some amazing and entertaining demonstrations. Rummaging in a huge briefcase, he would for instance pull out his famous red toy laser gun, into which his technician has skillfully installed a real small ruby laser. With serious voice at first, he would begin to explain: "We found the whole idea of the laser was some kind of a death ray. So Mr. Sherwin built us this ray gun. Having a weapon, we had to do a little hunting and went looking for some animals. Around Stanford, the only place you find animals is in the zoo. But at the zoo the animals were all rather big and fierce. So we just bought a balloon for the kids."

At this point he would begin to noisily inflate a large clear balloon. "But when we looked at the balloon there was something funny about it – there was something inside it." Gradually, a blue balloon begins to appear inside, with big ears like Mickey Mouse. "There was a mouse inside that balloon. You know it is terrible the way mice get into everything! So we had to get our more or less trusty laser and dispose of it." With a pull of the trigger, the ray gun flashes, and the inner balloon bursts with a loud pop while the outer balloon remains unharmed. This instant is captured in Fig. 1.

Figure 1. Art Schawlow using his ruby laser ray gun to dispose of a mouse.

"Now this is a very serious experiment. It works because the outer balloon is clear so that the red light flash of the laser passes through it without being absorbed. The inner balloon is dark blue and absorbs red light so that a hot spot was formed on the surface. This illustrates how, with lasers, light is no longer something to look with, it is something you can do things with, and you can do them at places where you can see but not touch, as for instance at the retina of the eye. One of the very first applications of lasers was for surgery inside the eye, to prevent blindness from either a detached retina or leaky blood vessels." Art Schawlow likes

4

to add that he had never heard of such diseases when he started his work on lasers. And if he had set out to find a treatment, he certainly would not have been fooling around with atoms and stimulated emission of radiation.

Sometimes the laser ray gun fails to work, but Art Schawlow is ready for such mishaps. By pulling another trigger, he can fire a small spring-loaded arrow with a rubber tip. "This is our second-strike capability."

Another of Art Schawlow's favorite demonstrations is his "laser eraser", a small flashlamp pumped Nd:glass laser which can be used to evaporate the ink from type-written or printed paper. With an impish twinkle in his eye, he would ask an unsuspecting visitor for a one-dollar bill. Pointing the eraser at the eye of George Washington, he innocently asks his victim to push a button. With a pop the eye vanishes so that Washington now appears to wear a monocle. "What have you done?", Art Schawlow would exclaim. "You know that it is illegal to deface US-currency!" Quite a few of these dollar bills are probably still being treasured as souvenirs, even though the future of the laser eraser may have become somewhat clouded by the success of personal computers and word processors.

Art Schawlow has little patience with abstract theory or tedious mathematical derivations. But he combines a vast range of knowledge and interests with a brilliant intuition. He has a unique gift of seeing the significance of new discoveries, and he can explain complex ideas in the most simple and lucid terms. To make a point in a public lecture, he can draw on all the skills of a good comedian. To give an example, I will never forget his explanation of "coherence":

"Now, nearly everybody knows what is meant by 'coherent'. But in case there is a chemist in the audience, let me explain: you see from the beginning of time until the last twenty years or so, all light sources have been essentially hot bodies, whether it is the sun, or a tungsten filament in a lamp, a flame, or the atoms in a neon sign. The atoms are jostled around by the thermal agitation, and as an atom gets struck, for a moment it stores a little bit of energy. But in a millionth of a second or so, it releases it and sends out another light wave. Now, there are a lot of these atoms and you can picture this process as being like raindrops falling into a still pond. As each drop lands - or as each atom emits - the waves spread out in ever widening circles." With movements of his arms and hands he would illustrate this spreading of waves. "So you have atoms here going ping, and ping - the big ones go pong - ..." Gesticulating with ever increasing speed, he bursts into a wild and hilarious jumble of "ping, ping, pong, pong, ping..." Suddenly he would stop and explain, after a measured pause: "Now, for some reason, people call this kind of light 'incoherent'..."

Art Schawlow has not only made immense contributions to the public understanding of lasers and optical science. With his keen interest in fundamental physics and his contagious enthusiasm, he has a rare ability to inspire students and co-workers to high achievements. Sometimes, he would visit a young graduate student in his laboratory and ask: "What have you discovered?" To most, the thought that they were there to discover something new came almost as a revelation. But how does one do that? Art Schawlow gives very important encouragement when he emphasizes that one does not have to study everything that is known about a subject in order to discover something new. One only has to find one thing that was not known.

Edible and Other Dye Lasers

After the "Sputnik-shock" in the sixties, quite a lot of money had flown into university laboratories in the United States, and Art Schawlow had managed to accumulate an enviable collection of instruments and expensive components. "I have been poor, and I have been rich," he sometimes quipped, "and let me tell you, rich is better! As experimentalists, we always can find something to do, even if we have to work with string and sealing wax. But then, a lot of talent, time and effort gets wasted. One problem is, one never knows what remains undiscovered simply because the right equipment is not there at the right time."

I soon started to enjoy my work in the laboratory tremendously, since Art Schawlow left me complete freedom in my research while giving me access to all his treasures. He even agreed to let me purchase an AVCO nitrogen laser which I was planning to use as pump source for a tunable dye laser. Laser action in dyes had been discovered a few years earlier by Peter Sorokin and independently by Fritz Schäfer. I felt that it should somehow be possible to make such a dye laser so highly monochromatic that it would permit us to study spectral lines of free atoms and molecules by the powerful new methods of Doppler-free saturation spectroscopy which Peter Toschek and I had begun to develop during my thesis work at Heidelberg.

The new nitrogen laser turned out to be a marvellous toy. By simply focusing its ultraviolet output beam with a cylindrical lens into a glass cell filled with an organic dye solution, we could produce spectacularly colorful intense beams of laser-like amplified spontaneous emission. By adding a diffraction grating and a mirror to form an optical cavity, the color of the output beam could be changed at will. At one time I focused the blue light of such a dye laser into a single drop of watery solution of fluorescein. This drop then became a dye laser all of its own, emitting an intense beam of green light. The laser cavity was simply formed by the surfaces of the liquid.

Observing this droplet laser with obvious delight, Art Schawlow postulated that "anything will lase if you hit it hard enough." Thinking about challenges to prove such a claim, we wondered if the colorful gelatine desserts popular with children would show laser action when pumped with the nitrogen laser. The next morning, Art Schawlow came to work waving a package with twelve different flavors of "Knox Jello". Using the hot water supply in the darkroom, we prepared two of the desserts in plastic cups, following the manufacturer's instructions. After they had begun to gel, we took them to the lab and focused the nitrogen laser beam to illuminate a line on the flat surface of the wobbly substance. There was distinct fluorescence, but no laser action. In resignation, Art Schawlow would return to his office and enjoy the obstinate experiment as a snack. This ritual was repeated every morning for a week until we had tried all twelve flavors without luck.

Determined to demonstrate an edible laser, we finally mixed up a packet of clear, flavorless gelatine and added some sodium fluorescein. This experiment was an instant success. With a knife we could cut the new laser material into rods or other shapes. The paper describing this laser would soon be posted on many bulletin boards [4]. A few people considered this experiment a frivolity, but it actually led to some rather important technical developments. Soon afterwards, Kogelnik and Shank [5] exposed a dichromated gelatine film to the interference pattern of two ultraviolet laser beams and demonstrated the first distributed feedback laser. Today, distributed feedback plays an increasingly important role in semiconductor diode lasers for optical communications.

From the beginning, Art Schawlow was very enthusiastic about my plans to develop a highly monochromatic tunable dye laser. Such a tool would open many exciting new possibilities for studying the structure of atoms, molecules, and solids. The nitrogen laser appeared as a particularly attractive pump source, because many dyes spanning the visible spectrum could be pumped with good efficiency at high pulse repetition rates. Past attempts to achieve narrow linewidths with a nitrogen-pumped dye laser had remained unsuccessful. But I was hopeful that it should be possible to isolate a single axial mode with the help of a holographic diffraction grating and a birefringent Lyot filter. After encouraging preliminary experiments, I submitted a paper for presentation at the APS Meeting at Stanford University in December 1970 [6].

As the conference approached, I became rather panicky, because the envisioned scheme did not work reliably. Almost in desperation, I tried an entirely different approach. I moved the grating far away from the dye cell, took out the Lyot filter, and inserted into the cavity a small Zeiss telescope which I happened to carry in my pocket so that I could read the slides during a lecture from the backrow. The idea was that a beam-expanding telescope would illuminate more lines at the grating and so improve the resolution. This little trick worked well beyond my expectations [7]. With a larger telescope, the linewidth could be reduced to a few hundredths of an angstrom. Even narrower lines could be achieved by inserting a tilted Fabry-Perot etalon into the cavity. With this suprisingly simple scheme, we could now produce coherent light of a spectral purity and brightness that was previously only available from gas lasers with very limited tuning range.

Spectroscopy without Doppler Broadening

During my thesis work at the University of Heidelberg I had become intrigued by the potential of lasers for high resolution spectroscopy. By exploiting the spectral hole-burning effect discovered by Willis Lamb, Bill Bennett and Ali Javan, I was able to develop an early form of saturation spectroscopy [8] which could circumvent the Doppler broadening of spectral lines. With Peter Toschek, we used this method to study collision processes and nonlinear optical phenomena in neon discharges. When I came to Stanford, I collaborated with Peter Smith, then on sabbatical at Berkeley. We demonstrated a simple new method of saturation spectroscopy, using just one single He-Ne laser and an external gas cell [9]. A very similar technique was developed independently by Christian Bordé in Paris [10].

In Art Schawlow's laboratory, Marc Levenson, then a graduate student, was working with a krypton ion laser. By placing a prism and etalon into the cavity, he could make the laser work in a single axial mode, manually tunable over a few gigahertz. During a visit of Peter Toschek in the summer of 1970, we decided jointly to try and use this laser for saturation spectroscopy of iodine vapor, since it was known that the diatomic iodine molecule has several absorption lines in accidental coincidence with krypton and argon ion laser wavelengths. The experiment was at first unsuccessful. A few weeks later, Marc Levenson found out that we had worked with a contaminated iodine cell. With a new cell, we soon obtained very pretty spectra, showing all 21 hyperfine components of one line completely resolved [11].

Art Schawlow was delighted with these results. Even then, he was so much in demand as a public speaker that he sometimes defined "genius" as "an infinite capacity to take planes." In his lectures, he would describe the new experiment with his unique clarity and simplicity by first reminding people of the Doppler effect: "If an object is moving towards you - like a train - the emitted sound goes up in pitch." While making this point he would walk towards the audience and his voice assumed a funny high pitch. Next he would walk backwards and explain in a deep bass voice: "...and if it is moving away from you, it goes down in pitch." Returning to his normal voice: "Now the same is true for light. So in a gas, where the molecules are moving in all directions, the spectral lines appear blurred and spread out by the Doppler effect."

Then he would go on to explain the method of Doppler-free saturation spectroscopy. Showing a drawing of the apparatus as in Fig. 2, he would point out: "The light from a tunable laser is divided by a beam splitter into a strong saturating beam and a weaker probe beam. These light beams are sent through the absorbing gas in nearly opposite directions. When the saturating beam is on, it bleaches a path through the cell, and a stronger probe signal is received at the detector. As the saturating beam is alternately stopped and transmitted by the chopper, the probe signal is

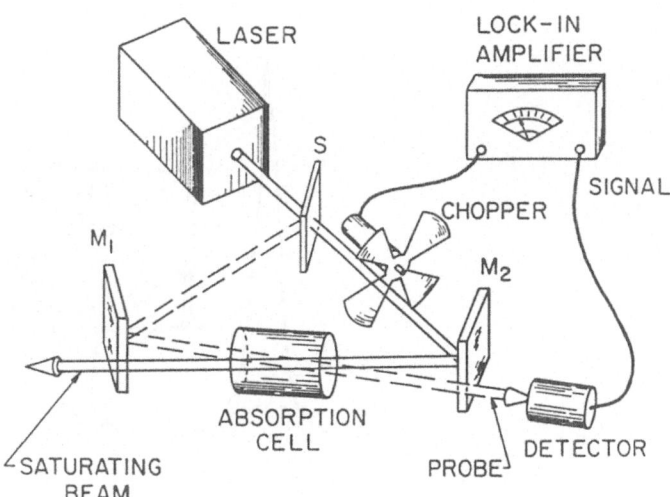

Figure 2. Apparatus for saturation spectroscopy without Doppler broadening.

modulated. However, this happens only for those wavelengths that interact with atoms that are standing still, or at most moving sideways. Because from a moving atom, the two light beams appear Doppler-shifted in opposite directions, so that they cannot both be in resonance at the same time."

Despite such powerful new techniques, the situation was at first quite frustrating to laser spectroscopists. Although there were hundreds of different lasers, they could only be tuned over very small wavelength ranges, and laser spectroscopy was limited to the laser transitions themselves or to a few molecular transitions in accidental coincidence with known laser lines. Lamenting this fact, Art Schawlow would recount how, as a graduate student, working in atomic spectroscopy, he discovered the true definition of a diatomic molecule: "It is a molecule with one atom too many!"

The whole situation changed completely with the advent of highly monochromatic tunable dye lasers. One of the first experiments with our new laser was the realization of an old dream. Jointly with Issa Shahin, one of Art Schawlow's graduate students, we recorded Doppler-free saturation spectra of the yellow D-lines of atomic sodium [12]. One such spectrum is shown in a little cartoon (Fig. 3) which I left on Art Schawlow's desk after an exhilarating night.

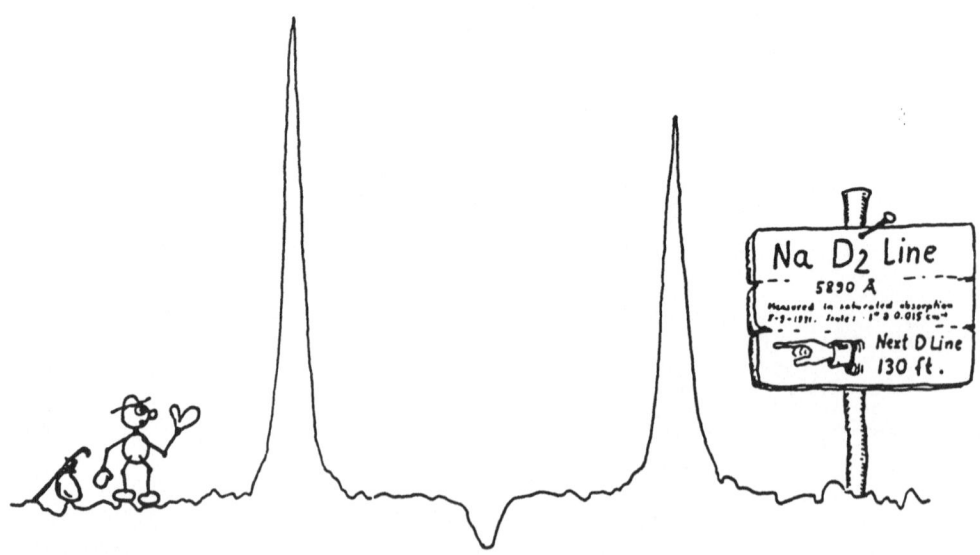

Figure 3. Cartoon dated August 9, 1971, with a Doppler-free saturation spectrum of the yellow sodium D_2 line

When Art Schawlow saw these results, he immediately urged: "You have to do the same with the red Balmer-α line of atomic hydrogen!" As the Rosetta Stone unlocked the secret of Egyptian hieroglyphics, the Balmer spectrum of the simple hydrogen atom opened up the laws governing atoms, and eventually molecules, liquids and solids [13]. It inspired the pathbreaking discoveries of Bohr, Sommerfeld, de Broglie, Schrödinger, and Dirac. For more than a decade, in the thirties and fourties, the line profile of the Balmer-α line presented one of the central problems of atomic spectroscopy. There appeared to be some discrepancies between observations and the predictions of the Dirac theory. But Doppler broadening is particularly large for the light hydrogen atoms, and classical spectroscopists never succeeded in resolving single fine structure components. In the 1940s, Lamb and Retherford used a radiofrequency method to reveal a new fine structure in the lower level of the transition that produces the Balmer-α line. The discovery of this Lamb shift led to the development of quantum electrodynamics by Feynman, Schwinger, and Tomonaga.

Within a few weeks, Issa Shahin and I set up an old-fashioned Wood-type gas discharge tube in order to produce hydrogen atoms in the excited n=2 state. With a simple change of the dye solution, our laser was ready for saturation spectroscopy of the Balmer-α line. Even the first Doppler-free spectra revealed single fine structure components [14]. We were thrilled that we could observe the Lamb shift directly in the optical spectrum, as illustrated in Fig. 4. This experiment set the beginning of an exploration of the simple hydrogen atom by high resolution laser spectroscopy that has grown into a fairly big enterprise in numerous laboratories [15]. Absolute wavelength measurements have meanwhile led to a 300-fold improvement in the accuracy of the Rydberg constant [16].

At Stanford, jointly with a group of bright graduate students and postdoctoral visitors, we went on to develop and demonstrate different variants of Doppler-free saturation spectroscopy. I remember a discussion with Art Schawlow at a conference in Esfahan in 1971, during which he invented intermodulated fluorescence spectroscopy, soon to be demonstrated by his student Mike Sorem [17]. Later, Art Schawlow and his group were very successful in applying saturation techniques to the simplification of complex molecular spectra by selective level labeling [18]. With the new tunable dye lasers, we could explore other fascinating coherent light techniques such as ultrasensitive detection of atoms and molecules by intracavity spectroscopy [19] or laser-excited fluorescence [20].

Much to our later chagrin, we long ignored a particularly elegant possibility for high resolution spectroscopy, the method of Doppler-free

11

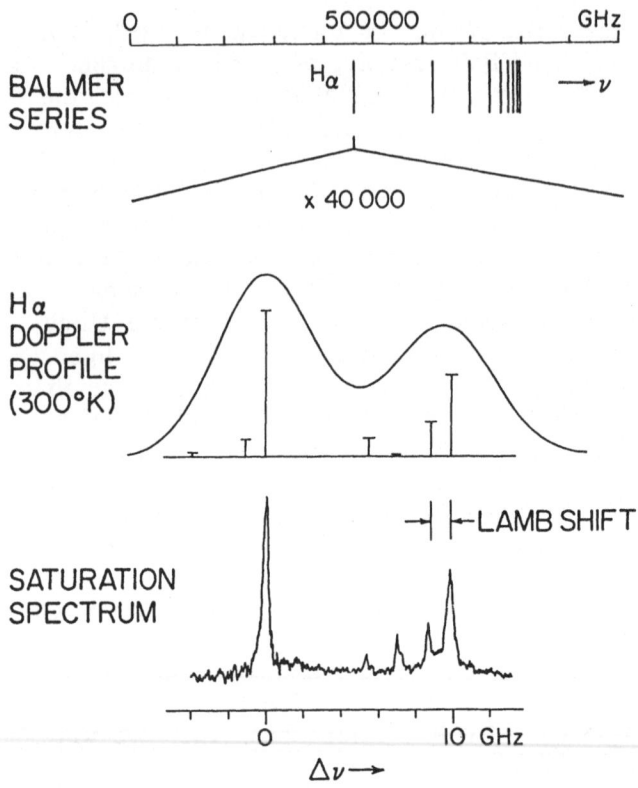

Figure 4. Balmer series of the simple hydrogen atom and fine structure of the red line H-α, resolved by saturation spectroscopy.

two-photon spectroscopy, as proposed in 1970 by Veniamin Chebotayev and coworkers [21]: If two counterpropagating light beams from a laser excite two-photon transitions in a gas, the two first-order Doppler shifts cancel, and a narrow resonance is expected, without any need to select slow atoms. This prediction was already implicitly contained in a theory of three-level gas atoms interacting with two laser beams which Peter Toschek and I had worked out [22]. However, we did not have widely tunable monochromatic lasers at that time, and we did not think seriously about exciting atoms with two equal photons.

At Stanford, Art Schawlow and I became aware of Chebotayev's proposal only in 1974 during a visit of Vladilen Letokhov. It was immediately obvious that such an experiment must work and that it should be very easy with our tunable dye lasers. Letokhov went on to visit Boston, where Marc Levenson was meanwhile working with Nico Bloembergen at Harvard. Soon afterwards, Marc succeeded in demonstrating pulsed Doppler-free two-photon spectroscopy in sodium vapor, almost simultaneously with an independent experiment of Bernard Cagnac's group in Paris. Our own paper [23] was unfortunately submitted a few weeks later, but we observed clean, well-resolved spectra with a cw dye laser of low power. This experiment was so simple that Art took delight in showing a color slide with a Mickey Mouse in front of the apparatus. "I have come to realize that I am really rather stupid", he would remark. "The only thing that saves me is that everybody else is rather stupid too - sometimes."

Doppler-free two-photon spectroscopy of atomic hydrogen holds particularly intriguing challenges, since the two-photon transition from the $1S$ ground state to the metastable $2S$ state has a natural linewidth of only about 1 Hz [21]. Unfortunately, such an experiment requires ultraviolet radiation in the difficult wavelength region near 243 nm. The first experiments had to resort to frequency-doubled pulsed dye lasers with rather large instrumental linewidths. But recent cw experiments at Stanford have measured the Lamb shift of the hydrogen ground state to within about 1 MHz [24]. Future experiments now being set up in our laboratories in Garching promise dramatic further improvements in spectral resolution [15]. They will permit tests of basic physics laws with unprecedented accuracy, and if past experience is any guide, the biggest surprise would perhaps be if we found no surprise.

Cooling with Laser Light

By 1974, our group at Stanford had become quite used to manipulating gas atoms with the help of intense, monochromatic laser light. In saturation spectroscopy, we were routinely selecting and exciting slow atoms with the pump beam. At about the same time, Art Ashkin and coworkers at Bell Laboratories reported on a series of interesting experiments levitating microscopic objects with laser radiation pressure. In a related earlier experiment, even before the advent of lasers, a beam of sodium atoms had been deflected by resonant radiation pressure.

Talking about these experiments, it suddenly became obvious to Art Schawlow and me that it should be possible to cool an atomic gas by resonant radiation pressure. The idea was to illuminate a gas sample with laser light tuned just below the atomic resonance frequency. Even if the light were propagating isotropically in all directions, it should exert a "viscous" braking force on any atom. Because of the Doppler effect, a moving atom will absorb preferentially those photons which are propagating in the opposite direction. Since the reemission is isotropic, each scattered photon will, on the average, reduce the momentum of the atom by an amount $\hbar k$. Some simple back-of-the-envelope calculations showed that atomic gases could be cooled very quickly to very low temperatures in this way.

Once we thought of it, the method appeared so obvious to me that I doubted we could publish it without demonstrating laser cooling in an actual experiment. But Art Schawlow called in a few colleagues and began to explain our idea. To my surprise, most shook their heads in disbelief and said something like "Cooling a gas by shining in laser light? You must be crazy!" After the visitors had left, Art said to me: "Now I am sure we have got something!" and we proceeded to write up our paper [25].

This scheme has been demonstrated recently with spectacular success by Steve Chu et al. [26] in their "optical molasses" experiments with sodium atoms. Some special tricks were required to avoid optical pumping between hyperfine levels. So Chu and his coworkers had the good sense to wait with this experiment until optical communications engineers had developed proper tools for modulating and controlling laser light. The ability to cool atoms by laser radiation pressure quickly to millikelvin temperatures and below is opening a whole new world to atomic physicists. It has become possible to trap cold neutral atoms magnetically or optically, to manipulate them, and to look for new quantum phenomena in their interactions with each other or with surfaces. Cold trapped atoms or ions may also lead to new incredibly precise atomic clocks.

More than once, Art Schawlow has uncovered new concepts which have opened up entire new fields of scientific endeavor, even though in hindsight some of his ideas may appear simple and almost obvious. The most dramatic example is of course the laser. Art Schawlow was the first to realize that, at optical wavelengths, just two mirrors would produce a laser resonator. With his role in the invention of the laser, he has reshaped modern science and technology to an extent not matched by many others. What a wonderful inspiration for future researchers when this great magician of science asserts that there must be many other beautiful and simple concepts lying around, ready to be discovered!

Acknowledgements

I am grateful to Peter Toschek for critically reading the manuscript and suggesting several changes. This article could not have been completed in time without the spirited help of Sigrid Oetjen, my secretary.

REFERENCES

1. A.L. Schawlow, J. Opt. Soc. Am. 67, 140 (1977)
2. T.W. Hänsch, Physics Today 30, 34 (1977)
3. A.L. Schawlow, Rev. Mod. Phys. 54, 687 (1982)
4. T.W. Hänsch, M. Pernier, and A.L. Schawlow, IEEE J. Quant. Electr. QE-7, 45 (1971)
5. H. Kogelnik and C.V. Shank, Appl. Phys. Lett. 18, 152 (1971)
6. T.W. Hänsch and A.L. Schawlow, Bull. Am. Phys. Soc. 15, 1638 (1970)
7. T.W. Hänsch, Appl. Opt. 11, 895 (1972)
8. T.W. Hänsch and P. Toschek, IEEE J. Quant. Electr. QE-4, 467 (1968)
9. P.W. Smith and T.W. Hänsch, Phys. Rev. Lett. 26, 740 (1971)
10. C. Bordé, C.R. Acad. Sci. Paris, 271, 371 (1970)
11. T.W. Hänsch, M.D. Levenson, and A.L. Schawlow, Phys. Rev. Lett. 26, 946 (1971)
12. T.W. Hänsch, I.S. Shahin, and A.L. Schawlow, Phys. Rev. Lett. 27, 707 (1971)
13. T.W. Hänsch, A.L. Schawlow, and G.W. Series, Scientific American 240, 94 (1979)
14. T.W. Hänsch, I.S. Shahin, and A.L. Schawlow, Nature 235, 63 (1972)
15. T.W. Hänsch, R.G. Beausoleil, B. Couillaud, C.J. Foot, E.A. Hildum, and D.H. McIntyre, Laser Spectroscopy VIII, S. Svanberg and W. Persson, Eds., Springer Series in Optical Sciences, Springer-Verlag, New York, Heidelberg, 1987
16. P. Zhao, W. Lichten, H.P. Layer, and J.C. Bergquist, Phys. Rev. Lett. 58, 1293 (1987)
17. M.S. Sorem and A.L. Schawlow, Opt. Comm. 5, 148 (1972)
18. R. Teets, R. Feinberg, T.W. Hänsch, and A.L. Schawlow, Phys. Rev. Lett. 37, 683 (1976)
19. T.W. Hänsch, A.L. Schawlow, and P. Toschek, IEEE J. Quant. Electr. QE-8, 802 (1972)
20. W.M. Fairbank, Jr., T.W. Hänsch, and A.L. Schawlow, J. Opt. Soc. Am. 65, 199 (1975)
21. L.S. Vasilenko, V.P. Chebotayev, and A.V. Shishaev, JETP Letters 12, 113 (1970)
22. T.W. Hänsch and P. Toschek, Z. Physik 236, 213 (1970)
23. T.W. Hänsch, K.C. Harvey, G. Meisel, and A.L. Schawlow, Opt. Comm. 11, 50 (1974)
24. R.G. Beausoleil, D.H. McIntyre, C.J. Foot, B. Couillaud, and T.W. Hänsch, Phys. Rev. A35, 4878 (1987)

25. T.W. Hänsch and A.L. Schawlow, Opt. Comm. 13, 68 (1975)
26. S. Chu, L. Hollberg, J. Bjorkholm, A. Cable, and A. Ashkin, Phys. Rev. Lett. 55, 48 (1985)

High-Power Solid State Lasers

J.F. Holzrichter

Lawrence Livermore National Laboratory, Livermore, CA 94550, USA

In 1958 SCHAWLOW and TOWNES [1] described the build-up of coherent light in an optical resonator, or laser. They noted that amplified coherent light would possess two desirable properties: precise, narrow wavelengths, and high brightness or focusability. The first laser, demonstrated by MAIMAN et al. [2] in 1961, produced a peak power of approximately 3 kW (3×10^3 W; i.e., 0.5 J of light in a pulse a few hundred microseconds long) and verified many of the Schawlow/Townes predictions.

Fig. 1. The Nova laser provides up to 100 TW (100 kJ in 10^{-9} s) at 1.05 μm. It can be harmonically converted to 0.53 or 0.35 μm with 60% efficiency. Most fusion experiments are conducted using 0.35-μm light

Today, high-power lasers (Fig. 1) are capable of more than 100 TW (10^{14} W). The high irradiances attainable with such lasers are making possible many technological advances, of which recent and important examples are the demonstration of inertial-confinement fusion (ICF) [3] and of laser gain at soft x-ray wavelengths [4].

Between 1959 and 1961, COLGATE et al. [5] suggested that high-power lasers could provide the irradiance required to compress small quantities of deuterium and tritium to high densities, initiating fusion and releasing thermonuclear energy.

In 1958, SCHAWLOW and TOWNES [6] wrote of the difficulty of producing lasers at x-ray wavelengths:

"...in the ultraviolet region at $\lambda = 1000$ Å, one may expect spontaneous emissions of intensities near ten watts.... Another decrease of a factor of 10 in λ would bring the spontaneous emission to the clearly prohibitive value of 100 kilowatts.... unless

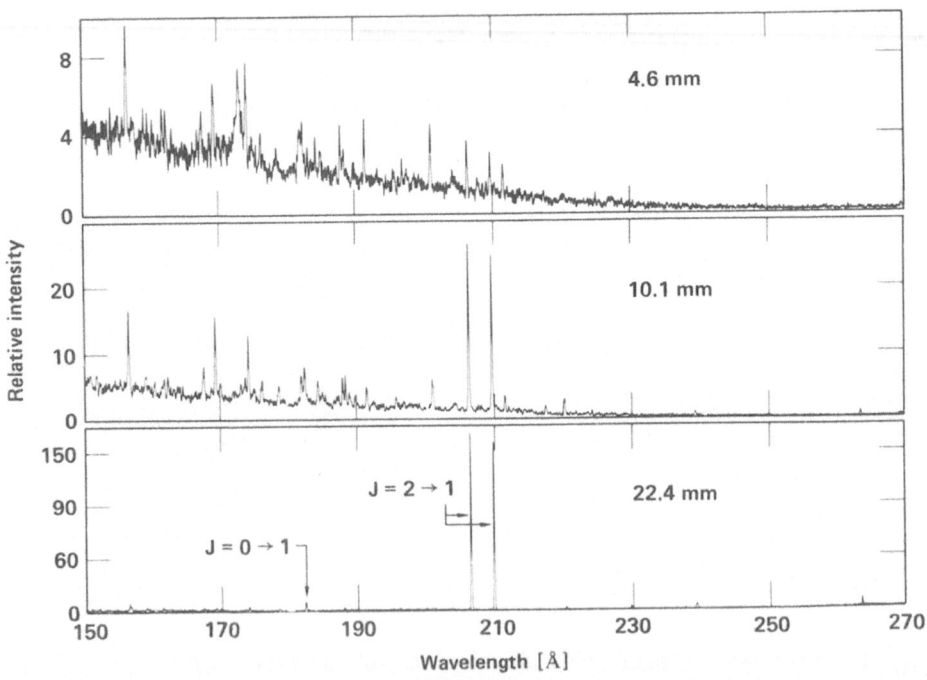

Fig. 2. Variation in line intensity with the length of the amplifying neon-like selenium plasma confirms lasing at 206 and 209 Å

some radically new approach is found, [maser systems] cannot be pushed to wavelengths much shorter than those in the ultraviolet region."

In 1985, MATTHEWS, ROSEN, et al. [4] observed laser gain in neon-like selenium at 206 Å (Fig. 2); recently, they observed gain at 66 Å in nickel-like europium [7]. In both experiments, the pumping source was the Novette laser, a 10-TW (10^{13}-W), harmonically converted, two-beam glass laser at the Lawrence Livermore National Laboratory (LLNL).

1. The 1960s at Stanford and Elsewhere

Schawlow and his associates and students demonstrated many of the laser technologies now being pursued in university, industrial, and national laboratory programs. These included the technology of high-power pulsed lasers with single temporal and spatial modes, frequency stabilization, wavelength tunability, the spectroscopy of laser media, and the focusing and diagnostic techniques necessary for the effective use of lasers in experimentation.

One of the more unusual experiments of the period—possibly one of the earliest ICF experiments—is shown in Fig. 3. The outer layer of a spherical object (a potato!) was irradiated with a 100-MW laser, one of the most powerful lasers available at the time. The ablation of the skin caused a shock wave to converge inward. Emmett and Schawlow are reported to have said at the time that if a powerful enough laser (the LLNL Nova laser would suffice) were to spherically irradiate a small enough potato, the skin would be instantly re-

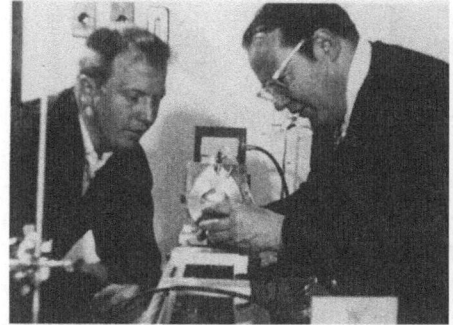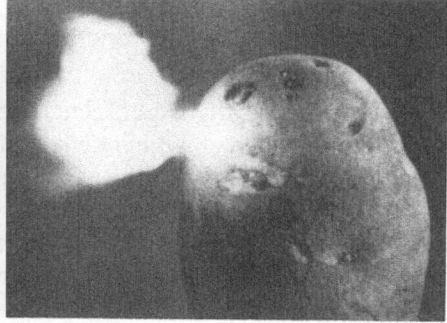

Fig. 3. J. L. Emmett and A. L. Schawlow align an early "fusion" experiment. Laser ablation of an Idaho potato "fusion-fries" the inside and removes the skin at the same time. Patent pending!

moved and the potato would be heated to near-stellar temperatures, creating a "fusion fry."

Experiments at Stanford and elsewhere (notably at Harvard and Michigan) showed that laser beams focused to high irradiance ($>10^9$ W/cm^2) produced electric fields strong enough ($>10^6$ V/cm) to modify the electron distribution around the nucleus. An intense laser field mixes high-lying electronic wave functions into the ground-state wave functions that are summed in calculating the susceptibility tensor of a material, thus making the optical response of the material to the laser field a nonlinear function of the field strength [8]. Through this interaction, an intense beam can increase the index of refraction of the medium (solid, liquid, or gas) in which the beam is propagating. This causes self-focusing of the laser beam, because wherever the beam is most intense the refractive index increases, causing a local lens-like action that further increases the intensity of that portion of the beam [9]. The nonlinear response to the electric field strength (a result of anharmonic electron motion) causes waves to be generated at harmonics of the frequency of the incoming beam [10]. This is used to efficiently ($>90\%$) convert the laser light to 2nd, 3rd, or 4th harmonic light [11].

In what are called three-wave interactions, incoming laser light can couple through virtual electronic transitions to elementary excitations in a material. The difference wave is Brillouin light if sound waves are the elementary excitation, or Raman light if vibrational transitions are the elementary excitations [12]. This difference wave can grow to high amplitude, and can ultimately carry more than 70% of the incoming laser energy.

High intensity can also cause complex multiphoton absorption processes, which can lead to ionization or other effects. A study [13] with a picosecond UV laser showed that at irradiances of roughly 10^{12} to 10^{13} W/cm^2, an atom such as uranium can lose its entire outer electron shell (8 electrons), corresponding to the absorption of nearly 100 photons from the laser field.

Most of these early experiments were performed with ruby lasers several millimeters in diameter, which delivered 10- to 100-MW beams (0.1 J in 10^{-8} s) that were focused (with some difficulty) to irradiances of 10^9 W/cm^2. A remarkable number of nonlinear processes were observed and quantified in the 1960s. Today, high-power fusion lasers such as Nova propagate 0.7-m-diameter beams at irradiances greater than 10^9 W/cm^2 over many meters before they are focused onto their targets. The irradiances of these beams are high enough without focusing to cause efficient (70%) harmonic conversion in 1-cm-thick plates of crystalline potassium dihydrogen phosphate (KDP); they are, in fact, high enough to push the limits of stable beam propagation in all media in which they travel. Propagation is limited by stimulated rotational Raman scattering and nonlinear self-focusing in air, and by transverse Raman or Brillouin scattering in transparent laser materials such as KDP crystals, fused silica, and even laser glass itself.

The 1960s were a time of great excitement in the laser field. Nonlinear optics was developed; in 1981 BLOEMBERGEN [14] received the Nobel Prize for his pioneering work in this field. The tunable dye laser was invented by SOROKIN [15], developed as a cw laser by SNAVELY [16], and developed to a high art form by HÄNSCH and his coworkers in Schawlow's laboratory at Stanford [17]. A renaissance in optical spectroscopy began in the 1960s and continues to this day. SCHAWLOW [18] received the Nobel Prize in 1981 for his leadership in this field. Being a student in Schawlow's laboratory during this period was a special experience. His attitudes and approaches to scientific investigation strongly influenced our work at Livermore and that of his students elsewhere.

The 1960s saw the development of a host of other important laser technologies. The invention of the semiconductor laser, together with glass fiber technology, has led to high-bandwidth communication by means of light waves. Arrays of semiconductor lasers are being used as efficient (>50%), high-power pump sources for solid state lasers and are leading to remarkable improvements in solid state laser technology. Laser-driven ICF and laser-induced isotope separation began; an important experiment by TIFFANY and SCHAWLOW [19] contributed to progress in isotope separation. Industrial and medical applications based on the welding, cutting, and photoselective properties of the laser were demonstrated (see Fig. 4 for an amusing

Fig. 4.   A. L. Schawlow demonstrates a laser technique relying on photoselectivity to eradicate a giant blue mouse using a red (ruby) laser beam. He also demonstrates the "action-at-a-distance" properties of the laser that make it so useful as a scientific probe, welding tool, etc.

example). Physicists, chemists, engineers, and others began to use the laser as a noninvasive probe for a wide variety of physical and chemical measurements that have contributed greatly to our understanding of the physical world.

2. Inertial-Confinement Fusion

In the 1970s, large national programs were begun to examine the possibility and practicality of ICF as a source of commercial electrical energy. This work grew out of pioneering work on high-power lasers in the 1960s at the Centre de l'Energie Atomique at Limeil by Bobin, Floux, and their coworkers; at the Lebedev institute by Basov and Prokhorov; and at LLNL by Nuckolls, Kidder, and others [20]. In 1970, laser power was about 0.01 TW (10^{10} W). Laser power of 10 to 100 TW (10^{13} to 10^{14} W) was needed to uniformly irradiate small fusion targets (0.1 to 1 cm in diameter) at $>10^{14}$ W/cm^2 [21]. This irradiance produces pressures greater than 10 Mbar (10 million atmospheres), enough to compress the most easily ignited fusion fuel, an equimolar mixture of deuterium and tritium, to fusion conditions—ion densities greater than 200 g/cm^2 (1000 times the density of solid DT) and ion temperatures greater than 5 keV (50 million degrees Celsius) [21]. Experiments at LLNL [3] recently confirmed that these conditions can be attained.

The development of the laser from its first demonstration in 1961 to the 100-TW fusion systems of today (see Fig. 5) has been made possible by con-

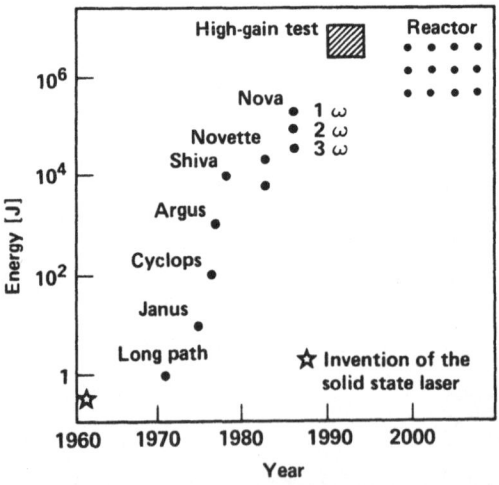

Fig. 5. **Laser performance has increased dramatically since the first demonstration by Maiman in 1961. The three "Nova" points represent (top to bottom) performance at 1.05, 0.52, and 0.35 μm. "High gain test" and "Reactor" are laser systems contemplated for future demonstrations**

tributions from many laboratories; see Ref. 22 for a review. This laser power gives us access in the laboratory to high temperatures and pressures that heretofore existed only in astrophysical environments (stars) or in nuclear explosions.

3. Physics of High-Power Lasers

A high-power laser designed for fusion and x-ray laser applications uses a small laser oscillator to generate a millijoule pulse in a single spatial and temporal mode, and then amplifies that pulse to thousands of joules (a gain of 10^6 to 10^8) [22]. Typical pulse durations are 10^{-9} s. The beam (a pulse of light about 30 cm long) propagates through the laser components at an irradiance greater than 10^9 W/cm^2.

Under these conditions, longitudinal and transverse nonlinear phenomena limit the irradiance of the pulse and the path length over which the pulse can propagate. Catastrophic breakdown of an optical material, associated with the formation of an absorbing, obscuring plasma, is the ultimate propagation limit and occurs at fluences of roughly 5 to 20 J/cm^2. The total power that can be carried by a single beam then depends on the irradiance and fluence of the beam and on the cross-sectional area to which the beam can be expanded before breakdown or nonlinear self-focusing occurs.

Nonlinear self-focusing [23] leads to near-field amplitude noise N and to far-field (focal) spread in a beam. The near-field noise grows as

$$N \simeq N_{init} \exp\left(\frac{2\pi}{\lambda} \int \gamma I \, dl\right) \equiv N_{init} \exp(B), \tag{1}$$

where N_{init} is the initial beam noise perturbation; λ is the laser wavelength; γ is the nonlinear retardation of the beam, caused by local increases in the refractive index of the material through which the beam propagates; and I is the beam irradiance. (Note that the beam noise grows more rapidly for shorter wavelengths.) The integration in (1) is carried out along the propagation path. The resulting so-called B-integral, giving the noise growth "gain," is also equal to the phase front retardation, in radians. (One wave of retardation corresponds to $2\pi = 6.28$ radians or to $B = 6.28$.) Typical near-field B limits are 2.5 to 3, which lead to near-field noise gain $\exp(B) = 12$ to 20. Far-field noise (beam spreading) grows as $\exp(2B)$, because both phase and amplitude noise contribute to the far field.

Given the cost of optical materials and the costs of construction generally, it is desirable to maximize the irradiance and fluence delivered by a given laser system. Progress in high-power lasers has resulted from efforts to minimize all parameters other than the irradiance (1), which we maximize subject to a fixed N:

- Lower-γ materials, e.g., fluoride-based glasses.
- Higher laser-gain coefficients and thinner optical components, which decrease the beam path length in the laser materials.
- Control of the spatial and temporal uniformity of I to prevent hot spots, which are exponentially amplified. Production of bandwidth-limited temporal pulses with no time spikes, and smooth, single-phase, noise-free transverse beams.
- Attainment of systems with low noise amplitude N_{init} by super-cleaning optical surfaces, by control of diffraction ripples with imaging apertures and apodizing apertures, and (most importantly) by periodically reducing the noise amplitude using spatial filtering.

Limits on the fluence ($\int I dt$) are different from irradiance limits because they occur when defects in optical elements absorb enough laser energy that an absorbing plasma develops, resulting in damage to the laser material. Low-loss, inclusion-free coatings, surfaces, and bulk materials permit optical elements to sustain higher fluences.

Higher fluences naturally occur for beams of fixed irradiance, defined in (1) above, as the laser pulse duration is increased from 1 to 10 ns. The current evolution of fusion lasers toward longer pulse duration for larger targets makes the development of high-fluence materials more and more important.

Large-area (high-Fresnel-number) systems are desirable because it is cost effective to maximize the power from a single beam aperture before paying the price of additional beams. Such systems have been made possible by advances in the production of low-cost, large-area laser materials, notably the development of new glasses, continuous polishing and lapping, and sol-gel coatings, and by the invention of imaging techniques (the Hunt relay [22]) that make it possibe to operate a laser with maximum fluence over the full optical aperture.

Beam area is ultimately limited by parasitic transverse laser oscillation in the gain medium or by transverse stimulated processes (e.g., Raman). If the transverse dimension of the medium is great enough, a sufficient laser gain-length or transverse stimulated gain-length can develop. This limits the gain coefficient of the laser host or the intensity of the laser beam.

To obtain laser power higher than that available from a single optimized beam, multiple beams are needed. The engineering and control of long, multibeam systems has been demonstrated. We have shown that multiple beams can be focused satisfactorily on a common target.

The output of glass laser systems can be harmonically converted from 1.05 μm to 0.52, 0.35, or 0.26 μm with efficiencies of up to 70%. This has made possible the generation of ICF plasmas free from suprathermal ("hot") electrons [3]. Hot electrons, which preheat the fusion fuel and greatly reduce the compression, are produced when plasma waves are stimulated by Raman and other processes. Laser light with wavelengths of 1 μm and longer is absorbed in a plasma at low electron densities ($n_e < 10^{21}$ cm^{-3}), at which electron-ion collision rates are too low to damp the plasma waves. But for $\lambda \leq 0.5\ \mu$m, absorption takes place at $n_e \geq 4 \times 10^{21}$ cm^{-3}; damping is then strong enough that at fusion irradiances ($>10^{14}$ W/cm^2), the plasma near the target is Maxwellian and free of hot electrons. Under these conditions, fusion implosions proceed satisfactorily [3].

4. Future of Solid State Lasers

Large investments in materials, research, computer models, and special fabrication technologies, together with new ideas and technologies, are leading to rapidly expanding applications of solid state lasers. Our experiments indicate that a demonstration of high-gain fusion in the laboratory will require a 5- to 10-MJ, short-wavelength laser. This represents roughly a fiftyfold increase in pulse energy with respect to the Nova system. Analysis of new laser architectures, new high-fluence materials, and lower-cost production technologies leads us to believe that such a system can be built at a cost only 2 or 3 times that of Nova, or roughly $300 to $600 million in 1987 dollars.

Recent calculations and experiments [24] show that crystal and (surprisingly) glass solid state laser media can be cooled well enough to generate high-optical-quality outputs of over 1000 W per laser plate. Multiplate, multibeam systems should be able to generate over a megawatt of average power. The ability of solid state laser systems to amplify nanosecond pulses makes it possible to convert the fundamental laser wavelength to almost any wavelength from the UV to the near IR using harmonic, parametric, or Raman processes. This can occur under average power conditions [24]. With the advent of GaAlAs laser diode arrays, which generate 810-nm light with an electricity-to-light conversion efficiency of over 60%, it is possible to contemplate solid state laser systems with overall efficiencies of 10 to 20%. A 1-cm^2 GaAlAs chip with 100 W/cm^2 pump output might be used to excite a 1-cm^2 solid state host at an efficiency of 30 to 40%; the 1-μm output of this system could be harmonically converted to 0.5 μm or shorter wavelengths with an overall system efficiency of 10 to 20%. Such a system, occupying a package a few cubic centimeters in volume, could operate on 110-V electrical energy, could be air-cooled, and could deliver up to 20 W of 0.5-μm light. Scaled up, this technology can be expected to lead to megawatt fusion systems capable of operating at efficiencies greater than 10%.

5. Summary

Work begun by Schawlow at Bell Laboratories in the 1950s on the optical properties of solids contributed to the demonstration of the laser, to a remarkable series of scientific achievements in the 1960s, and to applications in the 1970s. In the 1980s, solid state technologies are beginning to provide powerful, efficient, multiwavelength lasers for scientific, energy, defense, industrial, and medical applications, and they are likely to become the technology of choice for visible-laser applications in the 1990s.

Acknowledgments

While I have had the privilege of writing this article, which relates our work on solid state lasers at Livermore to Schawlow's work at Stanford, my Livermore colleagues have in fact done most of the work. In particular I acknowledge J. L. Emmett, Schawlow's second graduate student, who came to Livermore in 1972 to direct the Inertial Fusion Program and who began the Laser Isotope Separation Program in 1974. W. F. Krupke and A. C. Haussmann worked closely with Emmett from the beginning to develop these programs to their present levels of success. It has been a pleasure working with these people, and with my other Livermore colleagues; I have inadequately referenced their contributions, which can be found in the Lawrence Livermore National Laboratory *Laser Program Annual Reports* for 1972 to the present. I thank P. W. Murphy for editing this article.

Work performed under the auspices of the U.S. Department of Energy by the Lawrence Livermore National Laboratory under Contract W-7405-Eng-48.

References

1. A. L. Schawlow and C. H. Townes, "Infrared and Optical Masers," *Phys. Rev.* **112** (6), 1940–1949 (1958).
2. T. H. Maiman, R. H. Hoskins, I. J. D'Haenens, C. K. Asawa, and V. Evtuhov, "Stimulated Optical Emission in Fluorescent Solids, II. Spectroscopy and Stimulated Emission in Ruby," *Phys. Rev.* **4** (123), 1151–1157 (1961).
3. J. F. Holzrichter, E. M. Campbell, J. D. Lindl, and E. Storm, "Research with High-Power Short-Wavelength Lasers," *Science* **229**, 4718 (1985).
4. D. L. Matthews, P. L. Hagelstein, M. D. Rosen, M. J. Eckart, N. M. Ceglio, A. U. Hazi, H. Medecki, B. J. MacGowan, J. E. Trebes, B. L. Whitten, E. M. Campbell, C. W. Hatcher, A. M. Hawryluk, R. L. Kauffman, L. D. Pleasance, G. Rambach, J. H. Scofield, G. Stone, and T. A. Weaver, *Phys. Rev. Lett.* **54**, 110 (1985); M. D. Rosen, P. L. Hagelstein, D. L. Matthews, E. M. Campbell, A. U. Hazi, B. L. Whitten, B. MacGowan, R. E. Turner, R. W. Lee, G. Charatis, Gar. E. Busch, C. L. Shepard, P. D. Rockett, and R. R. Johnson, *Phys. Rev. Lett.* **54**, 106 (1985). In recent work, x-ray mirrors with >50% normal-incidence reflectivity near 200 Å have been made [T. W. Barbee, Jr., *AIP Conf. Proc.* **119**, 311 (1984); E. S. Spiller, *AIP Conf. Proc.* **119**, 312 (1984)].
5. S. Colgate, R. E. Kidder, J. H. Nuckolls, R. F. Zabawski, and E. Teller, Lawrence Livermore National Laboratory, Livermore, CA, unpublished calculations (1961).
6. [1], p. 1949.
7. D. L. Matthews, Lawrence Livermore National Laboratory, Livermore, CA, private communication (1987).
8. N. Bloembergen, *Nonlinear Optics* (Benjamin, New York, 1965); Y. R. Shen, *The Principles of Nonlinear Optics* (Wiley, New York, 1984).
9. R. L. Carman, R. Y. Chiao, and P. L. Kelley, "Observation of Degenerate Stimulated Four-Photon Interaction and Four-Wave Parametric Amplification," *Phys. Rev. Lett.* **17**, 1281 (1966); Y. R. Shen, "Self-Focusing: Experimental," and J. H. Marburger, "Self-Focusing: Theory," in *Progress in Quantum Electronics*, ed. by J. H. Sanders and Stenholm, Vol. 4, Part 1 (Pergamon, New York, 1975).
10. P. A. Franken, A. E. Hill, C. W. Peters, and G. Weinreich, *Phys. Rev. Lett.* **7**, 118 (1961).
11. D. Eimerl, "Thin-Thick Quadrature Frequency Conversion," Lawrence Livermore National Laboratory, Livermore, CA, UCRL-92087 (1984): also in *Proceedings of the International Conference on Lasers '84* (November 26–30, 1984); D. Eimerl, "Quadrature Frequency Conversion," Lawrence Livermore National Laboratory, Livermore, CA, UCRL-95424 (1986): to be published in *IEEE J. Quantum Electron.*

12. A. Denzhofer, A. Laubereau, and W. Kaiser, "High Intensity Raman Interactions," *Prog. Quantum Electron.* **6**, 55–140 (1979).

13. T. S. Luk, H. Pummer, K. Boyer, M. Sahidi, H. Egger, and C. K. Rhodes, *Phys Rev. Lett.* **51**, 110 (1983).

14. N. Bloembergen, "Nonlinear Optics and Spectroscopy," *Science* **216**, 4550 (1982).

15. P. P. Sorokin and J. R. Lankard, *IBM J. Res. Dev.* **11**, 148 (1967).

16. B. B. Snavely, O. G. Peterson, and R. F. Reithel *Appl. Phys. Lett.* **111**, 275 (1967).

17. T. W. Hänsch, "High Resolution Laser Spectroscopy," in *Advances in Laser Spectroscopy*, F. T. Arecchi, F. Strumia, and H. Walther, Eds. (Plenum Press, New York, 1983), p. 127; T. W. Hänsch, "Sub-Doppler Spectroscopy," in *Atomic Physics 8*, L. Lindgren, A. Rosen, and S. Svanberg, Eds. (Plenum Press, New York, 1983), p. 55.

18. A. L. Schawlow, "Spectroscopy in a New Light," *Science* **217**, 4554 (1982).

19. W. B. Tiffany, H. W. Moos, and A. L. Schawlow, "Selective Laser Photocatalysis of Bromine Reactions, *Science* **157**(2784), 40–43 (1967); W. B. Tiffany, "Selective Photochemistry of Bromine using a Ruby Laser," *J. Chem. Phys.* **48**(7), 3019–3031 (1968).

20. *Laser Interaction and Related Plasma Phenomena*, H. J. Schwarz and H. Hora, Eds. (Plenum Press, New York, 1971–1984), vols. 1–6.

21. J. H. Nuckolls, L. L. Wood, A. R. Thiessen, and G. B. Zimmerman, *Nature (London)* **239**, 139 (1972); J. L. Emmett, J. H. Nuckolls, and L. L. Wood, "Fusion Power by Laser Implosion," *Sci Am.*, pp. 24–37 (June 1974). For detailed descriptions of laser-plasma investigations, see the annual reports of the following laboratories: CEA, Limeil, France; KMS Fusion, Inc., Ann Arbor, MI; Naval Research Laboratory, Washington, DC; Osaka University, Osaka, Japan; Sandia Laboratories, Albuquerque, NM; the Lebedev Physical Institute, Moscow, USSR; the Max Planck Institute for Plasma Physics, Garching, Germany; the Rutherford Laboratory, Didcot, England; Lawrence Livermore National Laboratory, Livermore, CA; Los Alamos National Laboratory, Los Alamos, NM; University of Rochester (Laboratory for Laser Energetics), Rochester, NY.

22. J. F. Holzrichter, D. Eimerl, E. V. George, J. B. Trenholme, W. W. Simmons, and J. T. Hunt, *J. Fusion Energy* **2**, 5 (1982). Available as Lawrence Livermore National Laboratory, Livermore, CA, UCRL-52868-1. See also *Laser Program Annual Report*, Lawrence Livermore National Laboratory, Livermore, CA, UCRL-50021-73 (1974) to UCRL-50021-84 (1985).

23. J. B. Trenholme, in [22], p.28.

24. J. L. Emmett, W. F. Krupke, and J. B. Trenholme, *The Future Development of High-Power Solid State Laser Systems* (*Physics of Laser Fusion*, vol. 4), Lawrence Livermore National Laboratory, Livermore, CA, UCRL-53344 (1982).

From Micromasers to Antimasers: When One Photon in a Cavity May Be One Too Many...

S. Haroche

Ecole Normale Supérieure, Paris, France and
Yale University, New Haven, CT, USA

The direction of physics is sometimes unpredictable. When Arthur Schawlow and the other pioneers of lasers were trying to excite as many atoms as possible in an optical cavity to get the first lasers going, they would hardly have guessed that some twenty years later an active domain of quantum optics would be concerned with maser or laser systems in which the cavity is empty – or contains only one atom at a given time, or with cavities unable to sustain a single photon emitted by the atomic medium... This domain of quantum optics is now called "Cavity Quantum Electro-dynamics" [1, 2]. Although theory in this field can be traced back to articles [3] and short notes [4] published a long time ago, the development of cavity Q. E. D. experiments has recently come as a natural extension of the research on Rydberg atom radiative properties [1, 2, 5, 6]. In this paper, written as a token of my friendship with and admiration for Art Schawlow, I review some aspects of this research which I have been interested in over the last seven or eight years at Ecole Normale Supérieure in Paris, as well as more recently at Yale University.

1. Rydberg Atom Transient Masers : A Case Study in Superradiance

A sample of Rydberg atoms – i.e. of atoms excited in an energy level close to the atomic ionization limit – is an ideal active medium for a laser or a maser : there are a very large number of inverted transitions down to more bound states. The transition matrix elements towards nearby levels are huge (they scale as the size of the atom, i.e., as n^2 when n is the principal quantum number of the Rydberg state). Moreover, the relaxation times of these atoms are quite long (in the millisecond to microsecond range for n ~ 30, if the atoms are not too perturbed by external fields). As a result, superradiant emission without mirrors occurs with Rydberg samples containing typically about 10^6 atoms [7]. If the sample is prepared in a cavity with a moderate quality factor Q (in the range 10^3 to 10^4) a few hundred to a few thousand atoms are enough to reach the emission threshold [7, 8]. This emission usually occurs at microwave frequencies (centimeter or millimeter wavelengths) corresponding to transitions between nearby levels in the Rydberg spectrum.

In fact, we observed our first Rydberg masers somewhat by chance, when performing microwave spectroscopy in very excited states of sodium. The setup we were using is sketched in Fig. 1a : an atomic beam of Na was excited by a short laser pulse into a Rydberg state ns or nd (n ~ 20 to 40) inside a Fabry-Perot cavity into which we fed microwaves through a waveguide. After crossing the cavity, the atoms were analyzed with a Rydberg state selector (Fig. 1b) : a ramp of electric field was applied between two condenser plates. This field reached the ionization threshold for the two levels implied in the Rydberg-Rydberg transition at slightly different times (the lower more bound state requiring a somewhat larger ionizing field than the upper one). This resulted in two time resolved electron peaks detected by an electron multiplier. The resonance appeared as a change in the relative weights of the peaks associated with the initial and final states of the transition as the microwave frequency was scanned. We soon noticed that resonances could in fact be observed without applying any microwaves to the cavity. When the mirror distance L was tuned to an integer number of half-wavelengths of an atomic transition between the initially prepared Rydberg level and a lower state, we observed a strong and fast transfer of population to this latter state, demonstrating a transient maser or laser action in the system. In this way, we have observed hundreds of Rydberg transitions in Na, Cs [8] and more recently in lithium, corresponding to Δn = 1, 2, 3 transitions. Cascading transitions sharing a common level and masing successively or in competition have also been observed [8]. The thresholds - even with moderate Q cavities - were small (N ~ 10^2 to 10^4) and it was in fact difficult to avoid this maser action when it was a nuisance in other kinds of experiments. The first direct use of this effect has been to perform cheap and convenient Rydberg atom spectroscopy without a microwave source. The Rydberg-Rydberg frequency was merely determined by the lengths of the cavity corresponding to the maser action between the levels, a variant of the "spectroscopy with a ruler", type of experiment often advocated by Art Schawlow.

We then realized that these transient maser systems were ideal to test the theory of superradiance [9, 10], first developed by Dicke [11],

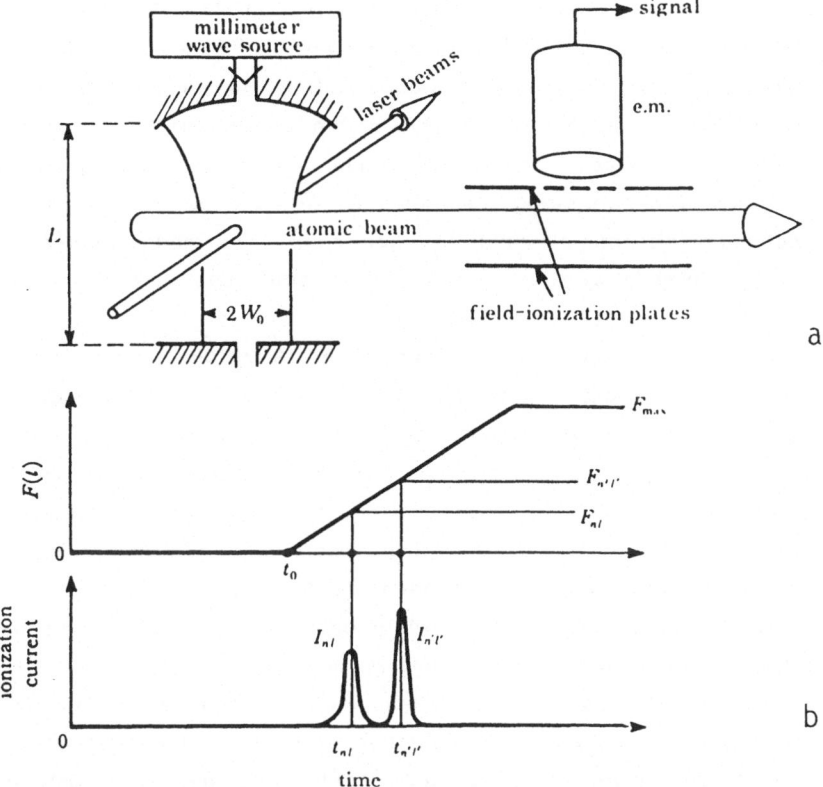

Figure 1 : Microwave spectroscopy of Rydberg states.
a) Sketch of setup showing atomic beam excited by laser pulses
inside the Fabry-Perot millimeter wave cavity. Rydberg atoms
interact with the microwave during their propagation through the
cavity waist (w_o), then drift into the detection zone (field ionization
plates). b) Time evolution of electric field produced by the plates
and of resulting ionization current. The delay of the field ramp is
set so that atoms excited at $t=0$ in the cavity have reached the
detector at time t_o. Times t_{nl} and $t_{n'l'}$ correspond respectively to
the ionization threshold fields F_{nl} and $F_{n'l'}$ of the Rydberg levels nl
and $n'l'$ involved in the transition. If the upper level nl is initially
prepared, resonant radiative transfer to level $n'l'$ results in a
change of the ionization peak shape from I_{nl} to $I_{n'l'}$. Maser action
is easily observed without any external microwave when cavity
length L is tuned into resonance.

which, until the development of Rydberg atom experiments, had been very
difficult to check experimentally in detail. The ideal textbook superradiance
situation deals with two-level atoms symmetrically coupled to a damped
mode of the electromagnetic field. The symmetrical coupling entails a fast
radiative deexcitation of the atomic system, on a time scale inversely

proportional to the number N of atoms involved in the process. This situation is quite easily achieved with a sample of Rydberg atoms in a microwave cavity : only the two levels connected by the transition resonant with the cavity are relevant during the time of the experiment since the rate of emission to other states is usually very small; the symmetric coupling to the field is realized by preparing the atoms at an antinode position in the cavity (a millimeter size sample fulfils this condition). The mode structure of the field in the cavity is very simple and usually one mode only is coupled to the atoms.

In a series of experiments performed in 1982-1984, we made an extensive check of the Dicke theory. We studied the evolution of the average atomic energy during a superradiant pulse and the fluctuations around this average [8,12]. We showed also that the collective behavior of the atomic system was manifest not only in emission processes, but also in absorption and we demonstrated an effect of collective absorption of thermal radiation by a sample of Rydberg atoms in a resonant cavity [13].

A simple way to analyze these experiments consists in describing the symmetrical ensemble of N two-level atoms as an angular momentum J=N/2 evolving in an abstract space [1,8] (Bloch vector). This vector evolution obeys a pendulum-like equation, the radiation reaction of the atoms being analogous to the pendulum gravity restoring couple and the cavity damping playing the role of a viscous drag force. In this model, the potential and the kinetic energies of the pendulum are respectively associated to the atomic and field energies in the cavity and the vacuum fluctuations and thermal field noise correspond to Langevin-type forces responsible for the Brownian motion of the pendulum. In other words, transient Rydberg maser experiments can be viewed as exploring the dynamics of a pendulum starting either from its unstable equilibrium position (emission) or from its stable one (absorption) and being triggered away from this position by Brownian forces. The Dicke superradiant regime corresponds in this point of view to the overdamped evolution of the pendulum starting from its unstable position (pendulum falling with a strong viscous drag force). If the cavity damping is small enough, the pendulum evolution becomes oscillatory, corresponding to a quasi-reversible exchange of energy between the atom and the field with a pseudofrequency proportional to the square root of the atom number in the cavity. This regime too had been predicted long ago in superradiance theories [14], but had never been clearly observed before in quantum

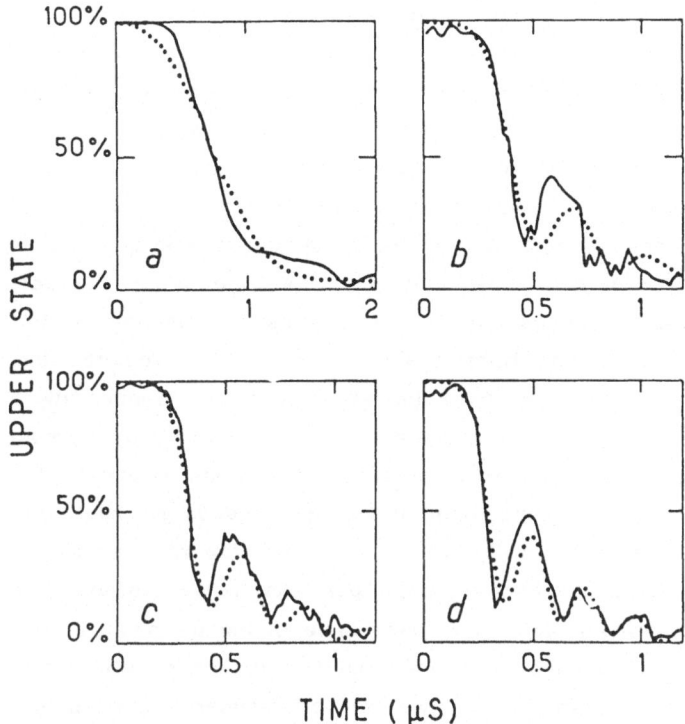

Figure 2 : *Time evolution of a transient Rydberg atom maser for increasing atom number N in the active medium (from [15]). The relative population of the upper Rydberg state, averaged over a few hundred pulses, is recorded as a function of the interaction time between the atoms and the cavity (t=0 corresponds to the initial preparation of the system). Traces a, b, c, d correspond respectively to N = 2000, 12000, 27000, 40000 atoms. The regime is overdamped in trace a. Traces b, c, d clearly exhibit the pendulum-like behavior of the system.*

optics. We were able to study it in detail with our Rydberg atom masers [15]. Figure 2 shows recordings of the mean atomic energy evolution, corresponding to increasing atom number in the sample. The pendulum-like oscillations of the system are clearly demonstrated in this experiment.

2. Microscopic Masers : One Rydberg Atom in the Cavity

In all the experiments mentioned above, N was in the range 10^3–10^5. A natural extension of these studies was to try to decrease the atom number in the resonator while increasing the cavity Q. The goal was eventually to achieve a Rydberg maser operating with a single atom in the cavity. This concept deserves some discussion since masers or lasers are usually supposed to operate as collective systems. In fact, a single atom maser starts as an atom-cavity coupled system in which the atom is induced to radiate a photon much faster than in free space. The cavity around the atom modifies the mode density of the electromagnetic field and, if it is resonant with the atomic transition, increases the density of photon final states open for the atomic decay, so that the spontaneous emission rate is enhanced [1,2]. Alternatively, one can say that the atom interacts collectively with its electric images in the cavity walls and radiates in this way faster than in free space. Here again, theorists had predicted this effect long ago, but Rydberg atoms gave the first experimental opportunity to observe it directly. Replacing our ordinary copper cavity mirrors by niobium superconducting ones, we were able to increase the Q value to about 10^6 and to observe in 1983 the enhancement of the spontaneous emission rate of a single Rydberg atom in the cavity [16]. The atomic flux was reduced so that most of the time only one Rydberg atom per laser pulse crossed the cavity. This atom was transferred to the lower state of the transition during the cavity crossing time when the cavity was tuned to resonance whereas it stayed excited in the initial level if the cavity was off-resonant (see Fig. 3). This experiment was the one-atom limit of the overdamped regime of a

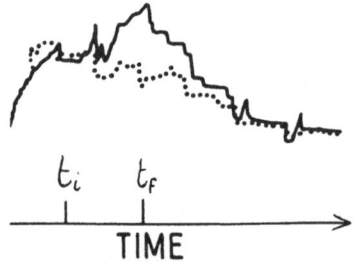

t_i t_f

TIME

Figure 3 : *Time resolved ionization signal demonstrating the one-atom transient maser action. The signal is the ionization current recorded during the ramp of ionizing field following each laser pulse (see Fig. 1b). The signal is averaged over a few hundred laser pulses. Times t_i and t_f correspond respectively to the ionization thresholds of the upper level i and lower level f of the transition ($23S_{1/2}$ and $22P_{1/2}$ states of Na). The laser intensity is reduced so that for most pulses, one or two atoms are excited in the cavity. The dotted and full line curves correspond respectively to off and on-resonance cavity. The increase of signal around t_f in the resonant case demonstrates the spontaneous emission enhancement effect (from [16]).*

transient Rydberg maser : the emitted photon was rapidly damped in the cavity walls.

In order to observe the one-atom maser oscillating regime – analogous to the oscillatory behavior mentioned above in the collective case – it was necessary to increase the cavity Q by at least two orders of magnitude. The emitted photon could then be stored long enough for the atom to be able to reabsorb it at a later time giving rise to a quasi reversible single photon exchange between the atom and the cavity mode. We described this ideal situation in several publications [5,8,16] and went on developing cavities for its experimental realization. But the Fabry-Perot resonators with open structures we were using at that time were mechanically too unstable to achieve the required high Q's. The solution to the cavity problem has been found by Walther, Meschede and Muller in Munich who have developed closed and rigid superconducting cylindrical microwave cavities operating in a low order with a Q in the 10^8-10^9 range. Using these cavities, in 1984 they observed for the first time a single atom maser with a long photon storing time [17]. In their experiment, the atoms (rubidium in 63P state) were prepared by continuous wave laser excitation, with a flux of the order of 10^3-10^4 atom per second, so that successive radiators entered the cavity before the photons radiated by preceding atoms had disappeared. In steady state operation, this system thus realizes a maser with less than one atom on average in the medium and only a few photons in the cavity. Although it has not yet been observed, the one atom-one photon exchange regime mentioned above is now within reach of experimental investigation.

Rydberg atom micromasers are now a class of quantum optics systems on their own. The statistical properties of the electromagnetic field radiated by these systems are quite interesting. Non-classical field statistics, squeezing, generation of pure Fock states corresponding to a well defined number of photons in the cavity have been predicted theoretically [18,19] and are now being checked in difficult experiments involving very high Q cavities at very low temperatures and well controlled atomic beams (velocity selection of atoms is very important in these experiments) [20].

3. Two-Photon Rydberg Atom Micromasers

The combination of Rydberg atom and high Q cavities also offers ideal opportunities to realize other kinds of new quantum optics devices. At Ecole Normale Supérieure, we are presently trying to build a Rydberg atom micromaser operating in a cw regime on a two-photon transition. Two photon lasers or masers have been for a long time the most elusive systems in quantum optics. Although the possibility of amplifying the electromagnetic field on a two-photon degenerate or non-degenerate transition between two atomic levels of the same parity was pointed out in the early days of lasers [21,22] and considered since in a large body of theoretical papers [23], these devices have never been demonstrated so far in cw operation (one report does exist however of two photon amplification in pulsed regime [24]). On ordinary transitions, the two photon gain is indeed very low. Furthermore, a two photon laser generally requires a triggering field in order to build up an oscillation large enough so that the two-photon quadratic gain overcomes the linear losses of the cavity. The triggering field is likely to produce unwanted competing effects (Raman processes, multiple wave mixing), depleting the pumped level before two-photon amplification can occur. All these difficulties vanish with Rydberg atoms in super-conducting cavities [25]. Very high gain is achievable due to the large intrinsic atomic dipoles of Rydberg levels and most importantly to the existence of near coincidences in the Rydberg spectrum with a relay level almost midway between the initial and final state of the two-photon transition (see Fig.4). Under optimum conditions (40S → 39S transition in ^{85}Rb

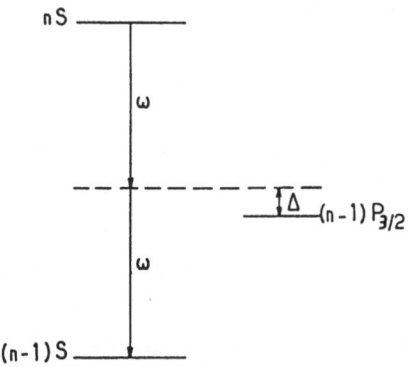

Figure 4 : Energy diagram relevant to the two-photon Rydberg atom maser. The cavity is tuned at frequency ω, corresponding to a two-photon degenerate transition between the $nS_{1/2}$ and $(n-1)S_{1/2}$ levels. The $(n-1)P_{3/2}$ level is very close to the middle of the energy interval, which considerably enhances the two-photon amplitude. The most favorable case is n = 40 in ^{85}rubidium ($\Delta \simeq 39MHz$ only). The two photon maser action will be demons-trated by recording the population transfer from $nS_{1/2}$ to $(n-1)S_{1/2}$, while the $(n-1)P_{3/2}$ population should remain zero.

atoms) with a niobium superconducting cavity having a Q ~ 2.10⁶ at ν = 68 GHz, we expect that the Rydberg maser should operate on the two photon transition with only about one atom at a time and a few tens of microwave photons in the cavity. Excitation of the atoms in the 40s state of rubidium is achieved by a four step cw process ($5S_{1/2} \rightarrow 5P_{3/2} \rightarrow 5D_{5/2} \rightarrow 40P_{3/2}$ transitions are induced by three laser diodes at λ_1 = 7802Å, λ_2 = 7759Å and λ_3 = 1.26 μm respectively and the final $40P_{3/2} \rightarrow 40S_{1/2}$ transition by a microwave source at 62 GHz). We have achieved in this way Rydberg atom fluxes in the range 10^6-10^7atoms/s, corresponding to tens of atoms in the cavity at a time. The fine cavity tuning will be achieved, as in [17], by mechanical pressure on the cavity walls and the maser action will be detected by monitoring the atomic populations in the 40S and 39S states in atoms emerging from the cavity.

We have performed a complete theoretical analysis of these systems [26]. Here again, the microscopic character of the device leads to interesting new effects to be studied. We have shown that the distribution of the photon number n_φ in this maser diffuses according to a Fokker-Planck equation, with an effective potential $V(n_\varphi)$ presenting typically several minima. Multistable operation and quantum diffusion between these minima are predicted and it will be interesting to investigate these experimentally. The passage time from one minimum to a neighboring one is shown to increase very quickly with the atom and photon numbers in the cavity, which means that such effects would be unobservable with "classical" macroscopic masers containing very large photon numbers. Even the initiation of the maser action is predicted to be different in the two-photon micromaser. In fact, spontaneous noise can be large enough to trigger the passage of the system in a finite time from the equilibrium around n_φ = 0 to stable operation with n_φ > 0, making external triggering unnecessary.

It is possible that the study of these new types of quantum systems will lead to situations analogous to the ones now being explored in the experiments monitoring single atoms or ions : the passage from one equilibrium state to another is in particular reminiscent of quantum jump behavior.

4. "Antimaser or Antilasers" : Inhibition of Spontaneous Emission in a Cavity

As discussed above, the basic process in the dynamics of a microscopic maser is the coupling of a single atom to a mode of the electromagnetic field sustained by a cavity. In this respect, micromaser physics is directly related to cavity Q.E.D., the domain of Q.E.D. studying how the cavity boundary conditions modify the spontaneous atomic radiative processes. In a micromaser, the cavity is resonant with an atomic transition, which corresponds to an enhancement of the field mode density at atomic frequency with respect to its free space value and to an acceleration of one-photon as well as two-photon spontaneous emission processes. The opposite effect – inhibition of spontaneous emission in an off-resonant cavity – has been predicted [27] and also observed in Rydberg atom physics : if the atom is excited in a cavity below cutoff for the atomic transition, there is no mode available for the photon decay and the atom should remain excited much longer than in free space.

The simplest cavity geometry corresponding to this situation is the biplanar configuration with two plane parallel metallic surfaces separated by a gap d. Figure 5a shows the mode density $\rho(\omega)$ versus frequency ω in this structure, for modes which at the midgap position $z = d/2$ are polarized either parallel (σ) or perpendicular (π) to the surfaces ($\rho_\sigma^{(cav)}$ and $\rho_\pi^{(cav)}$ respectively). The dotted curve indicates for comparison the free-space mode density $\rho_0(\omega) \sim (\omega^2/3\pi^2c^3)$. The striking feature of these distributions is the complete cutoff of field modes for $\omega < \omega_0 = (\pi c/d)$ in σ–

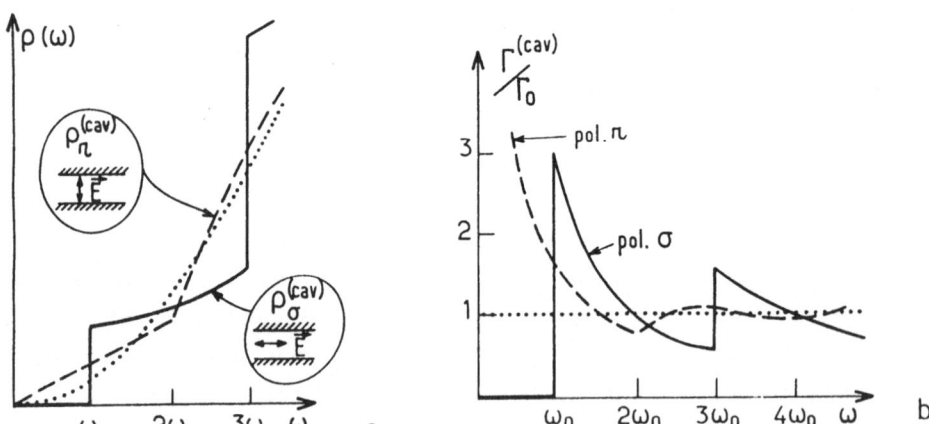

Figure 5 : a) Mode density $\rho(\omega)$ versus ω in a cavity made of two plane parallel mirrors separated by a gap d. Density evaluated at midgap position $z = d/2$. Full line : σ-polarization. Dashed line : π-polarization. For comparison, the free space mode density is represented by the dotted line. b) Ratio $\Gamma^{(cav)}/\Gamma_0$ versus frequency ω. Full line : spontaneous emission in polarization σ; dashed line : spontaneous emission in polarization π.

polarization whereas modes exist down to $\omega = 0$ in π-polarization (corresponding to the electrostatic limit with a static electric field perpendicular to the condenser plates). The spontaneous emission rate being proportional to the mode density, the rate in the cavity $\Gamma^{(cav)}$ is merely equal to the rate in free space Γ_0, multiplied by the ratio $\rho^{(cav)}/\rho_0$ (see Fig. 5b). In order to inhibit the atomic emission in such a structure, it is thus necessary to prepare an excited state corresponding to an electric dipole parallel to the mirrors and to have a plate separation $d < \lambda/2$. These conditions have been achieved by Hulet, Hilfer and Kleppner [28] who have sent Rydberg atoms in high angular momentum "circular" states between two aluminum mirrors separated by a gap $d = 0.23$ mm. The circular Rydberg states correspond to a toroïdal electronic orbital lying in a plane perpendicular to the atomic angular momentum. These states are prepared by feeding angular momentum into the Rydberg atoms by absorption of a large number of circularly polarized microwave photons [29]. The atoms are subjected to an electric field along the direction of this angular momentum and the microwave photons induce transitions between the Stark sublevels in this electric field. By aligning this directing field along the normal to the mirror, the Rydberg atoms are thus prepared with their orbit parallel to the conductor plates and the electric dipole corresponding to transitions to the adjacent lower circular state is parallel to the mirrors. The inhibition of spontaneous emission was observed by monitoring the flux of excited atoms having survived spontaneous decay in the structure, after a crossing time of the order of the natural lifetime of the transition (~ 1 ms). Evidence for a large suppression of spontaneous emission has been obtained in this experiment. This experiment had been in fact preceded by a demonstration of radiative decay inhibition in the cyclotron resonance of a free electron in a Penning trap [30]. The cavity was made by the electrodes of the trap. Due to the much more complex geometry of this cavity, the interpretation of this experiment and the quantitative check of cavity Q.E.D. theory were more delicate in this case.

In all these microwave experiments, the spontaneous processes to be inhibited were quite slow (natural spontaneous emission rates Γ_0 in the 10 s^{-1} to 10^3 s^{-1} range). We have very recently shown at Yale [31] that it is possible to extend these experiments in the optical domain, where spontaneous emission processes are intrinsically much stronger ($\Gamma_0 \sim 10^6$ s^{-1}). The atoms (which are now in strongly bound excited states and no longer in Rydberg levels) must be sent in micrometer-sized cavities. We have realized such microscopic cavities [32] by pressing two $\lambda/10$ flat mirrors against each other with 1μm-thick metal foils used as spacers between them. We were able in this way to build mirror tunnels with a gap $d\sim1\mu$ and a length $\iota=8$mm, subtending an angle of only $\sim10^{-4}$ rad.

The relevant energy levels for this experiment are shown in Figs. 6a and 6b and the experimental setup is sketched in Fig. 6c. Exciting the Cs atoms in the $5D_{5/2}$ level and sending these atoms across the mirror tunnel, we were able to observe the inhibition of the $5D_{5/2} \rightarrow 6P_{3/2}$ transition at $\lambda=3.49\mu$m. The natural lifetime of this transition – which has a branching ratio 1 – is $\tau=1/\Gamma_0=1.6$ μs. One laser system (laser 1) was used to prepare the atoms in the $5D_{5/2}$ level before they entered the cavity : the atoms were excited into the $7P_{3/2}$ state ($6S_{1/2} \rightarrow 7P_{3/2}$ transition at $\lambda_1=0.4555$ μm) from which $\sim10\%$ cascade down very rapidly (~150 ns) into the $5D_{5/2}$ state. This state has a hyperfine structure (F=1 to 6 where F is the total system angular momentum). Only levels F=4, 5 and 6 are prepared by our excitation scheme (see Fig. 6b). A second laser system (laser 2) was used to detect the atoms remaining in the $5D_{5/2}$ level after crossing the cavity (in a time $\sim20\mu$s). This laser was tuned across the transition from the $5D_{5/2}$ level into the 26f Rydberg state ($\lambda_2 \sim 6010$ Å). The Rydberg atoms were subsequently field-ionized. Counting the resulting electrons as a function of the frequency ν_2 of laser 2 provided a direct measure of the number of atoms remaining in the F=4, 5, 6 hyperfine levels of the $5D_{5/2}$ state after the cavity crossing (these numbers being proportional to the intensity of lines # 1, 2 and 3, respectively, in the energy diagram of Fig. 6b). We observed a large detection signal corres-ponding to peak # 3, providing a clear demonstration of spontaneous emission inhibition for atoms in the $5D_{5/2}$ F=6 state. The detected atoms had survived the gap crossing without decaying, remaining excited during about 13 natural lifetimes... The probability of such a long survival in free space would be quite negligible. The absence of peaks # 1 and 2 in the detection signal is easy to understand [31]. Among the atoms in the $5D_{5/2}$ level, only those which were in substates with electric dipoles radiating parallel to the surfaces had their lifetime lengthened to the extent that they could survive the cavity crossing. These substates correspond to the maximum possible value of the angular momentum component along the normal to the mirrors (F=6, $M_F=\pm6$ states). A mere inspection of Fig. 6b shows that only these substates radiate by pure σ-emission towards the $M_{F'}=\pm5$ substates of the $6P_{3/2}$ F'=5 final level. All the other levels – and in

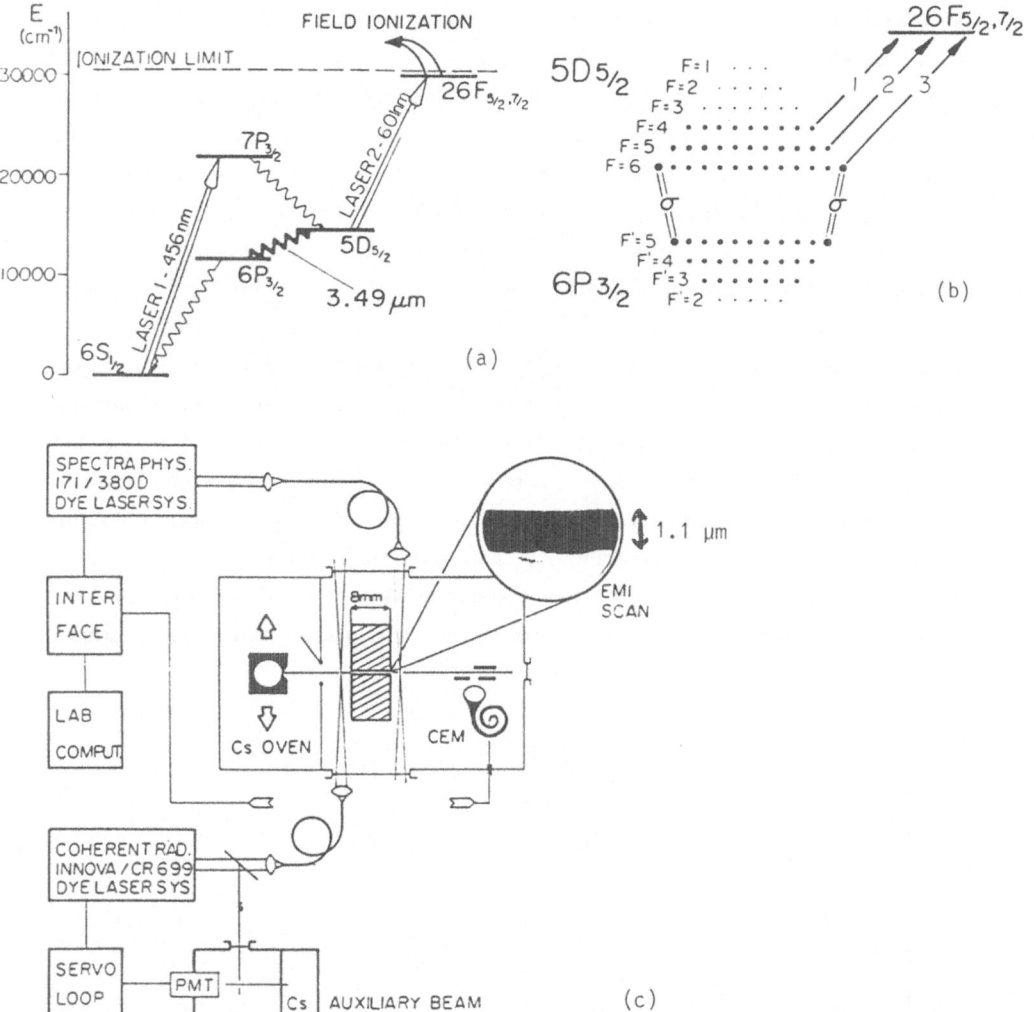

Figure 6 : a) Cesium energy level relevant to the optical spontaneous emission inhibition experiment. b) Close-up showing the hyperfine structure of the $5D_{5/2}$ and $6P_{3/2}$ states. The extreme $F = 6$, $M_F = \pm 6$ states of the $5D_{5/2}$ level are the only ones to radiate by pure σ-emission (to the $F' = 5$, $M_{F'} = \pm 5$ substates of the $6P_{3/2}$ level). Hence, only the hyperfine component # 3 remains in the absorption from level $5D_{5/2}$ after the atoms have crossed the cavity. c) Sketch of experimental setup. Laser 1 (at bottom) is locked on the $6S_{1/2} \rightarrow 7P_{3/2}$ resonance line of an auxiliary cesium beam. Laser 2 (at top) is scanned across the $5D_{5/2} \rightarrow 26F$ absorption line. Laser 1 excites the atoms upstream the mirror cavity; laser 2 detects them downstream. The Rydberg atoms are detected by field ionization and counting the resulting electrons with the channel electron multiplier (CEM). The insert shows an electron microscope picture of the mirror gap exit.

particular the F=5 and F=4 ones – can also decay via π-polarized ΔM=0 channels which are not cut-off in the cavity (see Fig.5). Hence, only peak # 3 appears in the absorption spectrum of the excited atoms after the cavity crossing. This provides evidence of the anisotropy of the spontaneous emission inhibition effect in the cavity. We have also demonstrated this anisotropy in a very dramatic way by studying the magnetic field dependence of the excited atom transmission signal. By applying a magnetic field with a component parallel to the mirrors, we were able to mix the $M_F = \pm 6$ long lived states with other $|M_F| \leq 5$ states able to radiate π-polarized photons in the cavity. This mixing thus resulted in a quenching of the excited state in the mirror structure and in a decrease of the excited state transmission. Figure 7 shows the result of this experiment : the excited atomic transmission through the tunnel is plotted as a function of the angle θ between the direction of an applied magnetic field $\vec{B}$ and the normal to the mirrors : the points are experimental and the full line curve is theoretical. The fair agreement between them demonstrates that the cavity Q.E.D. processes are well understood in this case. This experiment clearly shows

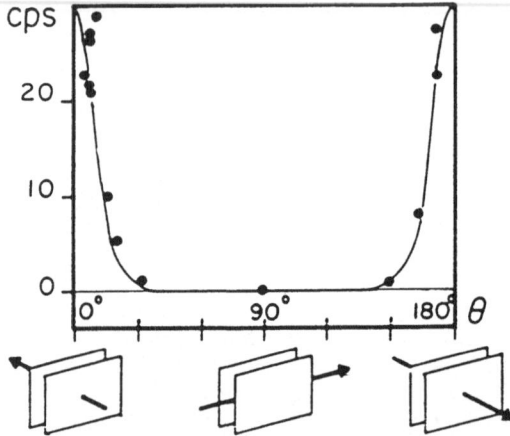

Figure 7 : Spontaneous emission inhibition experiment at optical frequency : Excited state transmission through the tunnel versus the angle θ between an applied magnetic field and the normal to the mirrors. For θ = 0 and θ = 180°, a large transmission is observed, demonstrating the spontaneous emission inhibition effect for F = 6, $M_F = \pm 6$ states. When θ is different from zero or 180°, the transverse field components mix the long lived $M_F = \pm 6$ states with shorter lived $|M_F| \leq 5$ ones and the excited states are quenched before emerging from the cavity. The points are experimental and the solid line curve is given by theory.

that it is possible to "manipulate" the vacuum fluctuations in a confined space and to radically alter the spontaneous emission processes induced by these fluctuations. The vacuum field, obviously isotropic in free space, becomes anisotropic in the cavity and this leads to an anisotropy of the spontaneous emission properties of the excited atomic state in this structure. It is tempting to call "antimaser" or "antilaser" such systems made of excited atoms crossing a cavity in which they are forbidden to emit a single photon... The reduction of the quantum noise in these devices bears some similitude with the effect known as "squeezing" [33]. In cavity Q.E.D., as in squeezing, one is able to decrease the effect on some carefully chosen physical observables of the vacuum field fluctuations.

The above discussion immediately raises the question of the possible use of this spontaneous emission inhibition effect for improving the ultimate resolution of spectroscopic experiments : since one can greatly enhance the natural lifetime of an excited atomic state, can one reduce in the same proportion the natural linewidth of a spectral line originating from this state ? The answer to this question is unfortunately not simple. In fact, the effect of the cavity is not only to modify the excited levels lifetime, but also to shift the atomic energies through the atom-metal surface Van der Waals interaction [34,35]. This latter effect cannot be separated from the lifetime enhancement phenomenon. The change of the radiative lifetime in the cavity is due to the atom coupling with the modes resonant with the atomic transition (dissipative atom-field interaction). The atomic system also interacts with the non-resonant modes (dispersive atom-field coupling). This coupling can be understood in terms of virtual photon exchanges and is responsible for Lamb-shift type of effects... The presence of the cavity changes these processes as it changes the resonant ones (virtual photons, as real ones, can only be emitted in modes compatible with the cavity geometry). The cavity-induced change in the atomic energy levels depends upon the atomic position in the cavity. It results in inhomogeneous level shifts which could be measured in spectroscopy experiments involving the atoms in the cavity. Furthermore, the derivative of these shifts with respect to the atomic position corresponds to the Van der Waals force pulling the atoms to the mirrors. In our 8mm long cavity, about 80% of the Cs atoms are deflected onto the walls by the force and the spontaneous emission inhibition effect is observed on the ~ 20% escaping collision with the mirrors. The lifetime enhancement ratio obtained in our experiment (~ 13) is thus a limit difficult to improve without loosing all the atoms to the walls of the cavity. The above discussion shows that while suppressing the atom coupling to the field resonant modes, we have in fact strongly coupled it to the cavity via the dispersive part of the atom-field interaction. Thus, if one tried to take advantage of the spontaneous emission inhibition to improve spectroscopic resolution beyond the natural width, one would be very quickly limited by the existence of inhomogeneous Van der Waals shifts. We would thus have replaced the "atom + vacuum field" system by the "atom + cavity" one. The experimental study of atomic energy level shifts in a cavity is certainly another interesting field of cavity Q.E.D. that we intend to study in the future.

The experiments in which I have been involved and which are reviewed here have been carried out in collaboration with colleagues and students in two research groups. In Paris, I have worked over a long period of time with C. Fabre, Philippe Goy, M. Gross and J.M. Raimond along with several graduate students (G. Vitrant, Y. Kaluzny, J. Liang, M. Brune, J. Hare) and foreign visitors (L. Moi). In Yale, my coworkers are E. Hinds, D. Meschede and again L. Moi with two graduate students, A. Anderson and W. Jhe... I was also very fortunate to be associated with Art Schawlow and his group at an early stage of my career. Many of the ideas and techniques I have developed later in the research reviewed in this paper owe much to the experience I gained in his laboratory. I am very glad to have the opportunity to acknowledge it in this article.

References

[1]   S. Haroche, J.M. Raimond: Adv. At. Mol. Phys. <u>20</u>, 347 (1985)
[2]   S. Haroche: in Proceedings of Symposium Alfred Kastler, Ann. Phys. (Paris) <u>10</u>, 811 (1985); P. Dobiasch, H. Walther: ibid. <u>10</u>, 825 (1985)
[3]   H.B. Casimir, D. Polder: Phys. Rev. <u>73</u>, 360 (1948)
[4]   E. Purcell: Phys. Rev. <u>69</u>, 681 (1946)
[5]   S. Haroche, P. Goy, J.M. Raimond, C. Fabre, M. Gross: Philos. Trans. R. Soc. London, Ser. A <u>307</u>, 659 (1982); S. Haroche: In New Trends in Atomic Physics, Les Houches Summer School, Session 38, ed. by G. Grynberg, R. Stora (North-Holland, Amsterdam 1984)

[6]   T. F. Gallagher: In Rydberg States of Atoms and Molecules, ed. by
        R. F. Stebbings, F. B. Dunning (Cambridge University Press, New
        York 1983); J. A. C. Gallas, G. Leuchs, H. Walther, H. Figger:
        Adv. At. Mol. Phys. 20, 413 (1985)

[7]   M. Gross, P. Goy, C. Fabre, S. Haroche, J. M. Raimond: Phys.
        Rev. Lett. 43, 343 (1979)

[8]   L. Moi, P. Goy, M. Gross, J. M. Raimond, C. Fabre, S. Haroche:
        Phys. Rev. A 27, 2043 (1983)

[9]   R. Bonifacio, R. Schwendimann, F. Haake: Phys. Rev. A 4, 302,
        854 (1971); F. Haake, H. King, G. Schröder, J. Haus, R. J.
        Glauber: Phys. Rev. A 20, 2047 (1979); F. Haake, J. Haus,
        R. J. Glauber: Phys. Rev. A 23, 3255 (1981)

[10]   M. Gross, S. Haroche: Phys. Rep. 93, 302 (1982)

[11]   R. H. Dicke: Phys. Rev. 93, 99 (1954)

[12]   J. M. Raimond, P. Goy, M. Gross, C. Fabre, S. Haroche: Phys.
        Rev. Lett. 49, 1924 (1982)

[13]   J. M. Raimond, P. Goy, M. Gross, C. Fabre, S. Haroche: Phys.
        Rev. Lett. 49, 117 (1982)

[14]   R. Bonifacio, L. A. Lugiato: Phys. Rev. A 11, 1507 (1975); ibid.
        12, 587 (1975)

[15]   Y. Kaluzny, P. Goy, M. Gross, J. M. Raimond, S. Haroche: Phys.
        Rev. Lett. 51, 1175 (1983)

[16]   P. Goy, J. M. Raimond, M. Gross, S. Haroche: Phys. Rev. Lett.
        50, 1903 (1983)

[17]   D. Meschede, H. Walther, G. Müller: Phys. Rev. Lett. 54, 551
        (1985)

[18]   P. Filipowicz, J. Javanainen, P. Meystre: Phys. Rev. A 34, 3077
        (1986)

[19]   P. Filipowicz, J. Javanainen, P. Meystre: Optics Commun. 58, 327
        (1986)

[20]   G. Rempe, H. Walther, N. Klein: Phys. Rev. Lett. 58,
        353 (1987)

[21]   P. P. Sorokin, N. Braslau: IBM J. Res. Dev. 8, 177 (1964)

[22]   A. M. Prokhorov: Science 149, 828 (1965); V. S. Letokhov: Pis'ma
        Zh. Eksp. Teor. Fiz. 7, 284 (1968) [JETP Lett. 7, 221 (1968)]

[23]   L.M. Narducci, W.W. Eidson, P. Furcinetti, D.C. Eteson: Phys. Rev. A 16, 1665 (1977); R.L. Carman: Phys. Rev. A 12, 2048 (1975); H.P. Yuen: Phys. Rev. A 13, 226 (1976); N. Nayak, B.K. Mohanty: Phys. Rev. A 19, 1204 (1979)

[24]   B. Nikolaus, D.Z. Zhang, P.E. Toscheck: Phys. Rev. Lett. 47, 171 (1981)

[25]   M. Brune, J.M. Raimond, S. Haroche: Phys. Rev. A 35, 154 (1987)

[26]   L. Davidovich, J.M. Raimond, M. Brune , S. Haroche: to be published

[27]   D. Kleppner: Phys. Rev. Lett. 47, 233 (1981)

[28]   R.G. Hulet, E.S. Hilfer, D. Kleppner: Phys. Rev. Lett. 55, 2137 (1985)

[29]   R.G. Hulet, D. Kleppner: Phys. Rev. Lett. 51, 1430 (1983)

[30]   G. Gabrielse, H. Dehmelt: Phys. Rev. Lett. 55, 67 (1985)

[31]   W. Jhe, A. Anderson, E.A. Hinds, D. Meschede, L. Moi, S. Haroche: Phys. Rev. Lett. 58, 666 (1987)

[32]   A. Anderson, S. Haroche, E.A. Hinds, W. Jhe, D. Meschede, L. Moi: Phys. Rev. A 34, 3513 (1986)

[33]   R.E. Slusher, L.W. Hollberg, B. Yurke, J.C. Mertz, J.F. Valley: Phys. Rev. Lett. 55, 2409 (1985); L.A. Wu, H.J. Kimble, J.L. Hall, H. Wu: Phys. Rev. Lett. 57, 2520 (1986)

[34]   J.E. Lennard-Jones, Trans. Faraday Soc. 28, 336 (1932)

[35]   C. Lütken, M. Raindal, Phys. Rev. A 31, 2082 (1985); Phys. Scr. 28, 209 (1983)

Laser Glass: An Engineered Material

S.E. Stokowski

Lawrence Livermore National Laboratory, Livermore, CA 94550, USA

Foreword

In 1958 SCHAWLOW and TOWNES [1] proposed that an optical maser could be made with the right combination of spectroscopic properties of an excited atom, ion, or molecule in a gas or solid, an optical pumping source, and a resonant cavity. The key to their proposal was their choice of a Fabry-Perot cavity as the resonator. This proposal stimulated several efforts to make an optical maser. In 1960 MAIMAN [2] demonstrated coherent emission from a ruby crystal, and the laser era had begun.

In the first years of its existence the laser was a curiosity; many people said that it was an interesting device looking for an application. Looking back from the perspective of a quarter century it is easy to deride such a comment. However, that is how a completely new and unexpected invention is usually received. Until people have a chance to store the unique characteristics of a new invention in their memories, they have difficulty in thinking of applications.

But applications for the laser have come in droves, ranging from the commonplace, such as supermarket scanners and surveying instruments, to the unusual, such as nuclear fusion drivers; from medical and surgical instruments for healing to weapons, such as missile designators and destroyers; from communications to compact disk players; from toys to scientific instruments used on the frontiers of science. We now take the laser for granted. In a little more than a quarter century the result of Schawlow's and Townes's proposal has made a substantial impact on our lives. Future laser applications are likely to make present ones pale in comparison.

Spectroscopy has played a key role in laser development. Schawlow, a spectroscopist by training and inclination, understood this from the beginning and has always been at the forefront of laser spectroscopy. My own interest in spectroscopy stems from Schawlow's excitement and enthusiasm for the new world of spectroscopy that opened up with the advent of the laser. When I was a serious graduate student, Schawlow, through his infectious joy and his ability to have fun with science, taught me that we probably do our best work when we approach nature with the curiosity and openness of a child. If we can work in this way, we are not burdened with prejudices that are obstacles to new insights.

This paper describes how glass has been made useful in lasers. My main theme is that laser glass has been "engineered" to meet a variety of applications. Engineering is defined in a dictionary as "a science by which the properties of matter and sources of energy are made useful to man in structures, machines, and products." In our search for better engineering materials, such as laser glass, we must understand materials science. Science and engineering are necessarily coupled; they are, in a sense, two sides of the same coin. When this coupling is effective, advances occur rapidly.

I acknowledge the many scientists who have contributed tremendously to laser glass development and with whom I have had the privilege of collaborating over the years, in particular, C. F. Cline, L. M. Cook, K.-H. Mader, T. Izumitani, S. D. Jacobs, H. E. Meisner, J. D. Myers, C. F. Rapp, E. Snitzer, H. Toratani, and M. J. Weber.

1. Introduction

The first laser material was a crystal, ruby. MAIMAN [2] used it to generate coherent light for the first time on May 15, 1960. Soon after that JAVAN et al. [3] demonstrated lasing action in a helium–neon gas mixture. Having realized that neodymium in a glass should produce laser light if the emission cross section was not too low because of the inhomogeneous line broadening (that is, broadening arising from inhomogeneity of the host glass), in 1961 SNITZER [4] demonstrated a neodymium-doped glass (Nd:glass) laser using barium crown glass. After this, other rare earth–doped glasses were made into lasers; dopants included Yb, Ho, Er, Tm, and Tb [5–9]. In 1963 YOUNG [10] made a cw laser from Nd:glass. By 1963 neodymium-doped yttrium aluminum garnet (Nd:YAG) had come into the marketplace, and rare earth-doped laser glass did not seem to be a competitor to crystalline YAG. Laser glass has lower thermal conductivity than YAG, which prevents it from being used in a cw or high-repetition rate mode. However, it was clear that laser glass has properties that complement those of laser crystals.

In their 1968 review, SNITZER and YOUNG [11] pointed out the principal advantages of laser glass: it can be made in a variety of sizes and shapes, with excellent homogeneity and low birefringence; it is less expensive than crystals; and it has good coupling to broadband pump sources, such as flashlamps, and is thus capable of high stored energy density.

Glass is described as an amorphous solid or as a frozen liquid. A crystal, on the other hand, has a unit structure that repeats itself over a long range. In a crystal this translational symmetry means that the laser-active ions reside in sites that are either identical throughout the crystal or can be reduced to a few equivalent sites. Thus transitions in crystals have narrow line widths. Glass by its nature has a variety of sites that differ from each other in the number and position of the surrounding anions. Thus transitions in glass are inhomogeneous and relatively broad. This basic structural difference between glasses and crystals leads to an important consequence: emission cross sections for ions in crystals are considerably higher than those for ions in glasses. The lasing threshold depends on the emission cross section, so why would laser glasses be of any interest?

Early researchers realized that emission cross section is not the whole story, because the laser gain coefficient g is equal to the product of the cross section σ and the excited state inversion density N^*:

$$g = \sigma N^* . \tag{1}$$

The broader emission lines and generally lower cross sections of glass are compensated by its broader absorption bands, which allow glass to couple more efficiently to broadband pump sources than do crystals.

Laser glass can store large amounts of energy because it can be produced in large sizes and because of its low cross section. Low cross sections are helpful in large laser systems because the obtainable gain is limited by amplified spontaneous emission and parasitic oscillations [12–14]. The rule of thumb is that the medium becomes harder and harder to pump when the product of the gain coefficient and the largest dimension of the laser material rises above 3 or 4. Cross sections in Nd:YAG crystals are about 10 times greater than in glasses: thus laser glasses can store energy at a density about 5 to 10 times that in Nd:YAG crystals.

Laser glass has an additional advantage over laser crystals in lasers operating in the short-pulse (picosecond) regime. The broad fluorescence line width of glass allows for shorter, transform-limited pulses. This property was first demonstrated in 1965 by MOCKER and COLLINS [15].

Glass has always had a fascination for materials scientists because of its variety of compositions, which give them a certain amount of control over its properties. In a sense you have to take what you get with crystals, but glass can be "engineered."

Making glass has been likened to making soup. In fact, early glass melters treated glass melts as a soup and took the empirical approach to discovering new glasses or better methods for making glass. In creating a soup, you decide on the base—for example, chicken, beef, or pork. Then you add modifiers (vegetables)—corn, carrots, onions, etc. Some people like their soup thick, so you may add thickeners, such as corn starch. Spices added in small amounts flavor the soup, and make the difference between a culinary delight and a bland, unappetizing mix.

The analogy between soup-making and glass-making is as follows: the base, or glass former, is usually the major constituent; SiO_2, P_2O_5, B_2O_3, and BeF_2 are examples. The glass formers form the backbone structure of the glass, which is based on SiO_4, PO_4, or BeF_4 tetrahedral units or BO_3 planar, triangular units. The modifiers do not form glasses by themselves, but open up the glass structure by breaking bonds. Examples of modifiers are the alkali metals Li, Na, and K, and the alkaline earths Mg, Ca, and Ba. Thickeners are the marginal glass formers or intermediates, which are incorporated into the backbone structure of the glass and which change the viscosity and improve durability of the glass. Examples of these are Al_2O_3, AlF_3, TiO_2, and TeO_2. Mixes of intermediates commonly form glasses. The spices are the minor additions that can make the difference between a usable commercial glass and one that never gets out of the research laboratory. Such additives as Sb_2O_3, Nb_2O_5, F_2, and As_2O_3 keep the glass free of bubbles, lower the water (OH^-) content, or prevent radiation-caused absorption (solarization).

In keeping with the soup analogy, the best soups, except for consommés, usually have the greatest number of ingredients. Practical glasses have a great number of components, primarily because of what we call the "confusion factor." The confusion factor solves the greatest problem in glass manufacture, crystallization. The more components in the glass melt, the more difficult it is to crystallize, as it is less likely that a critical nucleus will form and grow because components not incorporated into the crystals must be rejected.

SUN [16] categorized oxide compounds as glass formers, intermediates, or modifiers on the basis of their calculated bond energies, which are shown in Fig. 1. Glass formers generally have bond energies above 350 kJ/mole; intermediates have bond energies of 250 to 350 kJ/mole; and modifiers have bond energies below 250 kJ/mole. Neodymium oxide has a bond energy of 213 kJ/mole for 8-fold coordination, which places it near the high end of the modifier category.

The first laser glasses were silicates because they were the most common and the easiest to make. Initially, small-diameter (12 mm) laser glass rods were made and tested by many laboratories; rods of this diameter and 150 mm long were capable of 70-J output per pulse, with pulse durations of 0.6 ms. As the glass melters learned how to melt laser glass with greater

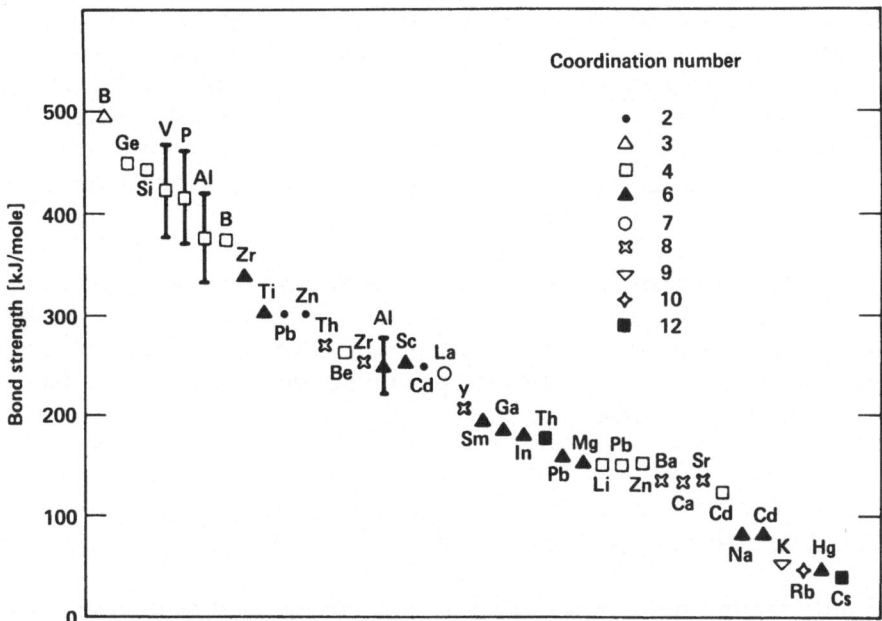

Fig. 1. Energies of cation-oxygen bonds calculated by SUN [16] for the coordination numbers indicated; in order of decreasing bond energy (abscissa arbitrary)

homogeneity and less stress, laser glass rods increased in diameter and length. By 1966, the energy output reached 5 kJ in a 3-ms-long pulse, from 30-mm-diam, 950-mm-long rods [17].

To increase coupling efficiency (the efficiency of coupling between the emission spectrum of the pump, typically a flashlamp, and the absorption spectrum of the Nd:glass), researchers from the beginning considered the use of sensitizers: that is, of an ion that absorbs flashlamp energy not absorbed by a lasing ion and transfers it to the lasing ion. Many ions have been tried for sensitizing Nd^{3+}, including Ce^{3+}, Mn^{2+}, UO_2^{2+}, Tb^{3+}, Bi^{3+}, Eu^{3+}, and Cr^{3+} [18–31]. Only Ce^{3+} proved to be of any practical use, and even Ce^{3+} is of marginal benefit because its absorption is in the ultraviolet.

Laser glasses based on glass formers other than silica were investigated: borates and germanates were first reported in 1963 [32], fluoroberyllates in 1966 [33], fluorophosphates in 1966 [34], and phosphates in 1967 [35]. The interest in laser glass studies increased significantly after BASOV and KROKHIN [36] initiated work on laser-produced plasmas, which was followed by suggestions that lasers could be used to compress and heat deuterium–tritium fuel for generating fusion energy [37,38]. In 1972 BASOV et al. [39] demonstrated neutron generation from a laser-irradiated target. The high energies and powers required for these lasers meant that larger laser glass components were needed [40]. The challenge of manufacturing large pieces of optically homogeneous laser glass was taken up by Corning, Hoya, Kigre, Owens-Illinois, and Schott. As larger and larger laser drivers for fusion were built, the melting technology developed by these companies kept pace with the requirements. By the late 1970s individual laser glass components had reached volumes of 7 liters [41,42].

In the last ten years, improved laser glasses have emerged from many compositional studies; these glasses are more efficient, have better energy storage, and are more robust. In this paper I describe some of the considerations that went into laser glass development. I hope I can give you a flavor of the research in this area.

2. What are the Important Laser Glass Properties?

Soon after the invention of the Nd:glass laser, researchers were measuring the variation in neodymium spectroscopic properties with glass composition [43,44]. However, it became clear very quickly that given the high number of possible glass compositions, one had better concentrate on those that had the greatest chance of providing the desired properties. But what properties made for a good laser glass? That question has several answers depending on the application.

In the early days of lasers the performance characteristics desired were high gain, high stored energy, and low loss. Gain and stored energy depend on stimulated emission cross section, fluorescence lifetime, and coupling efficiency.

In small oscillator systems, gain is the important parameter; therefore, according to (1), high stimulated emission cross section is desirable. On the other hand, in Q-switched systems both gain and stored energy are important. For large apertures the gain coefficient is limited by amplified spontaneous emission and perhaps by parasitic oscillations. Lower cross sections are favored to increase stored energy, but very low cross sections are undesirable because of the high fluences required to extract the stored energy and because of the need for very low material absorption. Energy extraction from a laser material becomes efficient when the laser fluence in the material reaches at least twice the saturation fluence

$$ E_s = \frac{h\nu}{(1 + g_u/g_l)\,\sigma}, \tag{2} $$

where g_u and g_l are the degeneracies of the upper and lower energy levels, respectively. High fluence is then needed for materials of very low cross section, in which case the damage-fluence threshold of the material may be exceeded. Thus the optimum cross section depends on the specific system.

In all laser systems the coupling between the pump and the active lasing ion must be maximized for best efficiency. Both a spectral and temporal match between the pump and material must be achieved. The most common pump source is the xenon flashlamp, which is an efficient (>80%), high-brightness source [45–47]. Xenon is primarily a broadband emitter with a blackbody temperature of about 10 000 K. The flashlamp pulse duration for most efficient lamp operation is 200 to 600 μs. The fluorescence lifetime of the active ion should be longer than the pump pulse. The lifetime of the $^4F_{3/2}$ state of Nd^{3+} varies between 10 and 1000 μs, depending on the glass composition and on the concentrations of neodymium and of quenching impurities.

Long fluorescence lifetimes are usually associated with low cross sections, through the relation

$$\sigma\tau_R = \frac{\lambda_p^4\beta}{8\pi cn^2\,\Delta\lambda_{eff}}, \tag{3}$$

where τ_R is the radiative lifetime, λ_p is the wavelength at the fluorescence peak, β is the branching ratio (the ratio of the fluorescence intensity for an individual transition to the total fluorescence intensity), n is the refractive index, and $\Delta\lambda_{eff}$ is the effective line width (determined by integrating the fluorescence intensity over wavelength and dividing by the intensity at the peak). For different neodymium glasses, λ_p and β do not vary substantially; therefore, the product $n^2\sigma\,\Delta\lambda_{eff}$ is linearly related to τ_R^{-1}, as can be seen in Fig. 2.

For maximum coupling to the flashlamp pump we want a material with long fluorescence lifetime and high absorption. For high absorption we need high neodymium concentrations and thick materials. However, high neodymium concentrations decrease the fluorescence lifetime and make the laser material harder to pump. The thickness of the laser material and the neodymium concentration are then determined by a competition between maximizing absorption and fluorescence lifetime. As an example of this phenomenon, Fig. 3 shows the calculated gain coefficient vs neodymium concentration and lamp pulse width [48]. For low neodymium concentrations the lifetime is long, but absorption is low, thus requiring thick laser glass; at high neodymium concentrations absorption is good, but the lifetime is too short. For short pump pulses, the flashlamp is not efficient because of lamp opacity, or reabsorption; for long pulses, too much energy drains out of the excited state during the pump pulse.

53

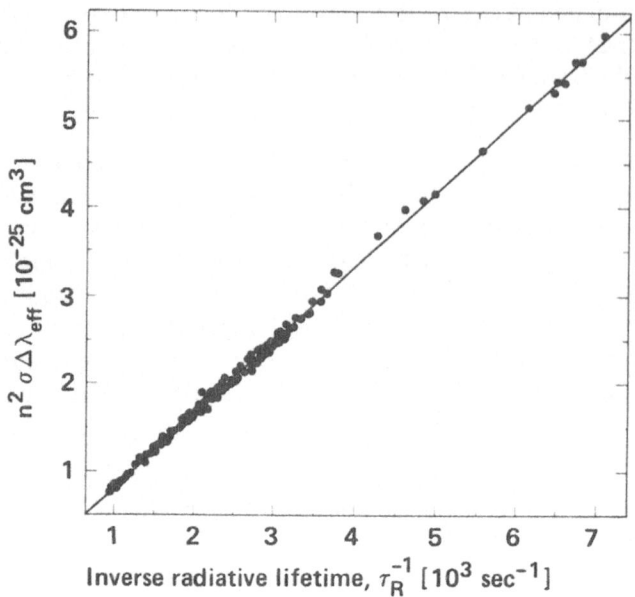

Fig. 2. The product $n^2\sigma\,\Delta\lambda_{eff}$ vs the inverse of the calculated radiative fluorescence lifetime, τ_R^{-1}, for 300 glasses

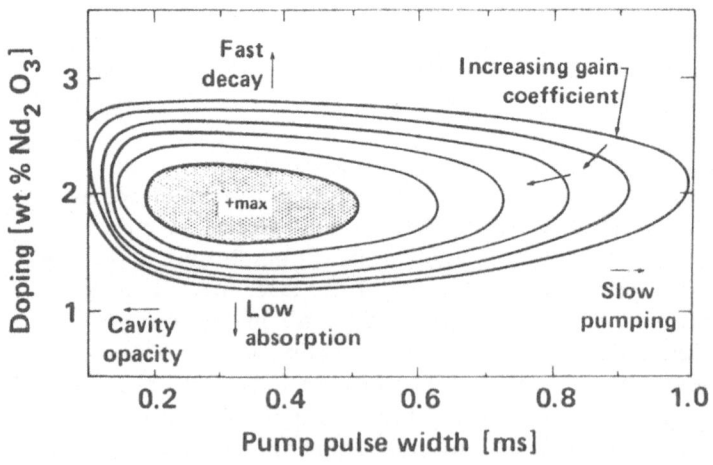

Fig. 3. Calculated gain coefficient contours for various neodymium concentrations and pump pulse durations

Soon after the advent of the laser, scientists found that its high fluence had an undesirable side effect, laser-induced damage of the laser material or of the optical material through which the beam passed [49]. Initially, filamentary "angel hair" tracks were seen in laser materials; these were later explained as due to self-focusing of the laser beam as a result of intensity-dependent refractive index changes [50]. This nonlinearity in the refractive index is critically important in designing laser oscillators and amplifiers. Even at laser fluences below those at which self-focusing occurs, refractive index nonlinearities lead to exponential growth of a small-scale irregularities in the beam profile. Thus diffraction fringes from small-particle obscurations in the beam grow as the laser beam passes through an optical material, ultimately leading to very large variations in the beam intensity. The fluctuations grow as $\exp(B)$, where [51,52]

$$B = \frac{2\pi}{\lambda} \int \gamma I \, dz , \tag{4}$$

where λ is the wavelength, γ is the nonlinear refractive index coefficient, and I is the optical intensity. The coefficient γ is defined by $\Delta n = \gamma I$, where Δn is the change in refractive index. Alternatively we have $\Delta n = n_2 \langle E^2 \rangle$, where n_2 is the nonlinear refractive index and E is the amplitude of the optical electric field; thus n_2 and γ are related by

$$n_2 = \frac{cn}{40\pi} \gamma , \tag{5}$$

where n_2 is in esu and γ is in m^2/W.

Nonlinear refractive indices are very difficult to measure because high-power lasers are required and the measurement samples must be of good optical quality [53–58]. Fortunately BOLING, GLASS, and OWYOUNG [59] formulated an empirical relation for predicting nonlinear refractive indices in optical solids from the linear index and the dispersion, which are relatively easy to measure. The dispersion is characterized by the Abbe number v, defined by

$$v = \frac{n_D - 1}{n_F - n_C} , \tag{6}$$

where n_F, n_D, and n_C are the linear refractive indices at 486.1, 589.3, and 656.3 nm, respectively. In terms of n_D and v, we have [59]

$$\gamma = \frac{K \left(n_D - 1\right)\left(n_D^2 + 2\right)^2}{n_D v \left[1.52 + \left(n_D^2 + 2\right)\left(n_D + 1\right) v/6n_D\right]^{1/2}} , \tag{7}$$

55

where $K = 2.8 \times 10^{-10}$ m²/W is obtained from fitting experimental values for γ at 1064 nm. Equations (7) and (5) are extensively used to predict γ and n_2, but (7) must be used with caution for high values of n_D, for which it overestimates γ.

Figure 4 is frequently used to illustrate the dependence of γ on n_D and Abbe number v; a similar figure was first published by GLASS [60] in 1975. Figure 4 also shows that fluoride glasses have much lower linear and nonlinear indices than oxide glasses, so that fluoride laser glasses are desirable for laser systems that deliver very high fluences.

Thus we see that emission cross section, the absorption spectrum and strength, the fluorescence lifetime as a function of neodymium concentration, and the nonlinear index of refraction are the important intrinsic properties for evaluating the utility of a glass for laser applications. Other properties are generally less important, but may be critically important in some applications. These properties include thermal expansion α and the change of refractive index with temperature $\delta n/\delta T$, which together affect the thermal change of optical path length s through the relation

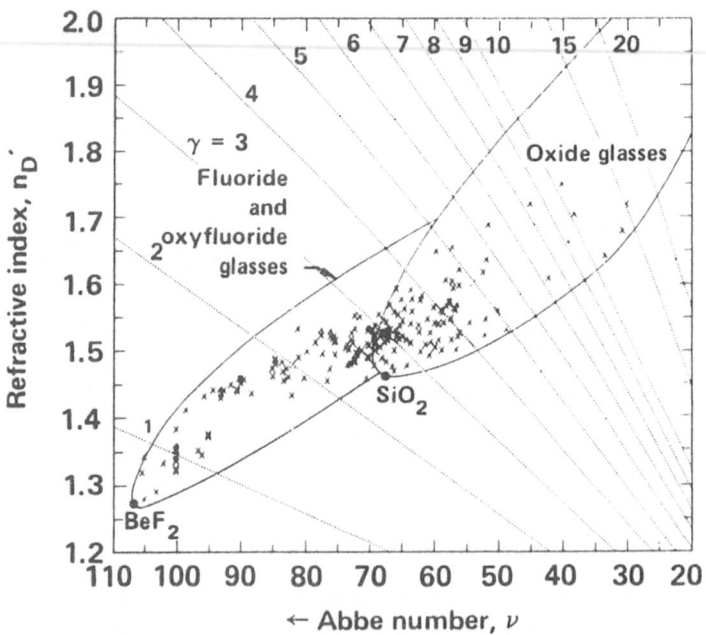

Fig. 4. Refractive index n_D and Abbe number v of optical glasses. Dotted lines of constant γ (intensity-dependent index change in 10^{-20} m²/W) calculated from (7)

$$\frac{\delta s}{\delta T} = \frac{\delta n}{\delta T} + \alpha(n - 1). \qquad (8)$$

Athermal glasses (those for which $\delta s/\delta T = 0$) are desired for applications in which large thermal gradients exist, particularly in laser oscillators or in amplifiers with high-average-power output.

The stress-optic coefficients of glasses determine the amount of birefringence induced by thermal stresses or by residual stresses from the melting process. Thermomechanical properties are important for high-average-power operation. The power available from a material is proportional to the thermomechanical figure of merit [61]

$$R_T = \frac{\kappa(1 - \nu)S_T}{\alpha E}, \qquad (9)$$

where κ is the thermal conductivity, ν is now Poisson's ratio, S_T is the tensile strength, and E is Young's modulus.

Some characteristics of laser glass are extrinsic, in that they depend on foreign or undesired material in the glass. These characteristics are susceptibility to laser-induced damage, absorption at the laser frequency, and solarization. Much of the utility of a laser glass depends on finding ways to avoid these effects.

Laser-induced damage in laser glasses is in most cases due to the presence of metallic inclusions [62]. Most glasses are melted and refined in platinum crucibles, so platinum particles of sizes 1 to 100 μm are commonly seen as inclusions. Laser fluences of about 2 J cm^{-2} can heat the surface of such small platinum particles above their vaporization temperature [63]. The pressure of the resulting platinum vapor is greater than the bulk strength of the glass, which therefore fractures. Glass manufacturers attempt to prevent the introduction of metallic platinum into the melt, which occurs as a result of vapor transport or of mechanical damage to the crucible. If it is present, platinum can be dissolved into the melt and its susceptibility to damage thus eliminated by refining the glass in an oxidizing environment. On the other hand, a high concentration of ionic platinum in the glass will absorb pump photons in the blue and ultraviolet and thereby lower pumping efficiency.

Improving laser glass thus depends on an understanding of the variation of spectroscopic, optical, and thermomechanical properties with composition. Because of the large number of possible glass compositions, this approach ensures job security for those who search for the Holy Grail of laser glass. I now describe these properties of Nd:glasses and give some background on the physical mechanisms for the observed variations.

3. What Determines Laser Glass Properties?

3.1 Spectral Inhomogeneity

Glass has no long-range spatial order. Thus, each neodymium ion in glass sees a different environment, resulting in substantial spectral inhomogeneity.

Within a given glass, what are the ranges of values of energy levels, transition strengths, electron-phonon coupling coefficients, and homogeneous line widths? Fluorescence line-narrowing (FLN) experiments [64,65], in which a spectrally narrow region can be explored by excitation with a narrow-band laser, have answered these questions for a variety of glasses. Typically the homogeneous width of the $^4F_{3/2} \rightarrow {}^4I_{11/2}$ transition in Nd^{3+} is 20 to 30 cm^{-1} [66–68], whereas the inhomogeneous width is about 250 cm^{-1}.

One indication of the spectroscopic inhomogeneity of a glass is in the deviation of the fluorescence decay from a pure exponential. The different decay rates of individual ions result in a nonexponential decay, even in the absence of energy migration or quenching. Typical e-folding times for fluorescence decays of Nd^{3+} in different glasses are given in Table 1. Of particular note is the substantial nonexponential nature of silicate glass decays compared with the nearly exponential decays of phosphates, which thus appear to be more homogeneous than silicates. This difference was confirmed, using FLN techniques, by BRECHER et al. [64,65], who found decay

Table 1. $Nd^{3+4}F_{3/2}$ fluorescence decay e-folding times in various glasses for low neodymium concentration, where neodymium self-quenching and Nd-Nd energy migration are absent. Percentage changes in these times indicate the inhomogeneity of the transition strengths

Glass	Glass type	e-folding times (μs)[a]			% change from τ_1 to τ_3
		1st	2nd	3rd	
LG-670	Silicate	370	385	395	7
LG-660	Silicate	420	560	580	38
LG-650	Silicate	690	860	1035	50
LG-750	Phosphate	390	390	385	1
UP-16	Phosphate	343	360	368	7
P-101	Phosphate	376	399	430	14
LG-810	Fluorophosphate	540	576	583	8
B-101	Fluoroberyllate	666	685	702	5
K-1261	Tellurite	175	176	181	3

[a] The nth e-folding time is the time taken for the fluorescence intensity to decay from e^{-n+1} to e^{-n} of its initial intensity.

rates in a silicate glass between 2400 and 3300 s^{-1} (a 38% variation) depending on the excitation wavelength within the $^4I_{9/2} \rightarrow \, ^2P_{1/2}$ transition. For a phosphate glass the rates vary from 2600 to 3050 s^{-1} (a 17% variation).

Spectral inhomogeneities can significantly decrease the extractable energy of Nd:glass amplifiers and oscillators. In a homogeneous system the saturation fluence E_s is given by (2). In an inhomogeneous system the satura-

Table 2. Measured saturation fluences of neodymium-doped laser glasses [69,70]

Glass	Supplier	Cross section (pm^2) 1053 nm	1064 nm	Saturation fluence (J/cm^2)[a] ±0.3 J/cm^2	$h\nu/\sigma$ (J/cm^2)
Phosphates					
P-101	Hoya	3.0	—	5.3	6.2
LHG-8	Hoya	4.0	—	4.3	4.6
Q-88	Kigre	4.0	—	4.1	4.6
Q-94	Kigre	3.8	—	4.2	4.9
Q-98	Kigre	4.1	—	4.0	4.5
LG-750	Schott	4.0	—	4.4	4.6
EV-4	Owens-Illinois	4.2	—	3.8	4.9
Silicates					
ED-2	Owens-Illinois	—	2.6	4.5	7.1
LSG-91H	Hoya	—	2.3	4.9	8.1
LG-56	Schott	0.84	0.8	9.0[b]	23
OI-H9	Owens-Illinois	—	1.45	6.7	12.8
Fluoro-phosphates					
LG-810	Schott	2.5	—	5.1	7.4
LG-800	Schott	2.7	—	4.0	6.9
LHG-10	Hoya	2.6	—	4.9	7.1
E-309	Owens-Illinois	2.4	—	5.0	7.7
Fluoro-beryllate					
B-101	Corning	2.6	—	5.5	7.1

[a] For 20-ns pulse length and an output fluence of 5 J/cm^2; saturation fluence for 1-ns pulses is about 5% lower.
[b] 8.2 ± 1.0 J/cm^2 at $E_{out} = 10$ J/cm^2, 1053 nm.

tion fluence will vary with the input energy because the excited ions with high cross sections will give up their energy more rapidly than those with low cross sections. Spectral hole-burning can also occur. MARTIN and MILAM [69] and YAREMA and MILAM [70] measured E_s for phosphate, silicate, fluorophosphate, and fluoroberyllate glasses. Their values, listed in Table 2, show that in phosphates the E_s values are close to $h\nu/\sigma$. However, in silicates E_s is about 50% lower than $h\nu/\sigma$, indicating that many ions are not in resonance with the extracting laser pulse.

3.2 Emission Cross Section and Absorption Spectra

The emission cross section, the absorption spectrum, and the absorption strength are needed to predict the performance of a laser glass. The absorption strength can be determined once the neodymium ion density is known. However, the emission cross sections for the $^4F_{3/2} \rightarrow {}^4I_{11/2}$ and $^4F_{3/2} \rightarrow {}^4I_{13/2}$ transitions cannot be determined from the absorption transitions because the terminal states are 2000 cm^{-1} and 4000 cm^{-1}, respectively, above the ground state. These states have very low populations and very weak, usually unobservable, absorptions.

Several spectroscopic means of estimating the emission cross section have been used; the most convenient and consistent way is through the application of the JUDD-OFELT [71–73] treatment of spectral intensities of the rare earths. This theory was first applied by KRUPKE [74] to calculating emission cross sections in 1974. In the Judd-Ofelt treatment, the line strength S of a transition between two J states is given by

$$S(aJ:bJ') = \sum_{t=2,4,6} \Omega_t |\langle aJ| |U^{(t)}| |bJ'\rangle|^2 \qquad (10)$$

where a and b denote other quantum numbers specifying the eigenstates. The reduced matrix elements of the unit tensor operator $U^{(t)}$ are calculated in an intermediate coupling approximation; KRUPKE [74] gives their values.

The Judd-Ofelt parameters vary from glass to glass and are determined from a least-squares fit of the integrated per-ion absorption band intensities. Once the Judd-Ofelt parameters are known, the peak emission cross section can be calculated from

$$\sigma(\lambda_p) = \frac{8\pi^3 e^2}{27hc(2J+1)} \frac{\lambda_p}{\Delta\lambda_{eff}} \frac{(n^2+2)^2}{n} S(aJ:bJ'), \qquad (11)$$

where λ_p is the peak fluorescence wavelength, $\Delta\lambda_{eff}$ is the effective line width, given by

$$\Delta \lambda_{eff} = \int \frac{I(\lambda) \, d\lambda}{I(\lambda_p)},$$

and n is the refractive index. The cross section thus calculated is accurate to $\pm 10\%$ if the Nd^{3+} concentration is accurately measured.

The Judd-Ofelt formalism predicts the strengths of individual absorption bands to better than 10%, except for a systematic 10 to 20% overestimation of the $^4I_{9/2} \rightarrow {}^4F_{3/2}$ transition strength. Measurements [75,76] of the branching ratios of the $^4F_{3/2} \rightarrow {}^4I_J$ transitions (J = 9/2, 11/2, 13/2, 15/2) have shown that they are very close to those predicted with the Judd-Ofelt formalism. Therefore, the calculated cross sections of all the $^4F_{3/2} \rightarrow {}^4I_J$ transitions may be too high by 10 to 20%.

Evidence for this overestimation comes from comparisons of the calculated radiative lifetime and the measured fluorescence lifetime at low neodymium concentrations, particularly for phosphate glasses. I find that the average calculated lifetime is about 20% lower than the measured lifetime in glasses. To maintain consistency in comparing different glasses, I have used the calculated Judd-Ofelt parameters in estimating emission cross sections, keeping in mind that some adjustment may be necessary if an improved calculation of rare earth spectral intensities becomes available.

3.3 Fluorescence Lifetimes

For good efficiency, fluorescence lifetimes of Nd:glasses should be longer than the pump pulse. Flashlamps driven at highest efficiency have pulse durations of 100 to 600 μs. In glass, Nd^{3+} has fluorescence lifetimes in the same range. POWELL, MURRAY, and JANCAITIS [77] found that the efficiency of pumping neodymium with flashlamps increases approximately as the square root of the lifetime if the fluorescence lifetime is about the same as the pump pulse duration.

The maximum fluorescence lifetime is set by the radiative lifetime. Non-radiative processes, if present, will shorten, or quench, this lifetime. These fluorescence-quenching processes, shown in Fig. 5, include (a) Nd–Nd self-quenching, (b) quenching induced by impurities (e.g., transition-metal ions), (c) nonradiative multiphonon relaxation, and (d) multiphonon energy transfer processes involving glass matrix vibrations or molecular impurities such as water.

Radiative lifetimes are related to the emission cross section through (3). Longer radiative lifetimes therefore generally mean lower cross sections (Fig. 2). Narrower effective line widths can offset this trend. The glass refrac-

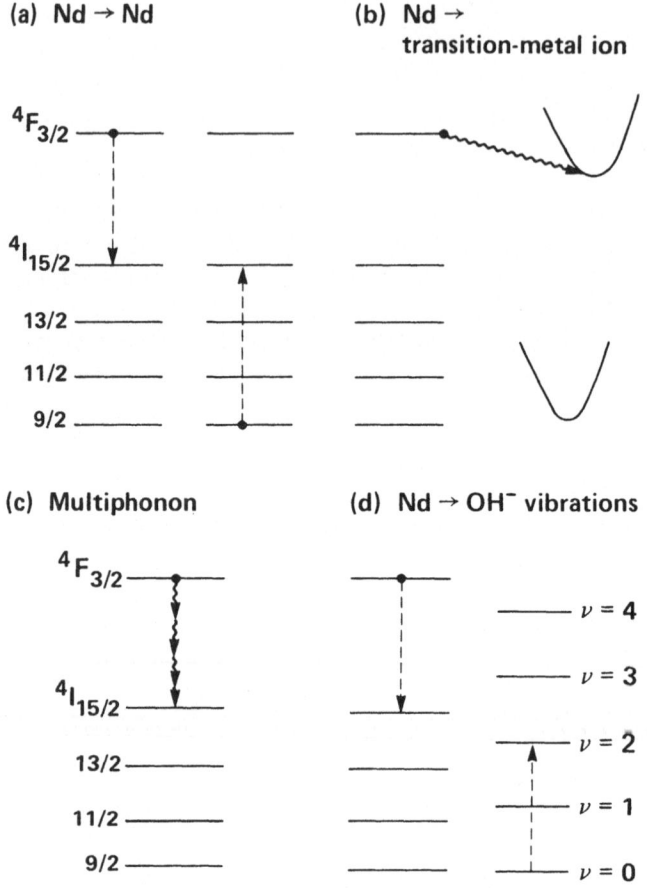

(a) Nd → Nd

(b) Nd → transition-metal ion

(c) Multiphonon

(d) Nd → OH⁻ vibrations

Fig. 5. **Fluorescence quenching processes for Nd^{3+}: (a) Nd–Nd self-quenching, (b) quenching induced by impurities (e.g., transition-metal ions), (c) nonradiative multiphonon relaxation, and (d) energy transfer to OH⁻ (water) vibrational energy**

tive index also plays a role, in that fluoride glasses generally have longer lifetimes because of the n^{-2} dependence in (3).

For Nd^{3+} the large energy gap between the $^4F_{3/2}$ and $^4I_{15/2}$ states makes the multiphonon relaxation rate between these states generally small compared with the radiative rate. LAYNE, LOWDERMILK, and WEBER [78,79] measured multiphonon rates W_{NR} for rare earths in a variety of glasses and found that they can be empirically expressed as

$$W_{NR} = W_0(T) \exp(-\alpha \, \Delta E), \tag{12}$$

where $W_0(T)$ depends on the host glass and temperature, α depends on the host glass, and ΔE is the energy gap between neighboring states. The rates are proportional to the highest-energy vibrational frequency in the different base glass types. For the $^4F_{3/2}$ state of Nd^{3+}, the results [78,79] are summarized in Table 3. Borate glasses, because of their high vibrational frequency, have high nonradiative rates and thus low quantum efficiency.

Table 3. **Multiphonon decay rates of rare-earth ions in various glass types, expressed as $W_{NR} = W_0(T)\exp(-\alpha\,\Delta E)$ with the listed parameters. The highest-frequency optical phonon in each glass type is listed, along with the predicted multiphonon decay rate ω_{max} for the $Nd^{3+}\,{}^4F_{3/2} \rightarrow$ $^4I_{15/2}$ transition ($\Delta E \simeq 5200$ cm^{-1}) (data from [78, 79])**

Glass type	W_0 (s^{-1})	α (cm)	ω_{max} (cm^{-1})	W_{NR} (s^{-1})[a]
Borate	5×10^{12}	3.94×10^{-3}	1350	6300
Phosphate	4×10^{12}	4.61×10^{-3}	1300	150
Silicate	3×10^{12}	4.95×10^{-3}	1100	20
Germanate	2×10^{11}	4.89×10^{-3}	875	2
Tellurite	1×10^{11}	4.91×10^{-3}	775	<1
Fluoroberyllate	9×10^{11}	4.98×10^{-3}	1050	5

[a] $\Delta E = 5200$ cm^{-1}.

GAPONTSEV [80] has also studied nonradiative rates in many glasses and relates them to the measured infrared multivibrational spectrum. Although the nonradiative rates follow the general exponential trend described by Layne, Lowdermilk, and Weber, the rates can increase significantly if ΔE is in resonance with overtone or combination vibrational frequencies of the host glass matrix.

An intrinsic nonradiative quenching mechanism is associated with neodymium pairs [Fig. 5(a)]. One excited Nd^{3+} ion can nonradiatively decay to the $^4I_{15/2}$ state and raise a neighboring neodymium ion to the $^4I_{15/2}$ or $^4I_{13/2}$ state with the emission or absorption of vibrational energy. This self-quenching increases as the average neodymium ion distance decreases with increasing neodymium concentration. This so-called concentration quenching is intrinsic to a given glass host. Many researchers have studied this phenomenon in an attempt to establish empirical rules that might help in finding glasses with low concentration quenching.

Energy transfer between ions is generally described on the basis of a dipole-dipole interaction, as first put forward by FÖRSTER [81] and DEXTER

[82,83]. The transfer rate W_{DA} from a donor D to an acceptor A can be written as

$$W_{DA} = \frac{9\chi^2}{128\pi^5 N_A n^4 \tau_{0D} R_{DA}^6} \int \frac{g_D(v) K_A(v)\, dv}{v^4},$$ (13)

where

χ^2 = a factor that takes into account the orientational averaging of the dipoles,
N_A = acceptor concentration,
n = refractive index,
τ_{0D} = donor radiative lifetime,
R_{DA} = donor-acceptor distance,
$g_D(v)$ = shape factor of the fluorescence band,
$K_A(v)$ = absorption coefficient of the acceptor,
v = frequency on a wave number scale.

Reduced energy transfer, and thus reduced fluorescence quenching, can occur under the following (independent) conditions: (a) the average distance between the ions is large, (b) the transition strengths are low, or (c) the overlap between the fluorescence and absorption spectra is small.

Concentration quenching of the neodymium system has been studied extensively. Figure 6 shows the fluorescence decay rates of several glasses as

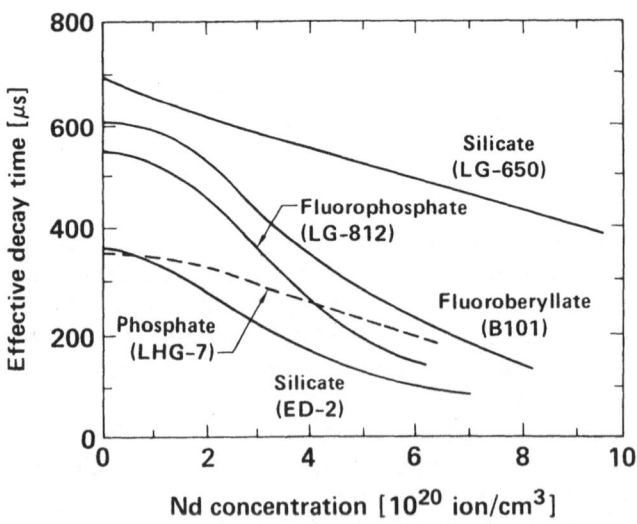

Fig. 6. Time required for the Nd^{3+} $^4F_{3/2}$ fluorescence to decay to $1/e$ of its initial value vs Nd^{3+} ion density (from STOKOWSKI [84])

a function of the neodymium concentration. In general, phosphate glasses (and particularly ultraphosphates) do not have strong concentration quenching. Most researchers believe that this weaker quenching is due to the greater average distance between neodymium ions in the phosphates. This view is supported by the low concentration quenching observed in ultraphosphate crystals, in which the smallest Nd–Nd distances are greater than 6 Å. Thus phosphate glasses have a decided advantage over other types of glass, particularly when high neodymium concentrations are needed in small systems.

The effect of transition strength on quenching can be seen in the silicate glasses LG-670 (ED-2), LG-660, and LG-650. The neodymium concentration for which the lifetime is reduced by half is listed in Table 4 and compared with the calculated absorption strength of the $^4I_{9/2} \rightarrow {}^4I_{15/2}$ transition. Lower line strengths result in lower concentration quenching.

Table 4. Fluorescence decay characteristics of three neodymium-doped silicate glasses: fluorescence decay time T_0 (first e-folding time), neodymium concentration $\rho_{1/2}$ at which the first e-folding time is one-half that for low neodymium concentrations, and peak emission cross section σ for the $^4F_{3/2} \rightarrow {}^4I_{11/2}$ transition.

Glass	T_0 (μs)	$\rho_{1/2}$ (10^{20} cm^{-3})	σ (pm^2)
LG-670	370	4	2.7
LG-660	420	6	1.8
LG-650	690	10	1.1

The neodymium self-quenching rate increases as the square of the neodymium concentration, because of the R^{-6} dependence of the energy transfer rate given by (13). However, in some ultraphosphate glasses the quenching rate increases linearly with neodymium concentration; in LG-650, the dependence is cubic (Fig. 7).

To investigate the effect of the overlap of the fluorescence and absorption transitions on self-quenching in the neodymium system, we measured the $^4F_{3/2} \rightarrow {}^4I_{15/2}$ fluorescence transition and the $^4I_{9/2} \rightarrow {}^4I_{15/2}$ absorption transition for different glasses and calculated the overlap integral. The overlap varies by less than 5% from glass to glass, and thus is not an important factor.

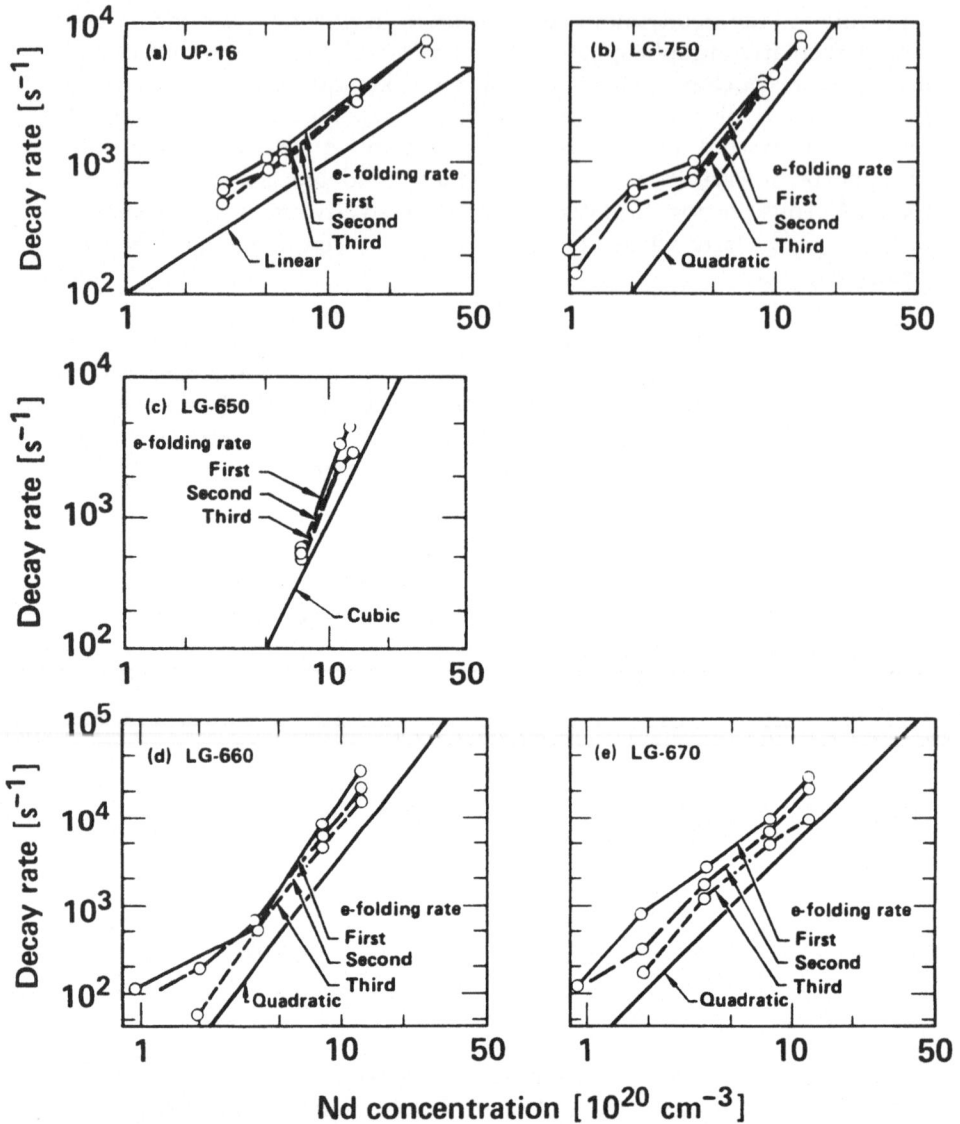

Fig. 7. Nonradiative decay rates vs neodymium concentration in (a) UP-16 phosphate, (b) LG-750 phosphate, (c) LG-650 silicate, (d) LG-660 silicate, and (e) LG-670 silicate laser glasses

Transition-metal and rare earth impurities can quench the neodymium fluorescence. STOKOWSKI and KRASHKEVICH [85] have measured the effect of these impurities on the neodymium fluorescence lifetime. Their results are listed in Table 5.

Molecular impurity vibrations can also lead to nonradiative quenching of the neodymium fluorescence. The only molecular impurity that has been studied to any extent is OH^-, whose high vibrational frequency gives it a strong effect on the nonradiative decay of neodymium. The quenching rates of OH^- are shown in Table 6. These rates are expressed in terms of the absorption coefficient at the OH^- fundamental vibrational frequency, which varies with glass type, as shown in Table 6. The water content of phosphate glasses can be very high because the OH^- enters the structure easily by converting a $P{=}O$ double bond to a single $P{-}OH$ bond. Silicate glasses are less susceptible to water contamination, but they sometimes contain enough water to affect the neodymium fluorescence lifetime. Commercial laser glass manufacturers have reduced the water content of their glasses to levels that result in an absorption coefficient of less than 2 cm^{-1} at the OH^- IR band peak.

Table 5. Nonradiative quenching rate of neodymium fluorescence for 300 wt. ppm of various transition-metal impurities for a neodymium concentration of 5×10^{19} cm^{-3}. Data from STOKOWSKI and KRASHKEVICH [85].

| Transition metal | Quenching rate (s^{-1}) | |
	Phosphate glass (UP-91)	Silicate glass (LG-660)
V	900	—
Cr	250	—
Fe	900	880
Co	1500	1740
Ni	1080	1050
Cu	1950	300

Table 6. Nonradiative quenching rate of neodymium fluorescence for various neodymium concentrations in three laser glasses assumed to contain water. Data from STOKOWSKI and KRASHKEVICH [85].

| Glass and glass type | Decay rate for pure sample (s^{-1}) | Vibrational frequency of OH^- (cm^{-1}) | Quenching rate (s^{-1}) per cm^{-1} coefficient at OH^- peak | | |
			0.5×10^{20} ions/cm^3	5×10^{20} ions/cm^3	10×10^{20} ions/cm^3
UP-91 phosphate	2650	2750	61	163	287
UP-16 phosphate	2565	2750	59	154	303
LG-660 silicate	2060	2880	70	162	319

3.4 Compositional Effects

The spectroscopic, optical, and thermomechanical properties of neodymium-doped glasses are sensitive to the glass composition. Variations in these properties are largely determined by the glass structure, which in turn depends on the glass former and on the glass modifiers.

To develop empirical rules to represent the variation of Nd:glass properties with composition, early researchers melted and characterized a large variety of glasses. Compositional studies of Nd:glass have been going on for over 25 years and will probably continue for a long time. The results of these studies are published in many articles [86–96]. The most extensive compilation of data on Nd:glass is that of STOKOWSKI, SAROYAN, and WEBER [97].

To summarize the results of a large amount of work, I describe below how the glass former affects Nd:glass properties in general, and then consider the effects of the typical glass modifiers, the alkali metals and alkaline earths.

Figure 8 shows the ranges of the spectroscopic properties found in the different major glass types. Phosphate glasses have higher cross sections than silicate or fluoroberyllate glasses. Differences in glass structure can account for this: silicate and fluoroberyllate glasses form a three-dimensional structure based on the SiO_2 or BeF_2 tetrahedra, whereas phosphate glasses consist of chains of PO_4 tetrahedra. In phosphate glasses, however, the average chain length depends on the P_2O_5 content. At low P_2O_5 content the chains are short. As the P_2O_5 content increases the chains get longer and theoretically are infinitely long at the metaphosphate composition (Fig. 9), assuming no cross-linking. In glasses with still higher P_2O_5 content (the ultraphosphates), cross-linking occurs and the structure is more three-dimensional. Borate glasses are based on the BO_3 planar molecular unit and thus constitute a third structural type.

Consider how a neodymium ion may integrate itself into the glass structure. The bonding of neodymium is primarily electrostatic, because the shielded f electrons do not participate strongly in covalent bonding. Given that the ionic radius of neodymium is about 1.08 Å, the closest packing that oxygen and fluorine ions can have with neodymium is about 8-fold coordination. The neodymium-oxygen bond energy of 213 kJ per mole with this coordination means, however, that neodymium is not a strong modifier of glass structure (Fig. 1). In fact, stable phosphate glasses can be made with neodymium concentrations of up to 4×10^{21} cm^{-3}.

The transition energies, transition strengths, and inhomogeneous line widths can tell us much about the local neodymium ion environment and its variation.

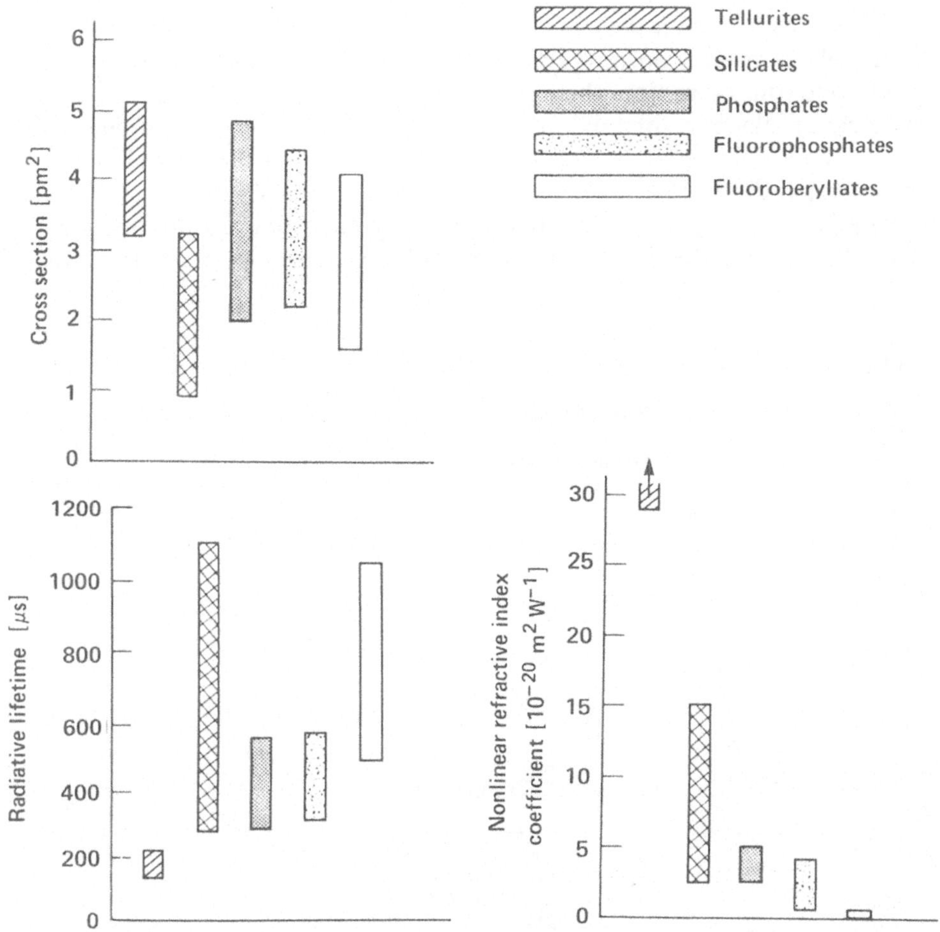

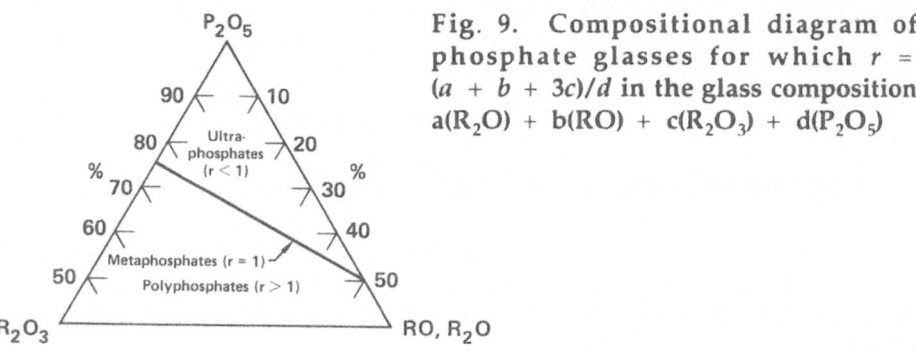

Fig. 8. Variation of Nd^{3+} $^4F_{3/2} \rightarrow {}^4I_{1/2}$ emission cross section, calculated $^4F_{3/2}$ radiative lifetime, and nonlinear refractive index coefficient γ for tellurite, silicate, phosphate, fluorophosphate, and fluoroberyllate glasses

Fig. 9. Compositional diagram of phosphate glasses for which $r = (a + b + 3c)/d$ in the glass composition $a(R_2O) + b(RO) + c(R_2O_3) + d(P_2O_5)$

The transition energies are determined by the splitting of the $4f^3$ states by the electron-electron interaction. The more covalent the neodymium-ligand bond, the more time the neodymium electrons spend apart; therefore, the electron-electron interaction is reduced and the energy differences and transition energies are reduced. This effect is known as the nephelauxetic or "cloud-expanding" effect. Thus, the $^4F_{3/2} \rightarrow {}^4I_{11/2}$ transition occurs at lower energy (i.e., at longer wavelengths) in oxide glasses than in fluoride glasses (Fig. 10). Further, the transition in phosphate glasses is at shorter wave-

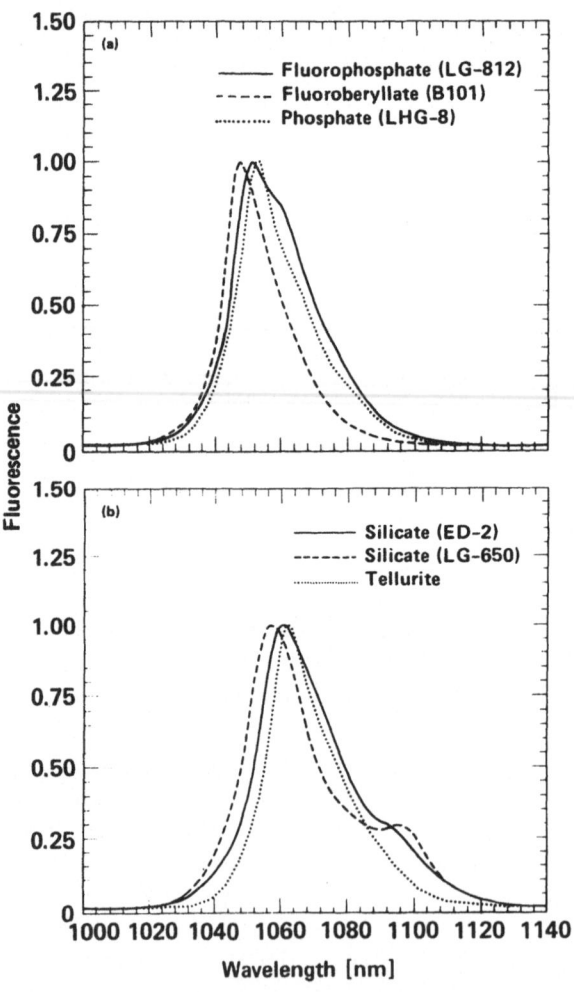

Fig. 10. Relative Nd^{3+} $^4F_{3/2}$ fluorescence spectra: (a) in a fluorophosphate, a fluoroberyllate, and a phosphate glass; (b) in two silicates and a tellurite glass (from STOKOWSKI [84])

lengths than in silicate glasses, implying that the neodymium in silicates participates more in bonding.

The transition strengths (or the Judd-Ofelt Ω parameters) are determined by the noncentrosymmetric local field surrounding the neodymium ion. This field mixes the higher-energy d-electronic states with the f states, thereby allowing electric dipole transitions to occur. The strength of this local field is determined by the distortion of the neodymium-anion ligands as a result of the bonding and structural requirements of the glass-former ion in the neighborhood of the neodymium ion. The transition strengths affect the radiative lifetime of the $^4F_{3/2}$ state. This lifetime depends on Ω_4 and Ω_6 and on the local field-correction factor $n(n^2 + 2)^2$. The calculated radiative lifetime is therefore a good indicator of the strength of the noncentrosymmetric field.

The inhomogeneous width of transitions is affected by variations in the local neodymium field that arise from structural variations in the glass. Large inhomogeneous line widths mean that the neodymium ion has less influence on its environment, which is determined more by the glass former. The site-to-site variations in glass structure are also determined by the temperature at which they are frozen in. Thus, we would expect greater variations in local structure for glasses that have a higher transformation temperature. The effective line width of the $^4F_{3/2} \rightarrow {}^4I_{11/2}$ fluorescence transition is determined by both the inhomogeneous width and the splitting of the six Kramer doublet states by the crystalline field.

Because of its importance in determining the emission cross section for the lasing transitions, we investigated the effective line width as a function of composition. Figures 11 through 14 illustrate the effects of different modifier ions on the calculated radiative lifetimes, emission cross sections, and effective line widths of neodymium in various glasses. Of particular note is the variety of dependences on the modifier. For instance, silicate and fluoroberyllate glasses, whose structures are both based on SiO_4 or BeF_4 tetrahedra, have opposite slopes with respect to changes in the alkali metal modifiers. The main difference between the glasses is that silicates are primarily covalent glasses, whereas fluoroberyllates are more ionic. The larger alkali ions have lower electrostatic field strengths, which explains the behavior of silicate glasses. We would expect the fluoroberyllate glasses to have the same trend; the fact that they do not indicates that the smaller alkali ions must be distributed around the neodymium ion in a more symmetrical way in the fluoroberyllates.

In general the effective line width depends little on the alkali ion and decreases somewhat as the alkaline earth varies from Mg to Ba in all the

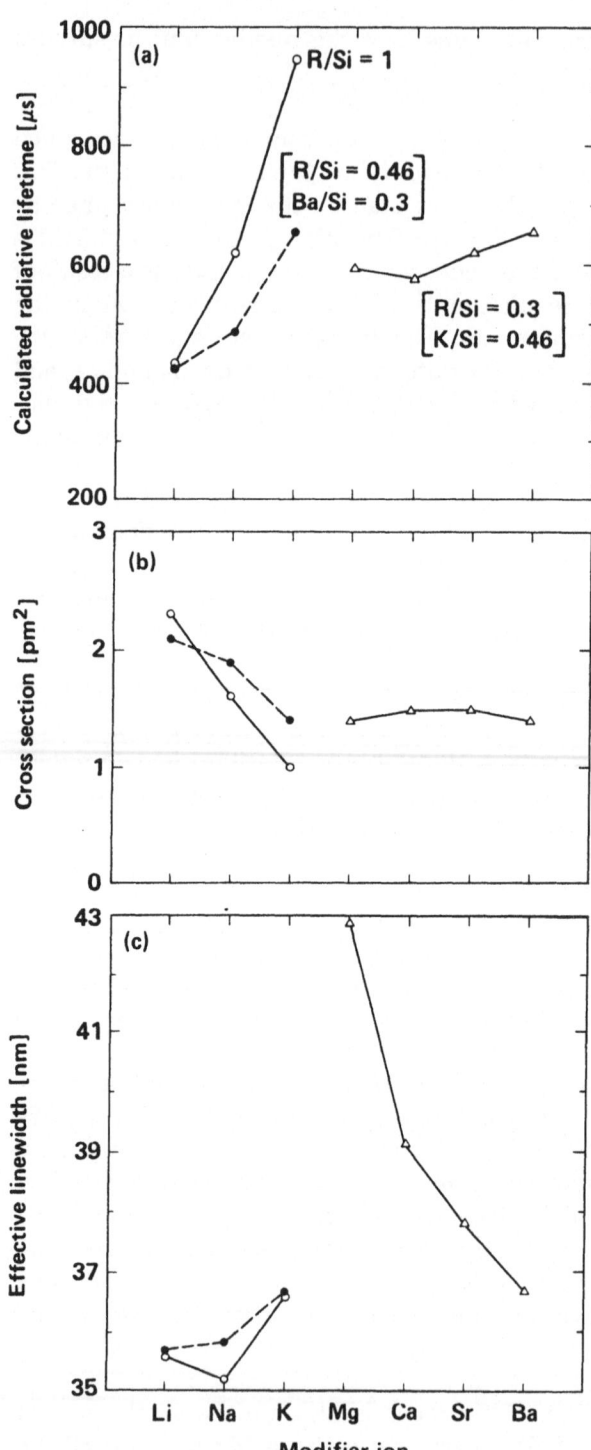

Fig. 11. Effects of modifier ions in silicate glasses on (a) calculated radiative lifetimes, (b) $^4F_{3/2} \rightarrow\ ^4I_{11/2}$ emission cross sections, and (c) effective line

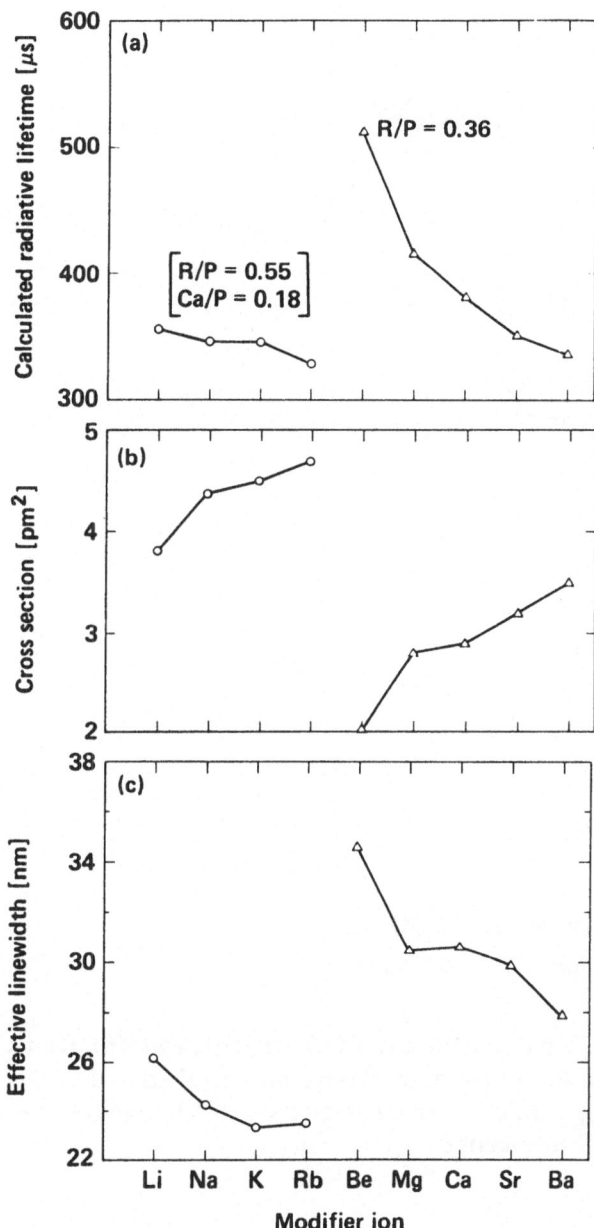

Fig. 12. Effects of modifier ions in phosphate glasses on (a) calculated radiative lifetimes, (b) $^4F_{3/2} \rightarrow {}^4I_{11/2}$ emission cross sections, and (c) effective line widths of the of the $^4F_{3/2} \rightarrow {}^4I_{11/2}$ fluorescence

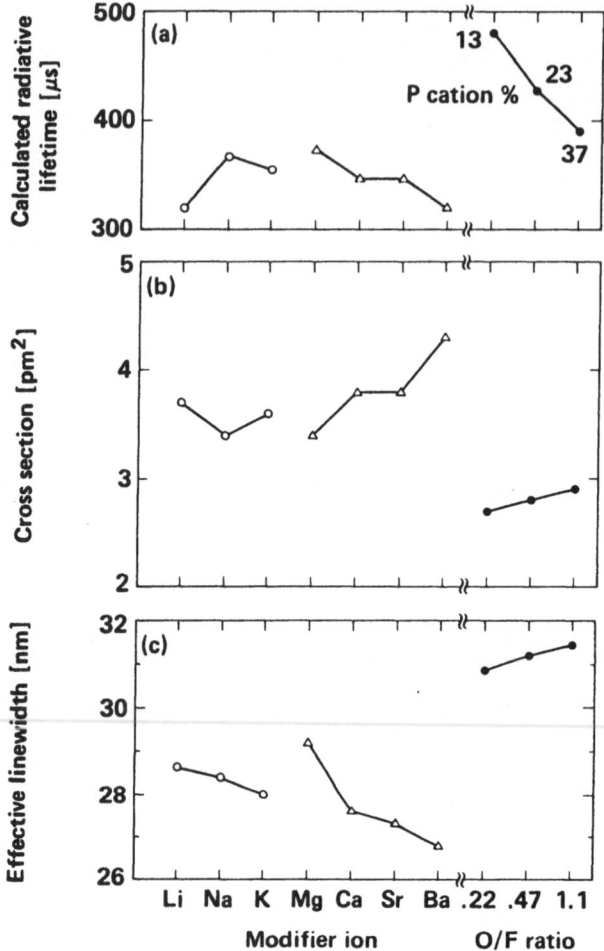

Fig. 13. Effects of modifier ions and the ratio O/F of ionic concentrations of oxygen and fluorine in fluorophosphate glasses on (a) calculated radiative lifetimes, (b) $^4F_{3/2} \to {}^4I_{11/2}$ emission cross sections, and (c) effective line widths of the $^4F_{3/2} \to {}^4I_{11/2}$ fluorescence

glass types. However, I noted an interesting correlation between the effective line width and the glass transformation temperature in a series of silicate glasses of silica content from 60 to 70 mole % with various modifiers. This correlation, shown in Fig. 15, lends support to the idea that the inhomogeneities in the local neodymium environment are greater when the glass structure is frozen in at higher temperatures, at which larger fluctuations in glass structure are present.

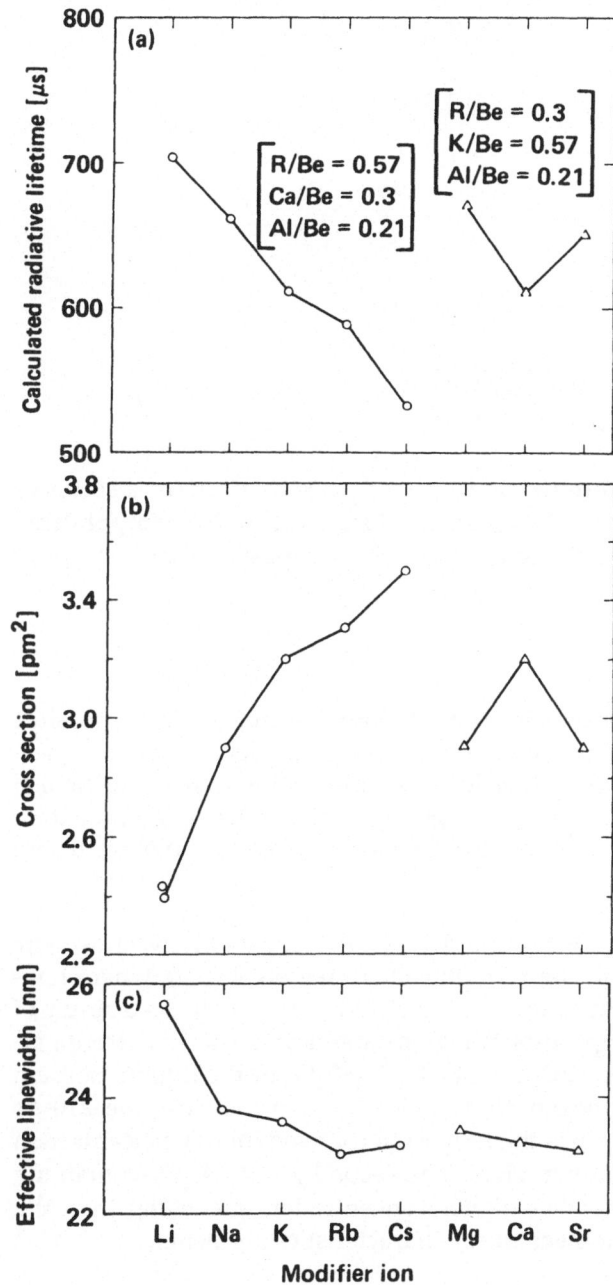

Fig. 14.   Effects of modifier ions in fluoroberyllate glasses on (a) calculated radiative lifetimes, (b) $^4F_{3/2} \rightarrow {}^4I_{11/2}$ emission cross sections, and (c) effective line widths of the $^4F_{3/2} \rightarrow {}^4I_{11/2}$ fluorescence

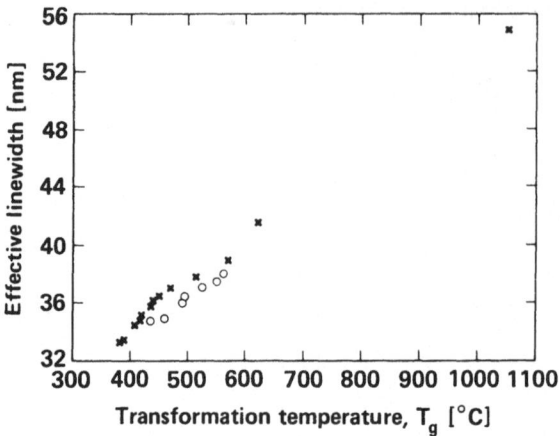

Fig. 15. Effective line width of the ${}^4F_{3/2} \rightarrow {}^4I_{11/2}$ transition for a series of Schott silicate glasses (crosses) vs measured transformation temperature. Open circles are for other silicate and borosilicate glasses. Cross in upper right corner is for fused silica

Phosphate glasses have proven to be the most useful of all the possible laser glass hosts. Their utility comes from their generally higher cross sections (which give higher gain), their lower concentration quenching (which allows more neodymium in the glass, giving higher efficiency), and their more homogeneous spectral characteristics (which allow for more efficient energy extraction).

In 1983 and 1984 LLNL, Schott, and Hoya investigated new phosphate compositions, particularly in the ultraphosphate region (Fig. 9), where low-quenching glasses were first found by VORONKO et al. [98]. As a result of these studies [99], two compositional regions stand out. The first group, al-kali-lanthanide-phosphates, such as $K_2O-La_2O_3-P_2O_5$ or the lithium-sodium equivalents, have lower concentration quenching. For instance; the 210-μs fluorescence lifetime of UP-16 is the longest, for a neodymium concentration of 10^{21} ions cm^{-3}, of any known glass. The second group, Al_2O_3-containing ultraphosphates, have somewhat higher concentration quenching than the first group, but their thermomechanical characteristics are better.

In the aluminum-free, alkali-ultraphosphate glasses, there is a range of compositions in which concentration quenching is minimal, as shown in Fig. 16. This compositional range seems to be correlated with the appearance of a stable glass structure, because, at high K_2O/P_2O_5 ratios, the glass melt crystallizes; at low ratios, the P_2O_5 volatilizes excessively. UP-16 is in the

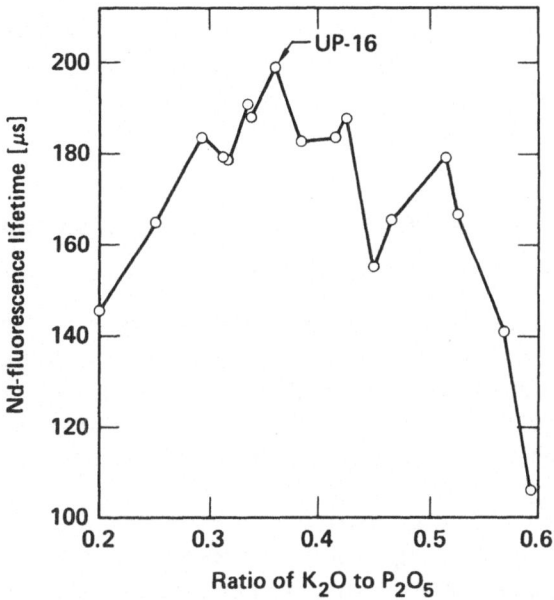

Fig. 16. Neodymium fluorescence lifetime for neodymium concentration of 13×10^{20} ions cm^{-3} in $K_2O–Ln_2O_3–P_2O_5$ glasses

center of this region and makes a good glass. Adding more than 5 mole % of alkaline earths (Mg, Ca, or Ba) or alumina to UP-16 results in increased concentration quenching, but the poor thermomechanical characteristics of UP-16 are improved by adding alumina.

We have found that adding up to 10 mole % of B_2O_3 can increase fluorescence lifetimes by 10% at high neodymium concentrations. Normally, B_2O_3 strongly quenches the neodymium fluorescence because of the high-frequency vibrational modes of the BO_3 group. However, B_2O_3 in concentrations of less than 10 mole % in ultraphosphates incorporates itself into the glass structure as BO_4 side groups on the phosphate chains, so that BO_3 chain terminators are absent.

3.5 Thermomechanical Properties

Thermomechanical properties of laser glasses are critically important in lasers for high-average-power operation. In such lasers, heat-removal rates and thermal stresses determine the maximum power that the laser medium can sustain. The thermomechanical figure of merit R_T [61] for a laser material designed to run at its fracture limit is given by (9). In brittle materials the fracture strength S_T, which appears in (9), depends on the critical-fracture toughness K_c, which characterizes the energy required to propagate a crack, and on the maximum surface flaw size a:

$$S_T = YK_c a^{-1/2}, \tag{14}$$

where Y is a dimensionless parameter that depends on the location and geometry of the flaw.

We have looked for correlations between thermomechanical properties and glass compositions. Figure 17 shows that thermal conductivity κ and the thermal expansion coefficient α are strongly correlated. The inverse relation between κ and α arises from the dependence of both on the anharmonics of glass structural vibrations. Large anharmonicities result in low thermal conductivities and high thermal expansions.

Young's modulus and the thermal expansion coefficient are also inversely related for phosphate glasses; their product αE has a value of 0.56 $\pm$ 0.14 MPa K^{-1}.

Poisson's ratio v for glasses is in the range of 0.25 $\pm$ 0.03, a relatively small variation.

The fracture toughness varies with the glass former and the modifier ion over the range 0.3 to 0.9 MPa m$^{1/2}$, with phosphate glasses at the lower end

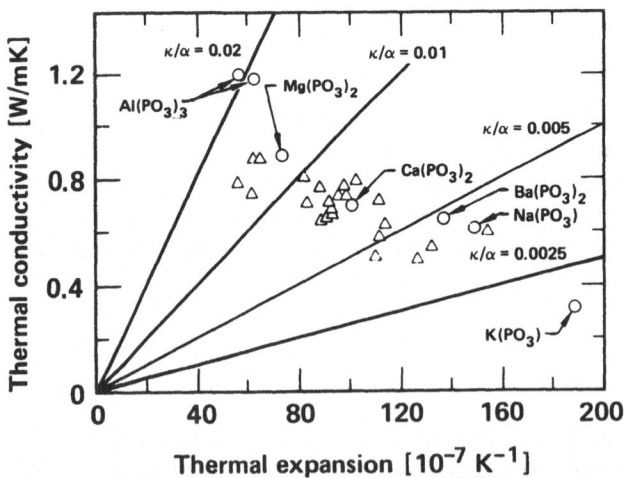

Fig. 17. Thermal conductivity and thermal expansion are inversely correlated in phosphate glasses. Lines of constant κ/α are shown. Unlabeled data points represent multicomponent glasses

and silicates at the upper end. Within a given glass type, K_c can vary by about a factor of 2.

As a result of variations, from glass to glass, in the material properties κ, v, α, and E in (9), and of variations in S_T insofar as it is affected by the material property K_c, thermomechanical figures of merit differ from one another by factors of no more than 2 or 3. In high-average-power lasers, we would like to have thermomechanical figures of merit an order of magnitude higher.

Such a large increase in R_T can be obtained only by going outside the material properties: this means increasing the glass fracture strength S_T through reductions in the maximum surface flaw size a, which is determined when the component is fabricated. Grinding and finishing are therefore critically important to obtaining high-strength laser glass components. MARION [100] has found that grinding leaves deep, but optically invisible subsurface cracks. This damage can be removed in subsequent grinding or polishing by removing more material than is typical in commercial processes. Factor-of-4 increases in S_T have been observed. Greater improvements (up to a factor of 15) can be obtained by etching the glass.

In use, the strength of a glass component will be degraded by handling, laser damage, and attack by atmospheric water vapor. To make the glass more durable, fractures generated by surface flaws under stress must be prevented from propagating by a compressive surface layer. Compressive layers can be formed by ion-beam–sputtered coatings [101] or ion-exchanged layers [102–104]. Ion-exchanged layers have made a factor-of-3 improvement in the fracture strength of laser glasses [104].

4. Future Possibilities

What further improvements might be made in laser glass? For large-aperture lasers such as those used in fusion research, the system cost per output joule is inversely proportional to the stored energy density in the material. Higher energy density will require higher neodymium concentrations. But high neodymium concentrations generally decrease fluorescence lifetimes, making flashlamp pumping more difficult. However, there is still some hope that glasses might be found that have very low concentration quenching. Our knowledge of the quenching mechanisms leads us to expect that these glasses will have large Nd–Nd distances and low transition strengths. The ultraphosphate compositional region should be investigated further—it may still have some surprises for us.

Increased energy density can also be obtained by improving pumping efficiency. Sensitization remains a possibility, because many sensitization schemes remain to be investigated, but I do not expect this approach to be successful.

The development of larger and more efficient arrays of laser diodes makes them increasingly attractive for pumping Nd:glass. The near-resonance pumping possible with laser diodes also means lower heat input to the laser glass, which makes diodes particularly desirable for pumping high-average-power lasers. However, the packing density of diode arrays must increase—and their cost must decrease considerably—before they will be used extensively.

Another way to increase energy density is to reduce the losses caused by amplified spontaneous emission. A saturable absorber in laser glass might be one way to reduce this loss. The search for such an absorber has so far been very limited; I believe that this area warrants more attention.

High laser fluences will be needed to extract energy stored at high densities. These high fluences cause damage in current optical and laser materials, primarily because of metallic inclusions. The only sure way to avoid these inclusions, which come from the metal crucibles commonly used to contain the melt, is to melt glass in nonmetallic containers. The optical glass manufacturers have begun work on using vitreous carbon, a highly promising future crucible material, for glass melts. The availability of metallic-inclusion-free laser materials will be critically important for future laser systems.

High laser fluences also mean that materials with lower nonlinear refractive indices are necessary to prevent self-focusing. Fluoride glasses have substantially lower nonlinear indices than oxide glasses. With the present high level of interest in fluoride glasses for optical fibers, research in this area will probably result in new practical laser glasses.

Research on composite laser materials promises to open up new territory. Marion and Stokowski have suggested that a composite laser material might be made consisting of a glassy phase and a crystalline phase. These composites can be formed by mixing small crystalline particles in a glass melt, which is then processed similarly to normal glasses. The two phases must match in refractive index and thermal strain to prevent high scattering losses. These composites are to be contrasted with glass ceramics, which are partially crystallized glasses in which the refractive-index mismatch between the two phases leads to high scattering losses. Marion and Stokowski have initiated an experimental program to find and make a suitable material.

5. Acknowledgment to Arthur Schawlow

Arthur Schawlow's scientific accomplishments are well known, particularly those connected with the laser and with high-resolution spectroscopy of the hydrogen atom. Those accomplishments are well documented in many publications. Schawlow's career as a university professor is not as well known, but his influence can be seen in the accomplishments of his students. I particularly remember the humor (irreverent at times) with which he emphasizes an insight into the physical world. His humor is an effective teaching technique; it's hard to forget the props, such as his "Mickey Mouse" balloon and his "Jell-o" laser, that he used to get across his points. But these amusing items are an example of his great imagination and intuitive thinking. In this way he taught me to step out of the constraining world of existing theory and calculations and to imagine new worlds of possibilities. I remember a discussion of the second-order phase transition in $SrTiO_3$, in which the crystal structure goes from cubic to tetragonal. We were comparing this transition to the ordering of the spin dipole (a first-rank tensor) in ferromagnets and antiferromagnets. Schawlow suggested that the transition we were observing in $SrTiO_3$ was analogous to an ordering in the orientation of a second-rank tensor (the lattice strain), and this was later confirmed.

I thank Arthur Schawlow for taking me on as a research student, for the opportunity to work with him, and for imparting to me his excitement about discovering the mysteries of the physical world.

Work performed under the auspices of the U.S. Department of Energy by the Lawrence Livermore National Laboratory under Contract W-7405-Eng-48.

References

1. A. L. Schawlow and C. H. Townes: "Infrared and Optical Masers," *Phys. Rev.* **112**, 1940–1949 (1958).
2. T. H. Maiman: "Stimulated Optical Radiation in Ruby," *Nature* **187**, 493–494 (1960).
3. A. Javan, W. R. Bennett, Jr., and D. R. Herriot: "Population Inversion and Continuous Optical Maser Oscillation in a Gas Discharge Containing a Ne–Ne Mixture," *Phys. Rev. Lett.* **6**, 106 (1961).
4. E. Snitzer: "Optical Maser Action of Nd^{+3} in a Barium Crown Glass," *Phys. Rev. Lett.* **7**, 444 (1961).
5. H. W. Etzel, H. W. Gandy, and R. J. Ginther: "Stimulated Emission of Infrared Radiation From Ytterbium-Activated Silicate Glass," *Appl. Opt.* **1**, 534 (1962).

6. H. W. Gandy and R. J. Ginther: "Stimulated Emission from Holmium Activated Silicate Glass", *Proc. IRE* **50**, 2113 (1962).

7. E. Snitzer and R. Woodcock: "Yb^{3+}–Er^{3+} Glass Laser," *Appl. Phys. Lett.* **6**, 45 (1965).

8. H. W. Gandy, R. J. Ginther, and J. F. Weller: "Stimulated Emission of Tm^{3+} Radiation In Silicate Glass," *J. Appl. Phys.* **38**, 3030 (1967).

9. S. I. Andreev, M. R. Bedilov, G. O. Karapetyan, and V. M. Likhachev: "Stimulated Radiation of Glass Activated By Terbium," *Sov. J. Opt. Tech.* **34**, 819 (1967): *Opt.-Mekh. Promst.* **34**, 60 (1967).

10. C. G. Young: "Continuous Glass Laser," *Appl. Phys. Lett.* **2**, 151 (1963).

11. E. Snitzer and C. G. Young: "Glass Lasers," in *Advances in Lasers*, ed. by A. Levine, vol. 2 (Dekker, New York, 1968).

12. J. M. McMahon, J. L. Emmett, J. F. Holzrichter, and J. B. Trenholme: "A Glass-Disk-Laser Amplifier," *IEEE J. Quantum Electron.* **QE-9**, 992 (1973).

13. D. C. Brown: "Parasitic Oscillations In Large Aperture Nd^{3+} Glass Amplifiers Revisited," *Appl. Opt.* **12**, 2215 (1973).

14. J. A. Glaze, S. Guch, and J. B. Trenholme: "Parasitic Suppression In Large Aperture Nd;Glass Disk Laser Amplifiers," *Appl. Opt.* **13**, 2808 (1974).

15. H. W. Mocker and R. J. Collins: "Mode Competition and Self-Locking Effects in a Q-Switched Ruby Laser," *Appl. Phys. Lett.* **7**, 270 (1965).

16. K. H. Sun: "Fundamental Condition of Glass Formation," *J. Am. Ceram. Soc.* **30**, 277 (1947).

17. C. G. Young: "Glass Laser Delivers 5000-Joule Output," *Laser Focus* **3**, 36 (February, 1967).

18. H. W. Gandy, R. J. Ginther, and J. F. Weller: "Energy Transfer in Silicate Glass Coactivated with Cerium and Neodymium," *Phys. Lett.* **11**, 213 (1964).

19. R. R. Jacobs, C. B. Layne, M. J. Weber, and C. Rapp: "$Ce^{3+} \rightarrow Nd^{3+}$ Energy Transfer in Silicate Glass," *J. Appl. Phys.* **47**, 2020 (1976).

20. S. Shionoya and E. Nakazawa: "Sensitization of Nd^{3+} Luminescence by Mn^{2+} and Ce^{3+} in Glasses," *Appl. Phys. Lett.* **6**, 117 (1965).

21. N. T. Melamed, C. Hirayama, and E. K. Davis: "Laser Action in Neodymium-doped Glass Produced Through Energy Transfer," *Appl. Phys. Lett.* **7**, 170 (1965).

22. N. T. Melamed, C. Hirayama, and P. W. French: "Laser Action in Uranyl-Sensitized Nd-Doped Glass," *Appl. Phys. Lett.* **6**, 43 (1965).

23. H. W. Gandy, R. J. Ginther, and J. F. Weller: "Energy Transfer in Triply Activated Glasses," *Appl. Phys. Lett.* **6**, 46 (1965).

24. J. C. Joshi, N. C. Pandey, B. C. Joshi, and J. Joshi: "Energy Transfer from $UO_2 \rightarrow Nd^{3+}$ in Barium Borate Glass," *J. Luminescence* **16**, 435 (1978).

25. A. Y. Cabezas and L. G. DeShazer: "Radiative Transfer of Energy Between Rare-Earth Ions in Glass," *Appl. Phys. Lett.* **4**, 37 (1964).
26. R. Reisfeld and Y. Kalisky: "Energy Transfer Between Bi^{3+} and Nd^{3+} in Germanate Glass," *Chem. Phys. Lett.* **50**, 199 (1977).
27. E. J. Sharp, M. J. Weber, and G. Cleek: "Energy Transfer and Fluorescence Quenching in Eu- and Nd-Doped Silicate Glasses," *J. Appl. Phys.* **47**, 364 (1976).
28. G. O. Karapetyan, V. P. Kovalyov, and S. G. Lunter: "Chromium Sensitization of the Neodymium Luminescence in Glass," *Opt. Spectrosc. USSR* **19**, 529 (1965); *Opt. Spectrosk.* **19**, 951 (1965).
29. G. Dauge: "Nonradiative Energy Transfer in Silicate Glass," *IEEE J. Quantum Electron.* **QE-2**, lviii (1966).
30. J. G. Edwards and S. Gomulka: "Enhanced Performance of Nd Laser Glass by Double Doping With Cr," *J. Phys. D.* **12**, 187 (1979).
31. A. G. Avanesov, Yu. K. Voron'kov, B. I. Denker, G. V. Maosimova, V. V. Osiko, A. M. Prokhorov, and I. A. Shcherbakov: "Nonradiative Energy Transfer from Cr^{3+} to Nd^{3+} Ions in Glasses with High Neodymium Concentrations," *Sov. J. Quantum Electron.* **9**, 935 (1979); *Kvantovaya Elektron.* **6**, 1583 (1979).
32. R. D. Maurer: "Nd^{3+} Fluorescence and Stimulated Emission in Oxide Glasses," in *Proceedings of the Symposium on Optical Masers*, Microwave Research Institute Symposia Series Vol. XIII (Polytechnic Press, Brooklyn, NY, 1963) p. 435.
33. G. T. Petrovksii, M. N. Tolstoi, P. P. Feofilov, G. A. Tsurikova, and V. N. Shapovalov: "Luminescence and Stimulated Emission of Neodymium in Beryllium Fluoride Glass," *Opt. Spectrosc. USSR* **21**, 72 (1966); *Opt. Spektrosk.* **21**, 126 (1966).
34. F. Auzel: "Emission Stimulée de Er^{3+} dans un Verre Fluorophosphate," *C. R. Acad. Sc. Ser. B* **263**, 765 (1966).
35. G. Deutschbein, C. Pautrat, and I. M. Svirchevsky: "Phosphate Glasses, New Laser Materials," *Rev. Phys. Appl.* **1**, 29 (1967).
36. N. G. Basov and O. N. Krokhin: "Conditions For Heating Up of a Plasma by the Radiation from an Optical Generator," *Sov. Phys. JETP Lett.* **19**, 123–125 (1964).
37. R. E. Kidder: "Applications of Lasers to the Production of High-Temperature and High Pressure Plasma," *Nucl. Fusion* **8**, 3–12 (1968).
38. J. Nuckolls, L. Wood, A. Thiessen, and G. Zimmerman: "Laser Compression of Matter to Super-High Densities: Thermonuclear (CTR) Applications," *Nature* **239**, 139–142 (1972).
39. N. G. Basov, Yu. S. Ivanov, O. N. Krokhin, Yu. A. Mikhailov, G. V. Sklizkov, and S. I. Fedotov: "Neutron Generation In Spherical Irradiation of a Target by High-Power Laser Radiation," *Sov. Phys. JETP Lett.* **15**, 417–419 (1972).

40. J. E. Swain, R. E. Kidder, K. Pettipiece, F. Rainer, E. D. Baird, and B. Loth: "Large-Aperture Glass Disk Laser System," *J. Appl. Phys.* **40**, 3973 (1969).

41. S. E. Stokowski, W. H. Lowdermilk, F. T. Marchi, J. E. Swain, E. P. Wallerstein, and G. R. Wirtenson: "Advances in Optical Materials for Large Aperture Lasers," in *Proceedings of Electro-Optics/Laser '81*, Anaheim, CA, Nov. 17–19 (Industrial & Scientific Conf. Management, Chicago, 1981), p. 203.

42. W. W. Simmons and R. O. Godwin: "Nova Laser Fusion Facility—Design, Engineering, and Assembly Overview," *Nucl. Technol. Fusion* **4**, 8–24 (1983).

43. D. W. Harper: "Assessment of Neodymium Optical Maser Glass," *Phys. Chem Glasses* **5**, 11 (1964).

44. C. Hirayama and D. W. Lewis: *Phys. Chem. Glasses* **5**, 44 (1964).

45. J. B. Trenholme and J. L. Emmett: "Xenon Flashlamp Model for Performance Prediction" in *Proceedings of Ninth International Conference on High Speed Photography*, ed. by W. G. Hyzen and W. G. Chase (Society of Motion Picture and Television Engineers, New York, 1970), p. 299.

46. A summary of the Trenholme-Emmett model and a review of flashlamp pumping of Nd:Glass lasers is found in: D. C. Brown: *High-Peak-Power Nd:Glass Laser Systems* (Springer, Berlin, 1981), Ch. 3.

47. H. T. Powell, A. C. Erlandson, and K. S. Jancaitis: "Characterization of High Power Flashlamps and Application to Nd:glass Laser Pumping," *Proc. SPIE Conf. on Flashlamp Pumped Laser Technology* **609**, 78 (1986).

48. J. B. Trenholme: Lawrence Livermore National Laboratory, Livermore, CA, private communication (1987).

49. M. Hercher: *J. Opt. Soc. Am.* **54**, 563 (1964).

50. R. Y. Chiao, E. Garmire, and C. H. Townes: "Self-Trapping of Optical Beams," *Phys. Rev. Lett.* **13**, 479 (1964).

51. J. B. Trenholme: "Small-Scale Instability Growth: Review of Small Signal Theory," in *Laser Program Annual Report 74*, Lawrence Livermore National Laboratory, Livermore, Calif., UCRL-50021-74 (1975), p. 179.

52. E. S. Bliss, J. T. Hunt, P. A. Renard, G. E. Sommargren, and H. J. Weaver: "Effects of Nonlinear Propagation on Laser Focusing Properties," *IEEE J. Quantum Electron.* **QE-12**, 402 (1976).

53. E. S. Bliss, D. R. Speck, and W. W. Simmons: "Direct Interferometric Measurements of the Nonlinear Refractive Index Coefficient n_2 in Laser Materials," *Appl. Phys. Lett.* **25**, 718 (1974).

54. D. Milam and M. J. Weber: "Measurement of Nonlinear Refractive Index Coefficients using Time-Resolved Interferometry: Application to Optical Materials for High-Power Neodymium Lasers," *J. Appl. Phys.* **47**, 2497 (1976).

55. D. Milam and M. J. Weber: "Nonlinear Refractive Index Coefficients for Nd Phosphate Laser Glasses," *IEEE J. Quantum Electron.* **QE-13**, 512 (1976).

56. D. Milam, M. J. Weber, and A. J. Glass: "Nonlinear Refractive Index of Fluoride Crystals," *Appl. Phys. Lett.* **31**, 822 (1977).

57. M. J. Weber, C. F. Cline, W. L. Smith, D. Milam, D. Heiman, and R. W. Hellwarth: "Measurements of the Electronic and Nuclear Contributions to the Nonlinear Refractive Index of Beryllium Fluoride glasses," *Appl. Phys. Lett.* **32**, 403 (1978).

58. M. J. Weber, D. Milam, and W. L. Smith: "Nonlinear Refractive Index of Glasses and Crystals," *Opt. Eng.* **17**, 463 (1978).

59. N. L. Boling, A. J. Glass, and A. Owyoung: "Empirical Relationships for Predicting Nonlinear Refractive-Index Changes in Optical Solids," *IEEE J. Quantum Electron.* **QE-14**, 601 (1978).

60. A. J. Glass: *Laser Program Annual Report 74*, Lawrence Livermore National Laboratory, Livermore, CA, UCRL-50021-84 (1975), p. 260.

61. J. L. Emmett, W. F. Krupke, and W. R. Sooy: "Future Development of High-Power Solid State Laser Systems," Lawrence Livermore National Laboratory, Livermore, CA, UCRL-53344 (1982): *Sov. J. Quantum Electron.* **13**, 1 (1983).

62. R. W. Hopper and D. R. Uhlmann: "Mechanism of Inclusion Damage in Laser Glass," *J. Appl. Phys.* **41**, 4023 (1970).

63. J. H. Pitts: "Modeling Laser Damage Caused by Platinum Inclusions in Laser Glass," Lawrence Livermore National Laboratory, Livermore, CA, UCRL-93249 (1985): to be published in *Proceedings of the 17th Annual Symposium—Optical Materials for High Power Lasers, Boulder, Colorado*, National Bureau of Standards, Washington, DC, NBS Special Publication.

64. C. Brecher, L. A. Riseberg, and M. J. Weber: "Line-Narrowed Fluorescence Spectra and Site-Dependent Transition Probabilities of Nd^{3+} in Oxide and Fluoride Glasses," *Phys. Rev. B* **18**, 5799 (1978).

65. C. Brecher, L. A. Riseberg, and M. J. Weber: "Site-Dependent Variation of Spectroscopic Relaxation Parameters in Nd Glasses," *J. Luminescence* **18/19**, 651 (1979).

66. V. I. Nikitin, M. S. Soskin, and A. I. Khizhnyak: "Influence of Uncorrelated Inhomogeneous Broadening of the 1.06 μm Band of the Nd^{3+} Ions on Laser Properties of Neodymium Glasses," *Sov. J. Quantum Electron.* **8**, 788 (1978): *Kvantovaya Elektron.* **5**, 1375 (1978).

67. V. I. Nikitin, M. S. Soskin, and A. I. Khizhnyak: "New Data About Internal 1.06 μm Luminescence Band Structure of Nd^{3+} in Silicate Glass," *Sov. Tech. Phys. Lett.* **2**, 64 (1976); *Pis'ma Zh. Tekh. Fiz.* **2**, 172 (1976).

68. V. I. Nikitin, M. S. Soskin, and A. I. Khizhnyak: "Uncorrelated Non-uniform Spreading—A Basic Reason for Narrow-Band Generation in Phosphate Glass with Nd^{3+}," *Sov. Tech. Phys. Lett.* **3**, 5 (1977): *Pis'ma Zh. Tekh. Fiz.* **3**, 14 (1977).
69. W. E. Martin and D. Milam: "Gain Saturation in Nd:Doped Laser Materials," *IEEE J. Quantum Electron.* **QE-18**, 1155 (1982).
70. S. M. Yarema and D. Milam: "Gain Saturation in Phosphate Laser Glasses," *IEEE J. Quantum Electron.* **QE-18**, 1941 (1982).
71. B. R. Judd: "Optical Absorption Intensities of Rare-Earth Ions," *Phys. Rev.* **127**, 750 (1962).
72. G. S. Ofelt: "Intensities of Crystal Spectra of Rare-Earth Ions," *J. Chem. Phys.* **37**, 511 (1962).
73. R. D. Peacock: "The Intensities of Lanthanide f↔f Transitions," *Struct. Bonding* **22**, 83 (1975).
74. W. F. Krupke: "Induced-Emission Cross Sections in Neodymium Laser Glasses," *IEEE J. Quantum Electron.* **QE-10**, 450 (1974).
75. T. S. Lomheim and L. G. DeShazer: "New Procedure of Determining Neodymium Fluorescence Branching Ratios as Applied to 25 Crystal and Glass Hosts," *Opt. Comm.* **24**, 89 (1978).
76. S. E. Stokowski: Lawrence Livermore National Laboratory, Livermore, CA: measurements made in 1978.
77. H. T. Powell, J. E. Murray, and K. S. Jancaitis: Lawrence Livermore National Laboratory, Livermore, CA, private communication (1987).
78. C. B. Layne, W. H. Lowdermilk, and M. J. Weber: "Multiphonon Relaxation of Rare-Earth Ions in Oxide Glasses," *Phys. Rev. B* **16**, 10 (1977).
79. C. B. Layne and M. J. Weber: "Multiphonon Relaxation of Rare-Earth Ions in Beryllium-Fluoride Glass," *Phys. Rev. B* **16**, 3259 (1977).
80. N. E. Alekseev, V. P. Gapontsev, M. E. Zhabotinskii, V. B. Kravchenko, and Yu. P. Rudnitskii: *Laser Phosphate Glasses* (Nauka, Moscow, 1980): Lawrence Livermore National Laboratory, Livermore, CA, UCRL-TRANS-11817 (1983), p. 3-97.
81. T. Förster: *Ann. Phys.* **2**, 55 (1948).
82. D. L. Dexter: "A Theory of Sensitized Luminescence in Solids," *J. Chem. Phys.* **21**, 836 (1953).
83. D. L. Dexter and J. H. Schulman: "Theory of Concentration Quenching in Inorganic Phosphors," *J. Chem. Phys.* **22**, 1063 (1954).
84. S. E. Stokowski: "Glass Lasers" in *CRC Handbook of Laser Science and Technology*, Vol. 1, *Lasers and Masers*, ed. by M. J. Weber (CRC Press, Boca Raton, FL, 1982), p. 215.
85. S. E. Stokowski and D. Krashkevich: "Transition-Metal Ions in Nd-Doped Glasses: Spectra and Effects on Nd Fluorescence," *Mat. Res. Soc. Symp. Proc.* **61**, 273 (1986).

86. V. F. Egorova, V. S. Zubkova, G. O. Karapetyan, A. A. Mak, D. S. Prilezhaev, and A. L. Reichakhrit: "Influence of Glass Composition on the Luminescence Characteristics of Nd^{3+} Ions," *Opt. Spectrosc. USSR* **23**, 148 (1967); *Opt. Spectrosk.* **23**, 275 (1967).

87. P. H. Sarkies, J. N. Sandoe, and S. Parke: "Variation of Nd^{3+} Cross Section for Stimulated Emission with Glass Composition," *J. Phys. D: Appl. Phys.* **4**, 1642 (1971).

88. R. R. Jacobs and M. J. Weber: "Dependence of the $^4F_{3/2} \rightarrow {}^4I_{11/2}$ Induced-Emission Cross Section for Nd^{3+} on Glass Composition," *IEEE J. Quantum Electron.* **QE-12**, 102 (1976).

89. H. G. Lipson, J. R. Buckmelter, and C. O. Dugger: "Neodymium Ion Environment in Germanate Crystals and Glasses," *J. Non-Cryst. Solids* **17**, 27 (1975).

90. N. B. Brachkovskaya, A. A. Grubin, S. G. Lunter, A. K. Przhevuskii, E. L. Raaben, and M. N. Tolstoi: "Intensities of Optical Transitions in Absorption and Luminescence Spectra of Neodymium in Glasses," *Sov. J. Quantum Electron.* **6**, 534 (1976).

91. G. O. Brachkovskaya, G. O. Karapetyan, A. L. Reishakhrit, and M. N. Tolstoi: "Luminescence of Neodymium in Alkali Silicate Glasses," *Opt. Spectrosc.* **29**, 173 (1970).

92. K. Hauptmanova, J. Pantoflicek, and K. Patek: "Absorption and Fluorescence of Nd^{3+} Ion in Silicate Glass," *Phys. Status Solidi* **9**, 525 (1965).

93. C. Hirayama: "Nd Fluorescence in Alkali Borate Glasses," *Phys. Chem Glasses* **7**, 52 (1966).

94. C. Hirayama, F. E. Camp, N. T. Melamed, and K. B. Steinbruegge: "Nd^{3+} in Germanate Glasses: Spectral and Laser Properties," *J. Non-Cryst. Solids* **6**, 342 (1971).

95. N. E. Alekseev, A. A. Izyneev, Yu. L. Kopylov, V. B. Kravchenko, Yu. P. Rudnitskii, and N. F. Udovenko: "Activated Nd^{3+} Laser Glasses Based on the Metaphosphates of Divalent Metals," *J. Appl. Spectrosc.* **24**, 691 (1976): *Zh. Prikl. Spektrosk.* **24**, 976 (1976).

96. N. E. Alekseev, A. A. Izyneev, Yu. L. Kopylov, V. B. Kravchenko, and Yu. P. Rudnitskii: "A Study of Neodymium Glasses Based on Alkali Metal Metaphosphates," *J. Appl. Spectrosc.* **26**, 87 (1977): *Zh. Prikl. Spektrosk.* **26**, 116 (1977).

97. S. E. Stokowski, R. A. Saroyan, and M. J. Weber: " Nd Doped Laser Glass Spectroscopic and Physical Properties," Lawrence Livermore National Laboratory, Livermore, CA, M-095 Rev. 2 (1981).

98. Yu. K. Voronko, B. I. Denker, A. A. Zlenko, et al.: "Spectral Lasing Properties of Li–Nd Phosphate Glass," *Opt. Commun.* **18**, 88 (1976).

99. L. M. Cook, A. J. Marker III, and S. E. Stokowski: in *Proc. SPIE*, Vol. 505 (Soc. Photo-Optical Inst. Engineers, Bellingham, Wash., 1984). pp. 102–111.

100. J. E. Marion: "Strengthened Solid-State Laser Materials," *Appl. Phys. Lett.* **47**(7), 694–696 (1985).
101. J. E. Marion: "Development of High Strength Solid State Laser Materials," *Proc. Amer. Inst. Physics* **146**, 234 (1986).
102. Owens-Illinois, Inc., Optical Products Division, Product Information on ED-2S Strengthened Glass (1978).
103. S. D. Stookey: *High Strength Materials*, ed. by V. F. Zakey (Wiley, New York, 1964), p. 669.
104. K. A. Cerqua, S. D. Jacobs, B. L. McIntyre, and W. Zhong: to be published in *Proceedings of the Boulder Damage Conference, 1985*, National Bureau of Standards, Washington, DC, NBS Special Publication.

One Is Not Enough: Intra-Cavity Spectroscopy with Multi-Mode Lasers

P.E. Toschek[1]* and V.M. Baev[2]†

[1]Joint Institute for Laboratory Astrophysics, University of Colorado and National Bureau of Standards, Boulder, CO 80309, USA
[2]I. Institut für Experimentalphysik, Universität Hamburg, D-2000 Hamburg 36, Fed. Rep. of Germany

1. Introduction

When the laser was discovered more than 25 years ago as the ultimate out-growth of spectroscopic work, it kept the physics community busy for more than a decade observing, surveying, and classifying all those surprising features of the new type of light and its interactions with matter. The early lasers, lacking technical sophistication, usually oscillated in a vast multitude of radiative modes. These systems were considered overly complex, and accordingly theoretical models and explanations were constructed for ideal single-mode lasers, which, however, barely could be demonstrated in experiments. It is perhaps not so surprising that it took ten years before emission spectra of a real multi-mode laser were sub-jected to detailed inspection. A close look revealed that the light flux of individual modes in their broad spectral comb did not vary smoothly -- as was anticipated -- across the full emission band. Instead, crests and crevices, rifts and ridges showed up in the recorded spectra. Soon these observed features were traced to either very weak absorption of matter or spurious interference inside the laser resonator [1]. This interpretation suggested that this ultra-sensitive response to weak extinction could be used for a novel spectroscopic technique.

An early example of an "intra-cavity" absorption (ICA) spectrum -- of HN_3 vapor [2] -- is compared in Fig. 1 with the analogous spectrum recorded by conventional differential absorption [3]. The new technique turned out to be 4×10^7 times more sensitive with the exposure time reduced from hours to 1 ms -- a considerable advancement!

The superior sensitivity of a multi-mode laser to intra-cavity fre-quency-selective extinction was noticed independently in some laboratories in the U.S. [4-6]. At Stanford, an experiment was devised that combined the demonstration of ultra-sensitive I_2 absorption inside a cw dye laser with a specific narrow-band technique of light detection [6]. The sche-matics is shown in Fig. 2. The rhodamine-6G laser with a Z-shaped cavity

---

*1986-87 JILA Visiting Fellow. On leave from I. Institut für Experimentalphysik, Universität Hamburg, D-2 Hamburg 36, F. R. Germany.

†On leave from Lebedev Physical Institute, Academy of Sciences of the USSR, Moscow, USSR.

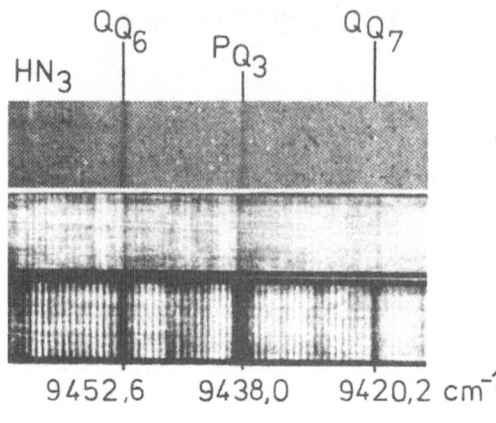

Fig. 1. (a) HN₃ absorption spectrum; 4 m absorber lengths;
0.4 bar [3]. ICA spectra
of HN₃ in a cell of 0.4-m
length, at 10^{-4} mbar (b),
and at 10^{-1} mbar (c)
(from Ref. 2).

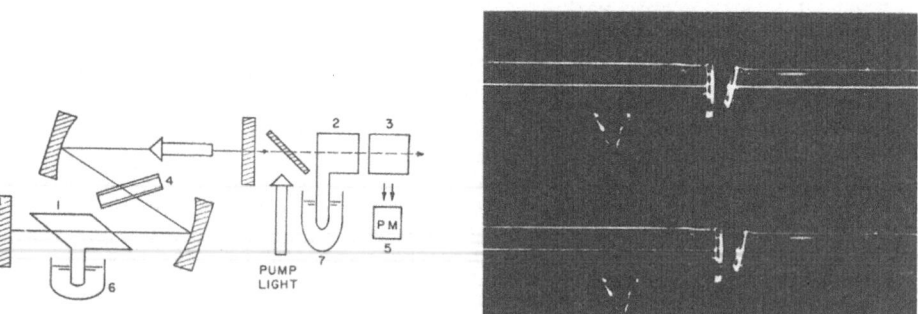

Fig. 2. Left: cw Rh 6G laser with ICA cell (I_2^{127}, 1) and fluorescence
cells (I_2^{129}, 2; I_2^{127}, 3). Dye cell (4), photomultiplier (5),
liq. N₂ (6,7). Right: Side view of fluorescence cells with
I_2^{127} in ICA cell (1) frozen out (upper portion), and at room
temperature (lower portion) (from Ref. 6).

contains a flowing dye cell and an absorption cell filled with isotop-
ically pure I_2^{127}, which can be frozen out in a side arm. The laser emission, covering the spectral range of some 3 nm, could have been analyzed
by a spectrograph in order to detect the quenching of the light at the
absorption lines. However, a simpler approach makes use of a fluorescence
cell external to the laser: If the cell's content matches the absorber,
little or no fluorescence is excited, since the laser modes required for
excitation are quenched (see Fig. 2). With another species (a different
isotope, for example) in the external cell, however, the fluorescence
appears due to the resonance mismatch of the absorber and fluorescent.
The enhancement in sensitivity over that of extra-cavity absorption was
estimated to be 10^5.

Such dramatic effects deserve further investigation. During the one-
and-a-half decades elapsed since then, many aspects of laser intra-cavity

spectroscopy (ICS) have been studied in quantitative detail and their results have been condensed in more than 500 papers. Today we understand rather well the fundamental dynamics of multi-mode lasers with spectrally selective perturbations, and also their intrinsic benefits and restrictions when used as analytic tools. Although some features are still being clarified, ICS today deserves application on a broader scale.

In Sec. 2, we outline the dynamics of multi-mode lasers with the help of a rate-equation model, and we describe the time evolution of the mean values of the mode amplitudes and the effects of fluctuations. The energy accumulated in the spectral bins corresponding to the modes defines line shapes for both the emission and the perturbation. The latter is dealt with in Sec. 3. In Sec. 4, examples are discussed for the application of high-sensitivity ICS to the detection of weak linear and nonlinear absorption, of gain, and of weak light. The limitations on spectral and time resolution of ICS are described in Sec. 5.

2. Dynamics of a Multi-Mode Laser

We use rate equations for the modeling of a multi-mode laser. Possible effects of coherence are briefly discussed later.

2.1 Rate equations for a multi-mode laser

The system of rate equations for the photon number in the field modes and for the inversion of the laser medium with homogeneously broadened gain is [7,8]

$$\dot{M}_q = -\gamma M_q + B_q N(M_q+1) - \kappa_q c M_q \quad , \tag{1}$$

$$\dot{N} = P - N/\tau - N \sum_{q=1}^{n} B_q M_q \quad , \tag{2}$$

where M_q is the photon number of mode q, N is the density of the inversion, $\gamma = T_{ph}^{-1}$ is the broadband cavity loss rate, $\kappa_q c$ is the narrow-band absorption rate, P is the pump rate, and τ is the decay time of the inversion. The mode number q extends up to the total number of modes, n. The first term on the right-hand side of Eq. (1) describes the cavity loss, the second one the gain, and the third one the narrow-band loss by intra-cavity absorbers. In Eq. (2), the second and third terms on the rhs describe spontaneous and stimulated decay of the inversion, respectively.

So far, Eqs. (1) and (2) give the mean values of the parameters that characterize the laser dynamics. Fluctuations of the light field due to its quantum nature require us to amend Eq. (1) by a Langevin term $F_q(t)$, normalized as

$$\langle F_q(t) \rangle = 0 \quad ; \quad \langle F_q(t) F_p(t') \rangle = \gamma \langle M_q \rangle \delta_{qp} \delta(t-t') \quad .$$

Interaction of field modes can play a significant role at elevated power levels. This interaction is taken into account by an additional term $\sum_i \Phi(M_q M_{q+i})$ on the rhs of Eq. (1), where $\Phi(M_q M_{q+i})$ gives the transfer of light to mode q from the i-th neighboring mode by nonlinear mode coupling, e.g., by Brillouin scattering.

2.2 Solutions of the rate equations

For some simple but important cases, Eqs. (1) and (2) are easily solved analytically:

(i) With lack of intra-cavity absorption ($\kappa_q = 0$), negligible fluctuations due to time averaging [$F_q(t) = 0$], same gain for all modes ($B_q \equiv B$), and with the inversion adiabatically following the field ($\gamma < \tau^{-1}$), the stationary solution of Eqs. (1) and (2) is

$$N^o = \gamma/B \quad , \tag{3}$$

$$M_q^o = (\eta-1) \, P_{th}/\gamma n \quad , \tag{4}$$

where $P_{th} = \gamma/B\tau$ is the pump threshold, and $\eta = P/P_{th}$ is the normalized pump rate.

(ii) If we add loss κ in one particular mode q, i.e., $\kappa_q = \kappa$, threshold and inversion are determined by the remaining modes with small and non-specific loss. After N and M_q have reached stationary values N^o and M_q^o for $t > \gamma^{-1}$, the light flux in mode q_0 decays according to

$$M_{q_o} = M_{q_o}^o \, e^{-\kappa ct} \quad . \tag{5}$$

The minimum detectable absorptivity is defined as

$$\kappa_{min} = (ct_\ell)^{-1} \equiv \ell_p^{-1} \, , \tag{6}$$

where t_ℓ is the laser pulse duration. Thus, an ICA line is equivalent to a line of conventional absorption with the optical path length in the absorber $\ell = \ell_p$. In Fig. 1b, e.g., this length is $\ell_p = 300$ km.

The minimum detectable IC absorption κ_{min} is estimated to be equal to spontaneous emission $\kappa cM_q^o = BN^o = \gamma$ [see Eq. (1)]; then

$$\kappa_{min} = \gamma/cM_q^o \quad . \tag{7}$$

With parameters of a typical laser,

$$\gamma = 3 \times 10^7 \, s^{-1} \quad , \quad M_q^o = 3 \times 10^7 \quad ,$$

we have

$$\kappa_{min} = 3 \times 10^{-11} \, cm^{-1} \quad .$$

The laser pulse duration which is required if we want to exploit this sensitivity is

$$t_\ell = M_q/\gamma = 1 \, s$$

with the equivalent pulse length 3×10^5 km.

(iii) With homogeneously broadened gain of bandwidth Γ, the laser modes differ in their amplification. Close to its center, the gain profile may be approximated by a parabola [9]:

92

$$B_q = B_0 \{1 - |2(q-q_0) \, \nu/\Gamma|^2\} \quad , \tag{8}$$

where ν is the mode separation, and B_0 is the gain of the central mode q_0, which controls threshold and inversion in the stationary state:

$$N_o = \gamma/B_o \quad , \quad P_{th,o} = \gamma/B_o\tau \quad .$$

For high enough pump rate, maintained over the time γ^{-1}, the laser starts oscillating all over the gain profile. Soon, the inversion stabilizes itself at N_o, and the emission narrows [9,10] according to Eq. (1):

$$M_q = M_q^o \, (\nu/\Gamma) \, (\gamma t/\pi)^{-1/2} \, \exp\left[-\gamma t \left(\frac{(q-q_o)\nu}{\Gamma/2}\right)^2\right] \quad . \tag{9}$$

The number of oscillating modes decreases in time as

$$n \approx \Gamma/\nu(\gamma t)^{1/2} \quad . \tag{10}$$

This "spectral condensation" is shown in Fig. 3 [10]. The spectrochronogram represents the spectrum of a multimode cw dye laser along the ordinate direction, which is time-scanned by a streak camera in the abscissa direction. After an interruption of its oscillation, the laser starts anew with a broad emission band whose width decreases in time, according to Eq. (9). In addition, the figure shows the decay of the light at wavelengths coinciding with atmospheric water absorption [compare Eq. (5)]. Similar phenomena have been observed under various conditions [9,11,12]. In a recent experiment [12], the three-mirror dye laser could be bandwidth-limited (<20 cm^{-1}) by a glass etalon. The pump light was switched on for 1 ms, and the laser spectrum was recorded with 10 µs gate time and variable decay, controlled by an acousto-optical deflector. Up to 500 µs delay, the validity of Eqs. (5) and (9) was confirmed with and without intra-cavity etalon, i.e., for different Γ.

In contrast with spectral condensation, which shows up with a homogeneously broadened gain medium, the emission band of a multi-mode laser widens in time with an inhomogeneously broadened gain profile: Spectral subgroups of molecules farther in the wings reach the threshold at a later time. The decay of the light flux at absorption lines has been proved valid, however, for at least 300 µs after start of the oscillation [13].

(iv) If the emission bandwidth of the laser is small compared with the absorption linewidth, we attribute the same absorptivity to all modes,

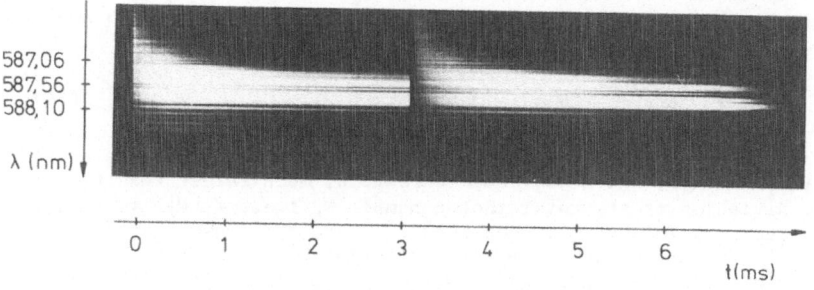

Fig. 3. Spectrochronogram of a multi-mode cw Rh-6G laser with water vapor ICA (from Ref. 10).

$\kappa_q = \kappa$. Now, the extra absorption is simply compensated by enhanced inversion,

$$N = (\gamma + \kappa c)/B \quad , \tag{11}$$

and the total photon number

$$M = P_{th} \frac{\eta - 1 + \kappa c/\gamma}{\gamma(1 + \kappa c/\gamma)} \tag{12}$$

is equally distributed among all modes. Then, the absorption coefficient is

$$\kappa = (1 - M/M_0) (\eta-1) \gamma/c \quad , \tag{13}$$

where M_0 is the photon number when no absorber is placed inside the laser resonator. We notice that the sensitivity of narrow-band lasers to ICA is high for small loss, and when operating close to the threshold ($\eta-1 \ll 1$). With the typical parameter values $\gamma = 3 \times 10^7 \text{ s}^{-1}$, $\eta = 1.5$, and $M/M_0 = 0.5$, we have $\kappa \simeq 10^{-4} \text{ cm}^{-1}$. The smallest detectable absorption could perhaps be pushed down by an order of magnitude closer to threshold. However, further improvement of sensitivity is hardly feasible in a narrow-band laser: Increasing the threshold by extra loss raises the steady-state gain, and that loss in turn has a less significant effect on the output power. In contrast, a broad-band laser acts differently upon narrow-band absorption: the pump threshold is controlled by the modes of minimum loss, whereas all other modes suffer exponential decay of their output. This decay is faster for modes subject to extra absorption. Only the total inversion couples the modes in the model used so far, and accordingly this contrast becomes more and more entrenched until the stationary state is reached after $t \simeq 1$ s. On the other hand, this is the time the spectrum would require to recover its initial shape if the extra absorption were suddenly removed [14].

2.3 Quantum fluctuations

So far we have dealt with a model that describes the dynamics of a multi-mode laser in terms of mean values of the light flux in individual modes q. However, the weakness of mode coupling and the concomitant sensitivity to perturbation leaves this system susceptible to large variation of the light flux M_q with mode number q. The values of M_q, initiated by spontaneous emission, also vary in time about their mean values. These "quantum fluctuations" are suitably modeled by an additional term on the rhs of Eq. (1) which describes a Langevin random force [15,16]. The analysis of the model thus modified shows that the values of M_q are exponentially distributed, and the fluctuations are large: The probability that mode q accumulates the photon number M_q is

$$P(M_q) = \langle M_q \rangle^{-1} \exp(-M_q/\langle M_q \rangle) \quad . \tag{14}$$

The mean absolute deviation of the photon number in mode q from its average is $\langle |\Delta M_q| \rangle \propto \langle M_q \rangle$, which is, above threshold, much larger than the corresponding deviation of the total photon number M, i.e., $\langle |\Delta M| \rangle \propto \sqrt{\langle M \rangle}$, or the fluctuations in a single-mode laser. Therefore those quantum fluctuations of individual modes are easily detectable [17]. A $LiF:F_2^+$ color center laser of 300 ns pulse length, pumped by a Xe ion laser, has been used for this purpose. Its spectral emission range was 899 nm through 907 nm, its cavity length 25 cm. With a 1-m spectrograph, individual modes have been resolved in its output. Figure 4 shows densitograms of

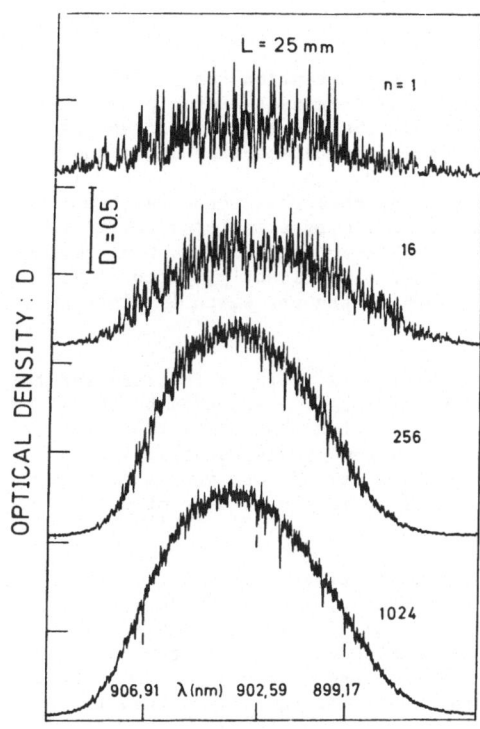

Fig. 4.
Densitograms of spectra of the output of a LiF:F_2^+ color center laser (n superimposed light pulses) (from Ref. 17).

recorded spectra with the number of superimposed light pulses varied. The top trace demonstrates a spectrum whose height distribution approximately obeys an exponential law. The deviations of the spectral flux from its mean value in <u>multi-pulse spectra</u> — or in spectra averaged by insufficient resolution — have a Gaussian distribution:

$$P(\Delta M_q) = (\sigma \sqrt{2\pi})^{-1} \exp|-(\Delta M_q)^2/2\sigma^2| \quad , \tag{15}$$

where $\sigma^2 = \sigma_0^2 \langle M_q \rangle^2 / jn$ is the dispersion, n is the number of superimposed pulses, and j the number of unresolved modes. Examples of spectra that are characterized by this distribution are shown in the lower traces of Fig. 4. A long laser cavity with its narrowly spaced modes suppresses the observation of the large quantum fluctuations in the frequency spectrum. In time domain, their characteristic duration is:

$$t_q \simeq \langle M_q \rangle \gamma^{-1} \quad . \tag{16}$$

For typical laser parameters [see evaluation of Eq. (7)], $t_q \simeq 1$ s. However, fluctuations on the order of milliseconds have been observed; their duration decreases with increasing laser output power. Thus, another mechanism is required to which the observed fluctuations can be attributed. This mechanism is a nonlinear coupling of the laser modes.

2.4 Mode coupling

There is much evidence for the existence of fluctuations on a millisecond time scale, whose mean period <u>decreases</u> upon increasing laser power [7,8, 10,18–25]. An example is shown in Fig. 5. The experimental setup for this recording was similar to the one used for the spectrochronograms of Fig. 3, but with spectral resolution being 10^6, in order to resolve individual modes. Satisfactory explanation of this phenomenon has been achieved recently on the basis of mode coupling by stimulated Brillouin scattering (SBS) [7,8]. To model this nonlinear interaction of modes, we assume that only neighboring modes interact significantly. Accordingly, Eq. (1) has to be amended by an extra term on the rhs which is given by $DM_q(M_{q+1}-M_{q-1})$.

When a realistic value on the order of 10^{-3} s^{-1} [7] is inserted for the coupling constant D, numerical solutions of Eq. (1), extended by this term, show strong fluctuations of mode power with characteristic times that are between tens of microseconds and a few milliseconds, in fair agreement with the observations. These time periods are much shorter than those inferred from quantum fluctuations in the absence of other dynamic perturbations.

An analytic estimate of the characteristic time of deterministic fluctuations is derived as follows [7]: Under stationary conditions, the first two terms on the rhs of the modified Eq. (1) cancel, and we set $\kappa_q = 0$, and $M_{q+1} - M_{q-1} \simeq \langle M_q \rangle$, as the amplitude of spectral fluctuations, according to Sec. 2.3, is on the order of the mean spectral power density. Since $F_q \simeq \sqrt{\langle M \rangle} \ll \langle M \rangle$ for large photon number M, Eq. (1) reduces to

$$\dot{M}_q = D \langle M_q \rangle M_q \quad , \tag{17}$$

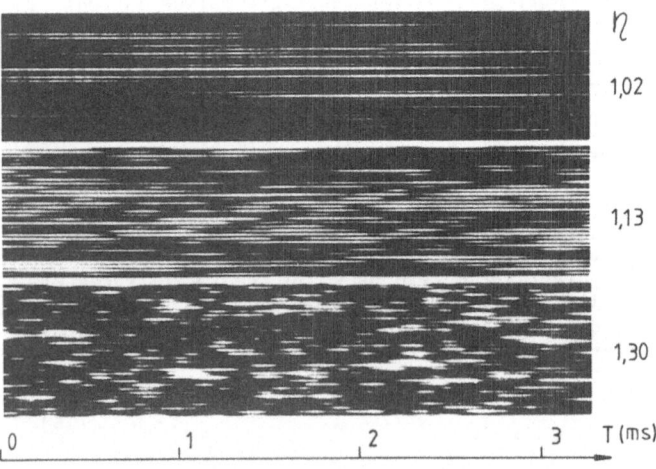

Fig. 5.   Spectrochronograms of the output of a cw Rh-6G laser at high spectral resolution, for various values of the relative pump rate η. The horizontal lines are individual modes separated by $\Delta\nu = 0.033$ cm^{-1} (from Ref. 25).

which gives, for the characteristic time of fluctuations in the output power of individual modes,

$$t_n \simeq (D \langle M_q \rangle)^{-1} \quad . \tag{18}$$

With $\langle M_q \rangle \simeq 10^6$, this expression gives the realistic value $t_n \simeq 1$ ms. Being the mean time for unperturbed oscillation in long enough dye laser pulses, t_n is the width of the intensity autocorrelation of individual modes and determines the sensitivity of ICAS. In general,

$$\kappa_{min} = (ct_m)^{-1} \quad ; \quad t_m = \min\{t_\ell, t_q, t_n\} \quad , \tag{19}$$

where t_ℓ and t_q are the laser pulse duration and the mean length of quantum fluctuations, respectively. Figure 6 shows how the sensitivity to absorption varies upon pump power and pulse length, when t_ℓ is either smaller or larger than t_n.

Numerical solutions of Eq. (1) for 100 modes and with different approximations discussed above are shown in Fig. 7 [7]. The spectrum on the top represents the simple case of an ICA spectrum described in Sec. 2.2(ii). The second spectrum includes quantum fluctuations (2.3), which are reduced by averaging over 10 laser pulses in the third spectrum (compare with the data of Fig. 4). The bottom spectrum includes strong mode coupling. An interesting feature of this last spectrum is the conspicuous asymmetry as a result of the nonlinear mode interaction. We present more detail on this problem in Sec. 3.

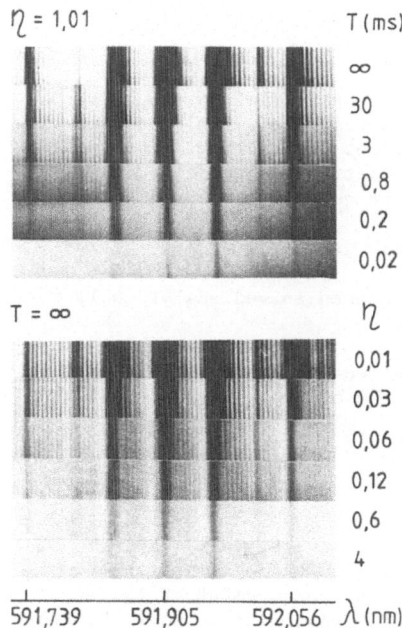

Fig. 6. ICA spectra of water vapor obtained with various pulse lengths of Rh-6G laser pulses at the relative pump rate $\eta = 1.01$ (top). Corresponding spectra at various pump rates, for cw operation (bottom) (from Ref. 25).

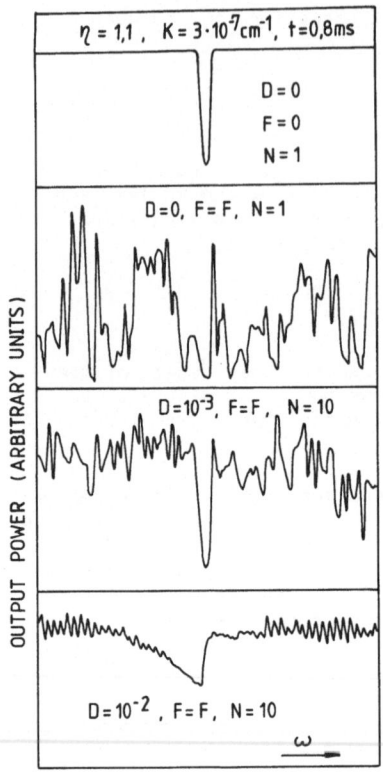

Fig. 7. Numerical solutions of rate equations for a 100-mode laser with T = 0.8 ms pulse duration, single-line ICA with κ = 3 x 10-7 cm-1 in center of emission spectrum, and η = 1.1, for various values of mode coupling constant D, Langevin random force F, and number of superimposed laser pulses, N (from Ref. 7).

2.5 Spatial hole burning

For some time in the history of ICAS, the role of spatial hole burning was controversial. It now appears that it does not severely limit the sensitivity of the method under most conditions met in practice. For quantitative evaluation, M_q and N must be written spatially dependent, and a Fourier decomposition is useful for the former quantity:

$$M_q \rightarrow M_q(z) = M_q^o \sin^2 q\nu z \quad ; \quad N \rightarrow N(z) \quad . \tag{20}$$

With this modification, Eqs. (1) and (2) have been solved analytically for some limiting cases [6,9,26,27].

(i) The gain medium is thought to fill the entire cavity. In the limit of a large mode number, n, the decrement of the output power of an attenuated mode is [6]

$$\frac{dM_q}{M_q} = -d\zeta \cdot 2n \cdot \frac{\alpha/\zeta}{\alpha-\zeta} \quad , \tag{21}$$

where α and $\zeta = \kappa L$ are the gain and extinction, respectively, taken to be equal for all modes unaffected by the absorber; L is the length of the resonator. The second and third factor show the enhancement of the attenuation due to ICA over the corresponding single-pass interaction. Enhancement according to the third factor, $(\alpha/\zeta)(\alpha-\zeta)^{-1}$, is expected for a single-mode laser with homogeneously broadened gain, if its output adjusts such that the saturated gain equals the loss. The second factor, $2n$, gives the genuine enhancement due to multi-mode oscillation. We estimate $\kappa_{min} = d_\zeta/L \simeq 10^{-7}$ cm^{-1} from this analysis. Necessarily, this situation is approximated, at best, since the absorber also requires resonator space.

(ii) The gain medium is concentrated in a thin slab of thickness $1 \ll L$. An analytical solution of Eqs. (1) and (2) including the spatial inhomogeneity shows [9] that the modification of the system is slight, and still absorptivity values as small as $\kappa \simeq 10^{-14}$ cm^{-1} can be detected. Under this condition, the previously discussed limitations have more severe effects on the sensitivity of the method.

If the gain medium fills only part of the laser cavity ($1 < L$), spatial inhomogeneities dominate the sensitivity of ICA only if the location of that medium is close to one of the cavity mirrors in a standing-wave configuration [26,27].

2.6 Sensitivity limits of ICA

We are now in a position to evaluate and compare the physical factors that can limit the sensitivity of ICA. With long enough laser pulses, or with cw operation, spontaneous emission and nonlinear mode coupling by SBS (providing the laser fluctuations) may effectively limit that sensitivity. The former effect fades with increasing laser power, the latter one grows. For best sensitivity the laser power is optimum when both mechanisms equally contribute. Equating t_ℓ and t_n [Eqs. (5a) and (17)] we find

$$\langle M_q \rangle_0 = \sqrt{\gamma/D} \qquad (22)$$

with $\gamma = 3 \times 10^7$ s^{-1} and $D = 10^{-3}$ s^{-1}, $\langle M_q \rangle_0 = 2 \times 10^5$. With 50 oscillating modes, this value corresponds to 0.1 mW of laser output power, which requires us to keep the laser closely above threshold, $\eta = 1.004$. Practically, we may achieve $P/P_{th} \gtrsim 1.01$ when the power stability of the pump laser is on the order of 1%. Thus, in most cw dye lasers, mode coupling by SBS determines the ultimate sensitivity.

From Eqs. (6), (18), and (22), we find the minimum absorptivity, which indicates the absolute ICA sensitivity of a cw dye laser:

$$\kappa_{min} = \sqrt{D\gamma}/c \qquad . \qquad (23)$$

This value is $\kappa_{min} = 6 \times 10^{-9}$ cm^{-1} for the example given above, and represents the maximum sensitivity reported so far with cw dye lasers [7,9,10,28,29]. Of course, this sensitivity can be accomplished only if fluctuations of technical origin are avoided, such as gas bubbles in the dye liquid, and, for a broader range of lasers, mechanical vibrations, dust in the optical path of the resonator, instability of the pump light level, and etalon effects due to reflection and/or scattering off resonator components.

In particular situations, other physical mechanisms not discussed so far can affect ICA sensitivity. In relevant studies, the saturation of the absorber [21], partial mode locking [30,31], self-focusing [32,33], and coherent contributions to the interaction of the light with the absorber [34] have been considered. Their effects are important under certain particular experimental conditions only.

3. Intracavity Line Shapes

When the multimode laser emission is analyzed by a spectrally selective detector with integration time τ_d, the light flux in individual modes, according to Sec. 2, is supposed to vary in time and frequency. The detection unavoidably includes time averaging over τ_d or the laser pulse length, t_ℓ, whichever is shorter. If the accumulated energy of the modes — or mode groups, with lower spectral resolution — varies slowly enough with frequency, we may define its envelope as a line shape. With the general condition $\gamma_0 \ll \Gamma$, this line shape is a superposition of the broad emission band and the narrow ICA line shape due to the perturber, of width γ_0.

Experimentally, symmetric and inconspicuous ICA line shapes have been observed [1,2,9,11-13,28]. Quantitative evaluation of ICA line shapes has been achieved with the recording of rotational overtones of O_2 in the visible [35]. Fitting of the lines using Voigt profiles with suitable contributions of Doppler and pressure broadening has been satisfactory, and no systematic line distortion was noticed. On the other hand, several observations have revealed complex asymmetric lines [7,8,23,30-34,36-41], sometimes with spectral light condensation on one or both wings of the line. These line distortions, if intrinsic to ICAS, would severely limit the applicability of the method. Fortunately, the distortions can be traced to two types of effects which can be avoided under most experimental conditions.

3.1 Mode stability of the cavity shifted by extra dispersion

The anomalous dispersion of the intra-cavity absorber modifies the diffraction loss of the cavity, κ_0, such that a frequency-dependent contribution develops [37]. For simplicity we assume that the cavity is formed by two spherical mirrors. We expand that contribution up to the first order in terms of the stability parameter $G(n) = g_1(n) \cdot g_2(n)$ of the cavity,

$$\kappa_d(\omega) = \kappa_0 + (d\kappa/dG)(dG/dn)\Delta n.$$

Here, $n-1$ is the dispersion of the absorber, $g_1(n) = 1 - L/r_i - \ell/nr_i$, and $L + \ell$, r_i, and ℓ are the distance and radii of curvature of the mirrors, and the absorber length, respectively. We combine this spectrally varying loss with the line absorption

$$\kappa(\omega) = k \, \mathscr{L}(\omega) \tag{24}$$

to form the overall extinction $\varepsilon(\omega)$. Here, $k = 2\pi\gamma_0 \cdot N \, c^2/\omega_0^2$, $\mathscr{L}(\omega) = \Gamma_0[\Gamma_0^2 + (\omega-\omega_0)^2]^{-1}$, N is the absorber density, γ_0 is the radiative absorption linewidth (HWHM), and ω_0 is the resonance frequency. This extinction is

$$\varepsilon(\omega) = k \, \mathscr{L}(\omega) + (d\kappa_d/dG) \times$$
$$[c_1 k \, \mathscr{L}(\omega) \cdot (\omega-\omega_0)/\Gamma_0 + c_2 k^2 \, \mathscr{L}^2(\omega) \cdot (\omega-\omega_0)^2/\Gamma_0^2] \quad, \tag{25}$$

where $c_1 = (g_1(1) - \ell/r_1)\ell/r_2 + (g_2(1) - \ell/r_2)\ell/r_1$, and $c_2 = 2\ell^2/r_1 r_2$.

In $\varepsilon(\omega)$, the first term in the brackets represents a line asymmetry, the second one a symmetric distortion. Extra contributions to ε by lensing may exist because of spatial inhomogeneity of the absorber (or of the light field) if nonlinear interaction is appreciable. To test this model, ICA spectra have been recorded with a pulsed LiF:F_2^+ color center laser, using Cs vapor in an intracavity heat pipe as absorber [37]. The observed spectral features are satisfactorily explained by the outlined model.

The analyzed type of line distortion is avoided when a resonator configuration is chosen which corresponds to minimum diffraction loss, i.e., $d\kappa_d/dG = 0$. Then, all the <u>spectrally varying</u> loss is genuine intracavity absorption.

3.2 Nonlinear interactions of the light

In certain experiments including cw dye lasers a characteristic line shape with enhanced red wing has been observed [7,8,23,33,36]. This distortion affects all ICA lines, and all narrow spectral structures, irrespective of their strength. Since it is related to the presence of light tightly focused into the dye jet, it is traced to the same nonlinear interaction of the light with the gain medium, which is responsible for the coupling of individual modes, i.e., SBS (see Fig. 7). To remedy this line distortion the laser pump power must be reduced, and the mode density increased by choosing a long laser cavity.

Enhancement of the light flux on those modes, which lie on one wing, or on both wings, of an ICA line, has been reported many times [30-34,38-41]. This phenomenon seems to develop upon modulation of the (inverted) population in the gain medium. If the period of modulation slightly exceeds that of a round-trip of the light in the cavity, mode coupling is maximum and loss is minimum <u>in the wings</u> of the absorption line, where the mode separation is smaller due to anomalous dispersion and matches the modulation. Stable mode locking somewhat off the transferred modulation frequency has been observed with a bichromatic helium-neon laser [42].

The above phenomenon has been demonstrated with external modulation of the laser gain [40,41]. This parametric interaction of the light can also show up, however, upon (partial) mode locking of the laser to a saturated IC <u>absorber</u> line [30,31,34,38,39]. Due to time-dependent phase shifts between neighboring spectral polarization components, the absorber imposes time-varying gain, rather than loss, upon light modes, which may time-average as net <u>gain</u> at a suitable frequency distance. Thus, this spectrally selective amplification gives rise to enhanced light flux in the wings of the line. The existence of parametric gain, which is similar in its orgin to this gain, has been pointed out in the context of saturation spectroscopy in simple systems that include only two light modes [43,44].

These nonlinear phenomena usually occur at elevated IC laser power levels, and with strong IC absorbers. They can be avoided in many situations of experimental interest.

4. Applications of ICS

In the following we describe the preconditions and several experiments for sensitive and time-resolved detection of linear absorption, of nonlinear absorption, of gain, and of light pulses.

4.1 Detection of linear absorption

From the very beginning, laser interactivity spectroscopy has been considered a technique for the detection of very weak optical extinction. In Sec. 2 we have shown that the spectral output, in mode q, of the broadband laser including loss obeys the Lambert—Beer law $J_q(\omega,t) = J_q(t) \times \exp(-\kappa_q(\omega)ct)$, where the optical path ct corresponds to the length of the uninterrupted wave of a particular laser mode. From the temporal decay of the light flux, the absorption can be determined from at least two measurements of the instantaneous spectral flux at the absorption line, which are separated in time by the interval θ, and corresponding normalization measurements outside the line:

$$\kappa_q = (c\theta)^{-1} \ln \left[\frac{J_q(t)}{J_{q+\delta q}(t)} \middle/ \frac{J_q(t+\theta)}{J_{q+\delta q}(t+\theta)} \right] \quad . \tag{26}$$

In this way, e.g., the absorption of rotational lines of the $000 \to 043$ transition of the CO_2 molecule was measured with a streak camera for time-resolved spectral recording [13]. With the laser light gated by an electro-optical switch for 10 μs, ro-vibrational overtone spectra of atmospheric H_2O [12] and O_2 [35] were measured. On the other hand, κ_q may be determined by a measurement of the time-integrated light flux, normalized to the flux outside the absorption line:

$$R \simeq \frac{\int_0^\theta J_q(t) \exp(-\kappa_q ct)dt}{\int_0^\theta J_{q+\delta q}(t)dt} \quad . \tag{27}$$

The relationship between κ_q and ln R is, in general, nonlinear. With constant flux during the pulse time t_ℓ, e.g., we have [13]

$$R = (\kappa_q ct_\ell)^{-1} [1 - \exp(-\kappa_q ct_\ell)] \quad . \tag{28}$$

As pointed out before, the light flux of a homogeneously broadened laser is not constant in time, but rather varies as $\sqrt{t}$ for modes in the central part of the emission profile. For these modes, the analytic result of Eq. (27) is [45]

$$R = \exp[-\kappa_q ct_\ell] \sum_{n=0}^{\infty} \frac{[\kappa_q ct_\ell]^n}{(\frac{3}{2} + n)!} \quad . \tag{29}$$

If $\kappa_q ct_\ell < 1$, we may restrict the sum to the leading term such that κ_q varies linearly with ln R. In this approximation, R is underestimated by less than 10%. Although accuracy and sensitivity of this version of IC measurements is slightly lower than that of time-resolved observation, it is much more convenient and has been used in most experiments so far.

Intracavity spectra are usually recorded photographically or photoelectrically with a spectrograph. Spectral resolution in these measurements is usually limited by the spectrograph. Since a laser beam is analyzed, resolving power need not be traded off for satisfactory transmission. In fact, resolution as high as 10^6 has been achieved with special designs [7].

The photographic recording of time-resolved spectra, or spectrochrono-grams, includes a streak camera for real-time observations, or some kind of gating of the laser light, for sampling techniques.

Photoelectric recording makes use of a vidicon, of photodiode arrays, or of a complete optical multi-channel analyzer. These photoelectric devices allow direct processing of the acquired data.

In various ways, the application of ICAS has extended the spectroscopic sensitivity limits of conventional absorption measurements. With a Nd-glass laser for example, extinction values as small as 10^{-9} cm^{-1} have been measured routinely. Consequently, extremely weak overtones of poly-atomic molecules have been detected, e.g., HN_3 (00000 $\rightarrow$ 30000) (Fig. 1), CO_2 (00°0 $\rightarrow$ 04°3), CH_4 (000 $\rightarrow$ 123), NH_3, C_2H_2, and C_2HD [2].

With the use of cw dye lasers and color center lasers, the number of accessible molecules and radicals is immense. The sensitivity of a cw dye laser to very weak absorption has been demonstrated by detecting the 3-0 band of the $b^1\Sigma_g^+ \leftarrow X^3\Sigma_g^-$ magnetic dipole transition of O_2 [15,36]. With 0.016 cm^{-1} resolution, absorption coefficients as low as 10^{-8} cm^{-1} have been measured, and line shapes have been recorded.

Other experiments were devoted to systematically recording atmospheric absorption [9,28,29,46]. The duration of undisturbed emission in indi-vidual modes in these experiments was several milliseconds, giving rise to 3×10^{-9} cm^{-1} minimum detectable absorptivity. Within the spectral range accessible with lasing of rhodamine 6G (584 nm through 603 nm), 717 lines have been recorded, 340 of which were not previously observed in the atmo-spheric absorption of the solar spectrum. Twenty-eight of these lines were identified as being caused by weak NO_2 pollution of the air in the laboratory [28].

4.2 Detection of two-photon absorption

The extension of ICAS to studies of nonlinear absorption is straightfor-ward. A two-photon absorption line, e.g., shows up in ICAS, if the ab-sorber is irradiated simultaneously by additional strong narrow-band light. A photon from the broad-band laser is chosen, which makes up for the energy defect between a photon of the narrow-band light and a two-photon resonance transition. Excitation of this line is characterized by a cross section that is resonantly enhanced when a real level, whose parity is opposite to the parity of the ground state, is close to the intermediate virtual level.

This situation can be met conveniently in alkali vapors. Two-photon absorption of sodium in a flame is easily detectable this way [47]. A measurement of the cross section for two-photon absorption in potassium vapor included a narrow-band ruby laser and a broad-band DOTS dye laser excited by the same ruby laser [48]. In fact, two states, the fine struc-ture levels $^2P_{1/2}$ and $^2P_{3/2}$, enhance the absorption probability but can give rise to a destructive interference [49]. This interference of in-distinguishable transitions has been demonstrated in two-photon absorption by monitoring the rate of excitation with the subsequent fluorescence [50]. Direct two-photon absorption observed recently in an experiment on sodium vapor is shown in Fig. 8 [51]. A pulsed broad-band dye laser was tuned to the frequency range of Na 3P-5S absorption, and a pulsed narrow-band laser close to the 3S $\rightarrow$ 3P resonance transitions, say, to the wavelength λ_{12}. One observes, in the spectrum of the broad-band laser, an

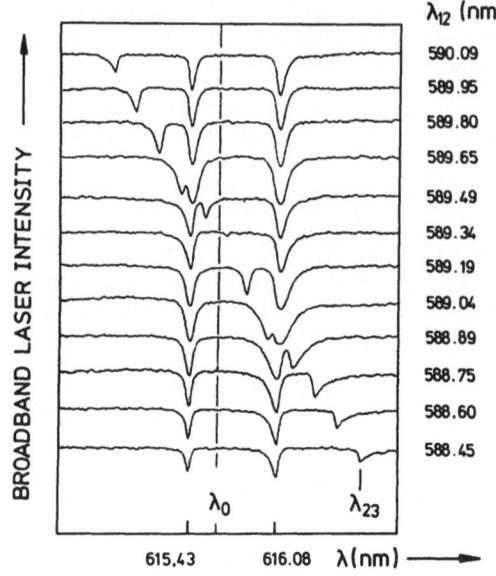

λ_{12} (nm)

590.09
589.95
589.80
589.65
589.49
589.34
589.19
589.04
588.89
588.75
588.60
588.45

λ_0 λ_{23}

615.43   616.08 λ(nm) ⟶

BROADBAND LASER INTENSITY ⟶

Fig. 8.
Two-photon absorption in Na, observed with broad-band dye laser, on the transition 3S-5S. λ_{12}, λ_{23} correspond to first and second photons, respectively. The lines at 615.43 nm and 616.08 nm represent single-photon absorption at the 3P-5S transitions (from Ref. 51).

additional absorption line of wavelength λ_{23}, which varies with λ_{12}, such that $\nu_{12} + \nu_{23} = \nu_{3s-5s}$. This absorption line corresponds to the second photon of the two-photon transition. For $\lambda_{12} = 589,40$; $\lambda_{23} = \lambda_0 = 615,64$ the amplitude of the two-photon line vanishes due to destructive interference of the two excitation channels.

4.3 Detection of amplification

Certain groups of modes of a multi-mode laser with a broad emission band may be enhanced in amplitude rather than quenched if frequency-selective gain is made available inside its resonator. In an ICAS experiment using a LiF:F_2^+ color center laser, which was pumped by the green emission of a pulsed xenon ion laser, the absorption of I_2 was to be studied [52]. Surprisingly, gain was observed on numerous ro-vibrational lines (see Fig. 9), since left-over pump light inverted the corresponding transition. In this way, pumping conditions and available inversible transitions can be studied systematically.

4.4 Detection of injected light

So far we have considered applications of ICAS where the gain of the broad-band laser was decremented by spectrally selective absorber loss or incremented by extra gain. There is a third way to exploit the stunning sensitivity of that system to frequency-selective perturbation: the injection of narrowband light pulses. In contrast to the previous variants, the light field is manipulated by placing an excess photon number in a particular mode, or group of modes. In a single-mode laser, the response of the inversion, characterized by time τ, is strong, and it quickly reestablishes the light amplitude. In multi-mode lasers, on the other hand, the instantaneously enhanced light amplitude of a particular mode (or mode group) does not much affect the total inversion. Stimulated emission funnels light into that mode whose amplitude is preserved over a

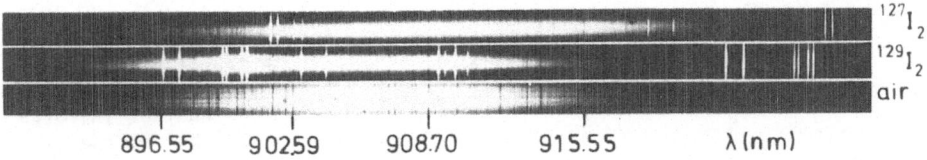

$^{127}I_2$

$^{129}I_2$

air

896.55   902.59   908.70   915.55 λ (nm)

Fig. 9.   IC gain spectra of I_2 isotopes, and atmospheric absorption (from Ref. 52).

much longer time. In practical situations, this time duration is determined by the time of uninterrupted oscillation of the modes in the broadband laser. In a recent experiment [53] (see Fig. 10), 2-ns pulses of a N_2-pumped coumarine-153 laser of 0.8 cm^{-1} spectral width were injected into a broad-band (30 cm^{-1}) jet-stream Rh-6G laser of 1,5-μs pulse length, excited by a xenon ion laser. The near concentric cavity of the broadband laser consisted of mirrors with 5 cm and 8 cm radius of curvature. The multi-mode light was analyzed by a 1-m spectrograph and a 1728-channel diode array. Simultaneously, the total instantaneous light flux was temporally resolved by a fast transient digitizer. Figure 11 shows spectra of the broad-band laser emission, and pulse shapes of the total light flux, both for various values of the time delay of the injection. Upon earlier arrival, the injected light pulses redistribute the light generation in the multi-mode laser more efficiently into the spectral channel subject to the injection. This light flux extension enables one to detect weak spectrally selective light pulses, since the time-integrated power of these pulses is amplified at the expense of the multi-mode laser gain. Although the sensitivity has not been pushed to its limits in this exploratory experiment, we stress the point that the prodigious sensitivity of multi-mode lasers to spectrally selective perturbation offers considerable prospects for the detection of light.

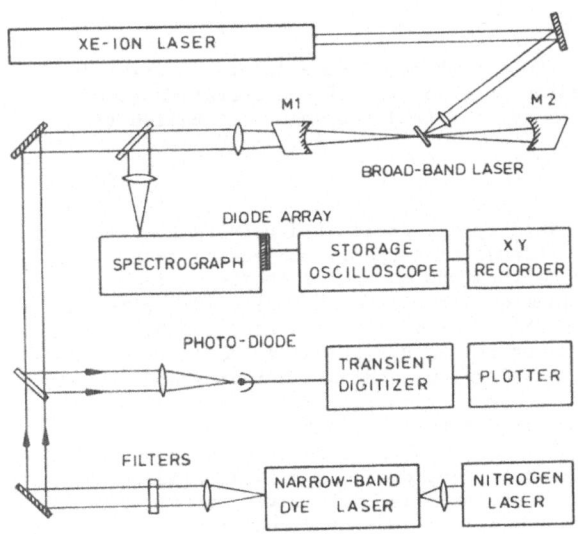

Fig. 10. Experiment for IC detection of narrow-band light emission (from Ref. 53).

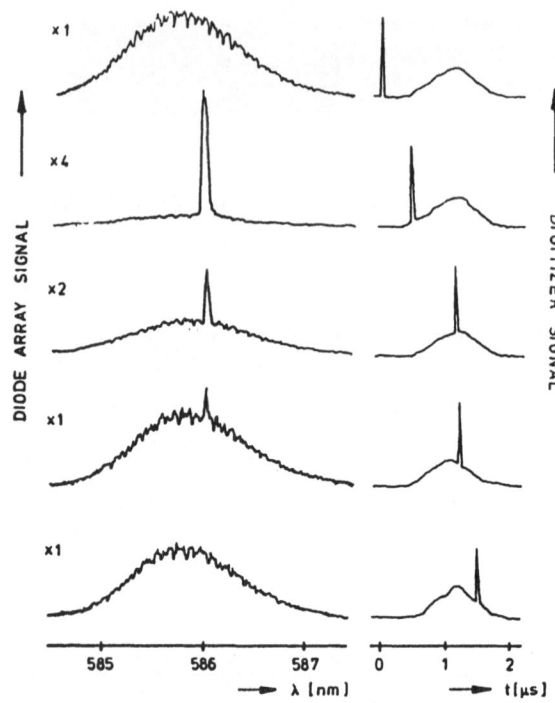

Fig. 11. Emission spectra of broad-band dye laser (left), and time-resolved superposition signal of injected light and spectrally integrated light of the broad-band laser. The injection delay increases from top to bottom (from Ref. 53).

5. The Ultimate Resolution of ICS

The high sensitivity of ICS is complemented by a considerable potential with respect to time and frequency resolution.

5.1 Time resolution

The sensitivity of the ICA technique with multi-mode lasers is determined, as shown above, by the laser pulse duration t_ℓ, if the operating conditions of the laser can be set so as to extend the mean fluctuation time beyond t_ℓ. Then, we have

$$\kappa_{min} \cdot t_\ell = 1/c \quad , \tag{30}$$

and the time resolution, determined by t_ℓ, is limited only by the required sensitivity of the detection. Indeed, time resolution can be traded for sensitivity within wide limits, with 10 ns being the shortest practical resolution time. With the minimum detectable absorptivity 10^{-5} cm^{-1}, e.g., the resolvable time is on the order of a microsecond [45,52]. Thus, ICS is applicable, if we want to track the variation of the macroscopic parameters of an absorber, as an afterglow, and/or the microscopic kinetics of its constituents. As an example, time-resolved absorption spectra of He$_2$ in the afterglow of a pulsed electric discharge are shown in Fig. 12 [45]. The multi-mode LiF:F$_2^+$ laser, with wavelength range between 900 and 930 nm and with 200 ns pulse duration, probed a pulsed helium discharge with variable delay. Each of the laser pulses was marked by spectral modulation from the absorption by two ro-vibrational bands of the

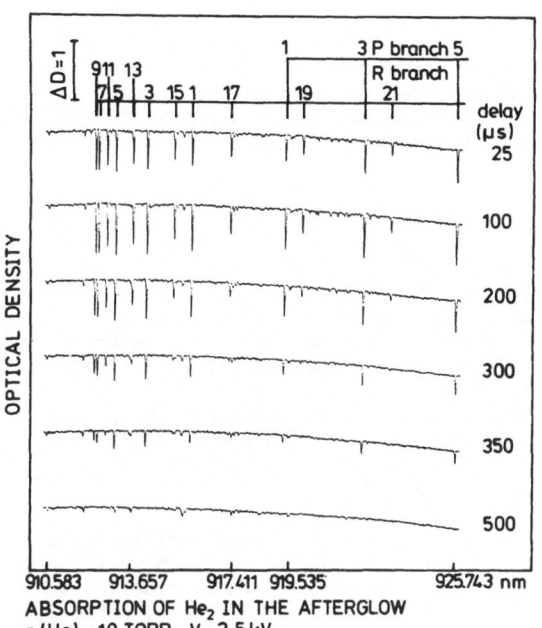

ABSORPTION OF He₂ IN THE AFTERGLOW
p(He) = 10 TORR, V = 2.5 kV

Fig. 12. ICA by He_2 in the afterglow of a He discharge. Excitation voltage 2.5 kV, He pressure 13.3 mbar (from Ref. 45).

electronic transition $c\Sigma_g^+ - a^3\Sigma_u^+$ starting in the lowest metastable state. Hoenl-London plots derived from these measurements enable one to determine the rotational temperature of the He_2 molecules, its deviation from thermal equilibrium, and the variation of temperature and metastable population as functions of the time delay in the afterglow.

High sensitivity along with time resolution is a prerequisite for studies of the kinetics of radicals [54,55]. In fact, ICAS has been shown to be the only analytic technique for measurements of the kinetic constants of reactions involving radicals such as NH_2, HCO, HNO, or PH_2 [55].

5.2 Spectral resolution

The spectral distance of modes in a multi-mode laser determines the spectral resolution in the IC spectrometers used to date. For 0.5 m resonator length, this spectral distance is 0.01 cm^{-1} or 300 MHz. The corresponding resolution is satisfactory for many Doppler-broadened lines and for molecular absorption at atmospheric pressure.

High spectral resolution, corresponding to the emission bandwidth of an individual mode, could be obtained if a variant of the detection technique described above is used. Instead of broadband-detecting the entire spectrum of the laser with a spectrograph, one would use a monochromator as a narrow-band filter for a particular mode. The frequency of this mode would be scanned across the laser spectrum in the conventional way, e.g., by piezo-electric drive of a cavity mirror. A measurement of the light flux in this mode would combine ultra-high sensitivity with high resolution. The latter is, in principle, limited by the Schawlow-Townes condition [56] for the ultimate laser bandwidth only, which is 30 mHz, or smaller. Of course, at present, technical noise keeps the limit at a level several orders of magnitude less favorable, on the order of 1 MHz for commercial laser systems. However, by use of intricate control circuits, cw dye lasers with a bandwidth of some hundred Hz, and He−Ne lasers with a bandwidth of 100 mHz have been demonstrated in the laboratory [57].

6. Conclusions

The scope of this presentation is necessarily limited: we have not aspired to offer a complete account of work on, or with, ICS. In particular, attempting to cover all the pertinent work that has been done in molecular spectroscopy and chemical kinetics would far exceed the limits of this article. Instead, we have tried to point out two facts that are, it seems, not fully appreciated by the broader spectroscopists' community: First, the considerable measure of understanding we now have of the fundamentals that underlie ICS, and second, the level of sophistication and maturity this field has acquired recently.

In adaptation of, but also — with due respect — challenging Arthur Schawlow's popular definition of a diatomic molecule, we claim that there indeed exist niches in physics where ONE is not enough! Eventually, what is lost in precious simplicity of the system under scrutiny is, perhaps, more than regained with inherent features, that allow the spectroscopist to sometimes extend spectacularly the limits previously put on his techniques.

P.E.T. is indebted to the JILA Visiting Fellows Program. V.M.B. appreciates support by the Alexander von Humboldt Foundation.

References

1. L. A. Pakhomycheva, E. A. Sviridenkov, A. F. Suchkov, L. V. Titova, and S. S. Churilov, ZhETP Pisma Red. 12, 60 (1970) [JETP Lett. 12, 43 (1970)].

2. T. P. Belikova, E. A. Sviridenkov, and A. F. Suchkov, Opt. Spektrosk. 37, 654 (1974) [Opt. Spectrosc. 37, 372 (1974)].

3. G. Herzberg, Molecular Spectra and Molecular Structure (Van Nostrand, New York, 1960), p. 427.

4. N. C. Peterson, M. J. Kurylo, W. Braun, A. M. Bass, and R. A. Keller, J.O.S.A. 61, 746 (1971).

5. R. J. Thrash, H. von Weyssenhof, and T. S. Shirk, J. Chem. Phys. 55, 4559 (1971).

6. T. W. Hänsch, A. L. Schawlow, and P. E. Toschek, J. Quant. Electron. QE−8, 802 (1972).

7. Yu. M. Ajvasjan, V. M. Baev, V. V. Ivanov, S. A. Kovalenko, and E. A. Sviridenkov, Kvantovaja Elektron. 14, No. 2 (1987).

8. H. Atmanspacher, H. Scheingraber, and V. M. Baev, Phys. Rev. A 35, 142 (1987).

9. V. M. Baev, T. P. Belikova, E. A. Sviridenkov, and A. F. Suchkov, Zh. Eksp. Teor. Fiz. 74, 43 (1978) [Sov. Phys.-JETP 74, 21 (1978)].

10. V. M. Baev, T. P. Belikova, S. A. Kovalenko, E. A. Sviridenkov, and A. F. Suchkov, Kvantovaja Elektron. 7, 903 (1980) [Sov. J. Quantum Electron. 10, 517 (1980)].

11. E. N. Antonov, V. G. Koloshnikov, and V. R Mironenko, Opt. Comm. 15, 99 (1975).

12. F. Stoeckel, M. A. Melieres, and M. Chenevier, J. Chem. Phys. 76, 2191 (1982).

13. T. P. Belikova, B. K. Dorofeev, E. A. Sviridenkov, and A. F. Suchkov, Kvantovaja Elektron. 2, 1325 (1975) [Sov. J. Quantum Electron. 5, 722 (1975)].

14. N. A. Raspopov, A. N. Savchenko, and E. A. Sviridenkov, Kvantovaja Elektron. 4, 736 (1977) [Sov. J. Quantum Electron. 7, 409 (1977)].

15. S. A. Kovalenko, Kvantovaja Elektron. 8, 1271 (1981) [Sov. J. Quantum Electron. 11, 759 (1981)].

16. V. R. Mironenko and V. J. Yudson, Opt. Comm. 34, 397 (1980).

17. V. M. Baev, G. Gaida, H. Schröder, and P. E. Toschek, Opt. Comm. 38, 309 (1981).

18. S. J. Harris, J. Chem. Phys. 71, 4001 (1979).

19. E. N. Antonov, A. A. Kachanov, V. R. Mironenko, and T. V. Plakhotnik, Opt. Comm. 46, 126 (1983).

20. S. J. Harris, Appl. Opt. 23, 1311 (1984).

21. H. Atmanspacher, H. Scheingraber, and C. R. Vidal, Phys. Rev. A 32, 254 (1985).

22. H. Atmanspacher, H. Scheingraber, and C. R. Vidal, Phys. Rev. A 33, 1052 (1986).

23. F. Stoeckel, G. H. Atkinson, Appl. Opt. 29, 3591 (1985).

24. Yu. M. Ajvasjan, V. M. Baev, A. A. Kachanov, and S. A. Kovalenko, Kvantovaja Elektron. 13, 1723 (1986).

25. Yu. M. Ajvasjan, V. M. Baev, T. P. Belikova, S. A. Kovalenko, E. A. Sviridenkov, and O. I. Yushchuk, Kvantovaja Elektron. 13, 612 (1986) [Sov. J. Quantum Electron. 16, 397 (1986)].

26. V. R. Mironenko, Kvantovaja Elektron. 7, 2069 (1980).

27. W. Brunner and H. Paul, Opt. Quantum Electron. 12, 393 (1980).

28. V. M. Baev, T. P. Belikova, M. B. Ippolitov, E. A. Sviridenkov, and A. F. Suchkov, Opt. Spektrosk. 45, 58 (1978) [Opt. Spectrosc. 45, 31 (1978)].

29. A. A. Kachanov and T. V. Plakhotnik, Opt. Comm. 47, 257 (1983).

30. Y. H. Meyer and M. N. Nenchev, Opt. Comm. 41, 292 (1982).

31. V. M. Baev, T. P. Belikova, O. P. Varnavskij, V. F. Gamalij, S. A. Kovalenko, and E. A. Sviridenkov, ZhETP Pisma Red. 42, 416 (1985) [JETP Lett. 42, 514 (1985)].

32. T. S. Zeilikovich, S. A. Pulkin and L. S. Gaida, Zh. Eksp. Teor. Fiz. 87, 125 (1984) [Sov. Phys.-JETP 60, 72 (1984)].

33. W. T. Hill III, T. W. Hänsch, and A. L. Schawlow, Appl. Opt. 24, 3718 (1985).

34. V. S. Egorov and I. A. Chekhonin, Opt. Spektrosk. 52, 591 (1982) [Opt. Spectrosc. 52, 355 (1982)].

35. M. Chenevier, M. A. Melieres, and F. Stoeckel, Opt. Comm. 45, 385 (1983).

36. W. T. Hill, R. A. Abreu, T. W. Hänsch, and A. L. Schawlow, Opt. Comm. 32, 96 (1980).

37. H. Schröder, K. Schultz, and P. E. Toschek, Opt. Comm. 60, 159 (1986).

38. Ya. I. Khanin, A. G. Kagan, V. P. Novikov, M. A. Novikov, I. N. Polushkin, and A. I. Shcherbakov, Opt. Comm. 32, 456 (1980).

39. Y. H. Meyer, Opt. Comm. 30, 75 (1979).

40. A. N. Rubinov, M. V. Belobon, and A. V. Adamushko, Kvantovaja Elektron. 6, 723 (1979) [Sov. J. Quantum Electron. 9, 433 (1979)].

41. V. M. Baev, V. F. Gamalij, E. A. Sviridenkov, and D. D. Toptygin, Kratk. Soob. po Fiz. No. 8, 6 (1986) [Sov. Phys. Lebedev Institute Report].

42. W. Neuhauser and P. E. Toschek, Opt. Comm. 11, 331 (1974).

43. S. Haroche and F. Hartmann, Phys. Rev. A 6, 1280 (1972).

44. M. Sargent III and P. E. Toschek, Appl. Phys. 11, 107 (1976).

45. B. Stahlberg, V. M. Baev, G. Gaida, H. Schröder, and P. E. Toschek, J. Chem. Soc. Faraday Trans. 81, 207 (1985).

46. V. M. Baev, S. A. Kovalenko, E. A. Sviridenkov, A. F. Suchkov, and D. D. Toptygin, Kvantovaja Elektron. 7, 1112 (1980) [Sov. J. Quantum Electron. 10, 638 (1980)].

47. J. P. Reilly and J. H. Clark, Proc. Int. Conf. on Advances in Laser Chemistry, California Institute of Technology, Pasadena, Calif., 1978, ed. by A. W. Zewail (Springer-Verlag, Berlin, 1978), p. 355.

48. V. M. Baev, T. P. Belikova, V. F. Gamalij, E. A. Sviridenkov, and
    A. F. Suchkov, Kvantovaja Elektron. 11, 2413 (1984) [Sov. J. Quantum
    Electron. 14, 1596 (1984)].

49. V. M. Baev, V. F. Gamalij, E. A. Sviridenkov, and D. D. Toptygin,
    Kratk. Soob. po Fiz. No. 8, 3 (1986) [Sov. Phys. Lebedev Institute
    Report].

50. J. E. Bjorkholm and P. F. Liao, Phys. Rev. Lett. 33, 128 (1974).

51. K. Boller, V. M. Baev, and P. E. Toschek, Opt. Comm., in press.

52. V. M. Baev, H. Schröder, and P. E. Toschek, Opt. Comm. 36, 57 (1981).

53. V. M. Baev, K.-J. Boller, A. Weiler, and P. E. Toschek, Opt. Comm., in
    press.

54. D. M. Sarkisov, E. A. Sviridenkov, and A. F. Suchkov, Chimicheskaya
    Fizika 9, 1155 (1982).

55. F. Stoeckel, M. Schuh, N. Goldstein, and G. H. Atkinson, Chem. Phys.
    95, 135 (1985).

56. A. L. Schawlow and C. H. Townes, Phys. Rev. 112, 1940 (1958).

57. J. L. Hall, private communication.

Spectroscopic Applications
of Frequency Modulated Dye Lasers

A.I. Ferguson, S.R. Bramwell, and D.M. Kane

Department of Physics, University of Southampton,
Southampton SO9 5NH, UK

1. INTRODUCTION (AIF)

My interest in frequency modulated lasers was stimulated when I was a visiting scholar at Stanford in the Schawlow/Hänsch group from October 1977 to February 1979. During the first part of my visit to Stanford I worked with Jim Eckstein on the use of a coherent train of mode-locked dye laser pulses as a source for Doppler-free coherent multiple pulse spectroscopy [1]. The idea behind these experiments was that a mode-locked laser could provide a source of coherent radiation with high peak power but yet could be used for high-resolution spectroscopy. At Jim's PhD viva, Tony Siegman asked if he thought that frequency modulated lasers could be used for this kind of spectroscopy. On hearing about this question, most of us smiled knowingly and then began to wonder exactly what is a frequency modulated laser! I can't remember how Jim managed to extricate himself from this question but the next day several of the people involved in this project went off to the library to find out about frequency modulated lasers.

It did not take long to discover that nearly all the important work on frequency modulated lasers had been done at Stanford. Steve Harris and Russel Targ had discovered in 1964 that, when a phase modulator crystal of KDP, inserted into the cavity of a He Ne laser operating at 633 nm, was driven close to the cavity mode spacing, a broadband frequency modulated output could be obtained [2]. In the time domain the amplitude was constant with the carrier frequency sinusoidally sweeping over the Doppler width. In the frequency domain this appeared as a series of equally spaced modes with Bessel function amplitudes. In 1965 Steve Harris and Otis McDuff developed a theory of frequency modulated lasers [3]. In this linear theory for an inhomogeneously broadened laser they assumed that each mode was coupled to its nearest neighbours by an intracavity phase perturbation. Several characteristic features came out of this theory. Firstly, they discovered that the modulation index, Γ of the FM oscillation was given by

$$\Gamma = \left(\frac{\delta}{\pi} \right) \left(\frac{\Omega}{\nu_m - \Omega} \right) , \tag{1}$$

where δ is the single pass phase retardation of the phase modulator, Ω is the cavity mode spacing and ν_m is the modulation frequency. We refer to $\Delta\nu = \nu_m - \Omega$ as the detuning. They also discovered that there were three regions of operation, called the FM, unquenched and phase-locked regions. The FM region corresponds to constant amplitude with Bessel function mode amplitudes and is characterised by a relatively large detuning. As the detuning is reduced the unquenched region is reached where several FM oscillations may occur. At the smallest detunings the mode amplitudes become Gaussian in shape and the laser consists of a periodic train of pulses. This is sometimes called FM mode-locking. Many of these predictions were confirmed by the work of Amman et al [4] using a He Ne laser.

In 1970 Dirk Kuizenga and Tony Siegman investigated the FM operation of a Nd:YAG laser [5]. They had already worked on the theory and experiment of FM mode-locking of a Nd:YAG laser [6,7]. They discovered that the FM bandwidth was much greater in Nd:YAG than the mode-locked bandwidth. They also emphasised that despite the large bandwidth, FM lasers are coherent and can be viewed in a generalised sense as 'single mode'.

As far as we could see in the Schawlow/Hänsch group, not much more had been done on frequency modulated lasers. The idea that larger bandwidths could be obtained in the FM region than in the mode-locked region was tantalising. Jim Eckstein and I had managed to generate bandwidths of up to 500 GHz coherently, with a synchronously pumped mode-locked dye laser [8]. Could an FM dye laser give bandwidth well in excess of a THz?

The spectroscopic aspects of FM dye lasers looked interesting. For example, lasers separated by a large frequency interval could be compared by heterodyning the reference and unknown laser frequency together with the FM laser. The low frequency beat between the reference laser and one of the modes of the FM laser and a similar beat with the unknown laser, together with a knowledge of the equally spaced FM laser mode frequency could be used to measure the laser frequency differences modulo the FM mode spacing. Less precise methods could establish the integer mode separation between the lasers unambiguously. The difference frequency between the lasers could be as large as the bandwidth but only low frequency beats (~ 100 MHz) would have to be measured.

Other intriguing possibilities came up. For example Heinz Weber and Ernst Mathieu [9] had predicted in 1967 that an FM beam could be divided into two halves, one beam being delayed by half the modulation period and the two beams being combined in a nonlinear crystal in such a way that only the sum frequency was generated. The resulting output would be a single frequency at twice the carrier frequency of the FM oscillation. In a similar vein, we at Stanford worked out that if an FM beam was used for Doppler-free two-photon spectroscopy, with the counter-propagating beam delayed by half the modulation period, a spectrum identical to the single frequency spectrum could be obtained as the carrier frequency was swept through the two-photon resonance. The theory of this process was worked out by Ted Hänsch and N.C. Wong in 1980 [10].

Our ideas about FM dye lasers went untested until I moved to Southampton in 1984 and was fortunate to obtain a Metrology Award and research contract from the National Physical Laboratory. What is described in the rest of this paper is the development of an FM dye laser together with some of the performance characteristics, experiments on single frequency UV generation using an FM laser and experiments on Doppler-free two-photon spectroscopy using an FM laser.

2. FM DYE LASER

When we started developing FM dye lasers it was not clear which would be the best configuration to pursue. Dramatic differences between the performance of single frequency lasers had been observed for standing-wave and ring dye lasers. We decided to purchase a commercial dye laser which could be operated in a ring configuration or standing-wave configuration and was versatile enough to enable a variety of tuning elements to be tested (Coherent 699-21). The dye was chosen to be Rhodamine 6G since it was thought to be typical and because it was efficient and easy to use. Detailed performance of ring and standing-wave lasers have been reported in previous publications [11, 12, 13]. The main difference between ring and standing wave was that the bandwidth of the standing wave laser was twice that for a ring and that the phase modulator had to be placed close to one of the end mirrors of the standing-wave laser. It was found that the bandwidth of the FM laser was well described by eqn. (1). In Fig. 1 we show the spectrum of a standing-wave FM dye laser restricted to run over a narrow range of frequencies by two intracavity etalons and a 3-plate birefringent filter as a function of detuning for a fixed single pass phase retardation. We compare this spectrum with the spectrum predicted for a pure FM oscillation. The observed and predicted spectra show excellent agreement.

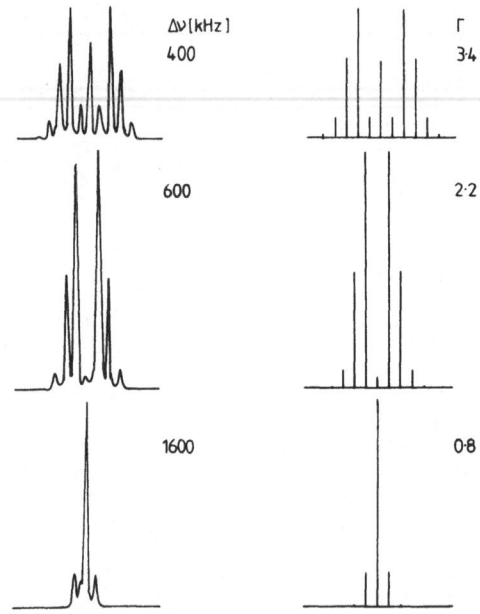

Fig.1 FM optical spectra obtained by modulating a single frequency dye laser as a function of detuning. The line spectra on the right are computer generated Bessel function spectra with the modulation index indicated. The mode spacing is 127 MHz

In order to attempt to develop a laser with as large a bandwidth as possible, we have studied dye lasers without etalons. In the standing-wave and bidirectional ring we found that it is difficult to obtain a single pure FM oscillation. We have now settled on the use of a unidirectional ring laser design. When such a laser is free-running it tends to operate on a single mode which is rather free to wander. The application of phase modulation tends to stabilise this mode and gives rise to a single FM oscillation.

A schematic diagram of this laser is shown in Fig. 2. The basic laser was unchanged from the commercial laser (Coherent 699-21) consisting of four mirrors, one of which was mounted on a piezoceramic. This could be used to make rapid adjustments to the cavity length when frequency stabilising. Slow changes in the cavity length were provided by a fused silica plate, mounted at Brewster's angle on a galvanometer, and inter-secting two beams. The laser was made to operate in one direction by use of a device based on the Faraday effect. Tuning was accomplished with a 3-plate birefringent filter. An etalon of 225 GHz free spectral range and 25% reflectivity could be incorporated for further spectral narrowing and control. Several different phase modulators have been tested and are located in the laser where etalons are usually placed.

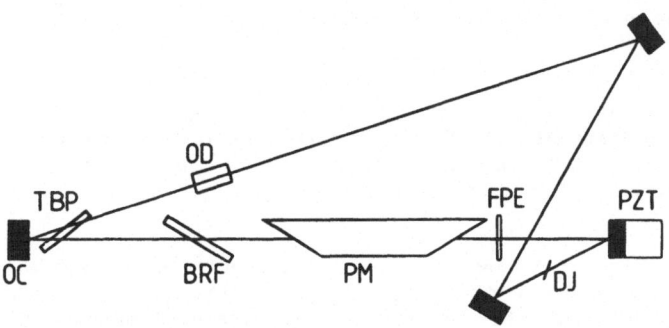

Fig.2 Schematic diagram of the FM dye laser. The symbols represent the following: OC, output coupler; OD, optical diode; BRF, birefringent filter; PM, phase modulator; DJ, dye jet, PZT, piezoelectric tweeter mirror; FPE, etalon; TBP, tipping Brewster plate.

Most of the phase modulators that have been tested consist of 45° y-cut ADP crystals. ADP is available in large sizes of excellent optical quality but has a rather small electro-optic coefficient. Several schemes for driving these crystals have been studied. These include inductively coupling to a coil which forms part of a resonant circuit with the capacitance of the crystal. One of the more successful phase modulators is a commercial unit (Gsänger) consisting of two Brewster angled ADP crystals configured to give no deflection or displacement of the laser beam. This modulator has rather a large capacitance and was normally directly driven. The usual source for driving the systems was a frequency synthesiser and 5W amplifier. For the ring laser the modulation frequency was in the region of 181 MHz. In Fig. 3 we show some typical spectra taken using this modulator at a single pass phase retardation of 1 mrad for different detunings. Bandwidths of greater than 30 GHz could easily be obtained.

In more recent tests we have obtained $MgO:LiNbO_3$ modulators. These tend to be smaller and of poorer optical quality than ADP but have much larger

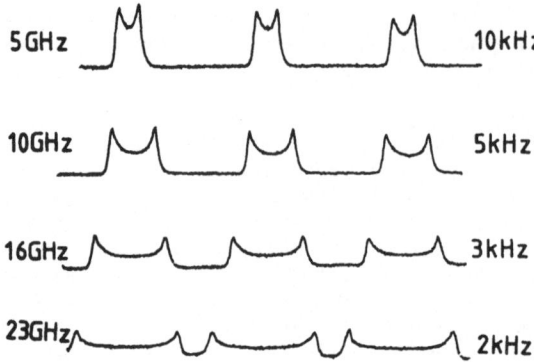

5GHz　　　　　　　　　　　　　　　　10kHz

10GHz　　　　　　　　　　　　　　　5kHz

16GHz　　　　　　　　　　　　　　　3kHz

23GHz　　　　　　　　　　　　　　　2kHz

Fig.3   Spectra of the FM dye laser as a function of detuning observed
with 30 GHz free spectral range interferometer.   The detuning $\Delta\nu = \nu_m - \Omega$
is given on the right and the measured bandwidth is given on the left.

electro-optic coefficients.   We have obtained a single pass phase
retardation of $\delta = 3$ radians at 181 MHz using this material.   This gives
a bandwidth of up to 6 THz, although this appears to be strongly
dependent on the bandpass of the tuning element.

At these exceptionally large bandwidths it is difficult to analyse the
FM spectrum to see how closely the output approximates to a pure FM
oscillation.   We have used the techniques of single frequency UV
generation and Doppler-free two-photon spectroscopy to help us analyse
the spectrum.   In both of these methods all modes contribute to the
signal.   Experiment and analysis show that amplitude distortions of the
FM output appear as sidebands around the UV or two-photon signal whereas
frequency distortions give rise to broadened signals.   We will now
describe these techniques in more detail.

3. SINGLE FREQUENCY ULTRAVIOLET GENERATION

The simplest way to understand the principle of single frequency UV
generation using an FM laser is to consider the instantaneous electric
field $E(t)$ of an FM oscillation at carrier frequency ω_0 given by

$$E(t) = \frac{1}{2} E_0 \exp\left[-i\left\{\omega_0 t + \Gamma \sin \Omega t\right\}\right] + c.c.,$$

where we have arbitrarily set the phase of the FM oscillation to zero.
If we now take this same FM oscillation but introduce a phase delay of θ
radians, the field is given by

$$E'(t) = \frac{1}{2} E_0 \exp\left[-i\left\{\omega_0 t + \Gamma \sin (\Omega t + \theta)\right\}\right] + c.c.$$

If these two beams interact by a second-order nonlinear process we can
pick out the sum-frequency oscillation given by

116

$$E''(t) \propto E(t) \; E'(t) + c.c$$

$$\propto E_0^2 \exp \left[i\{2\omega_0 t + 2 \; \Gamma(\cos(\theta/2)) \sin(\Omega t + \theta/2)\} \right] + c.c.$$

This represents an FM oscillation centred on a carrier frequency of $2\omega_0$ with modulation frequency Ω and modulation index $2\Gamma\cos(\theta/2)$. If we set the phase delay to be $\theta = \pi$ the modulation index becomes zero and we are left with a single frequency oscillation at twice the original carrier frequency. We can investigate the spectrum we would expect to see when we detune from $\theta = \pi$ by introducing a detuning phase given by $\Delta\theta = \pi - \theta$. The modulation index of the FM oscillation will then become $2\Gamma\sin(\Delta\theta/2)$.

The experimental arrangement for demonstrating its effect consists of our standard FM ring dye laser which is split into two components. One component is delayed and recombined, with the aid of a lens, at a crystal of ADP, phase-matched for sum frequency mixing at the laser wavelength. The sum frequency is generated at the bisector of the two beams and then enters a UV confocal interferometer for spectral analysis. A typical pair of traces is shown in Fig. 4. This shows the spectrum of the dye laser observed with a 30 GHz free spectral range confocal interferometer at the bottom together with the spectrum of the sum frequency UV observed with a 2 GHz free spectral range interferometer. The bandwidth of the dye laser in this case was 7.7 GHz whereas the UV linewidth was less than

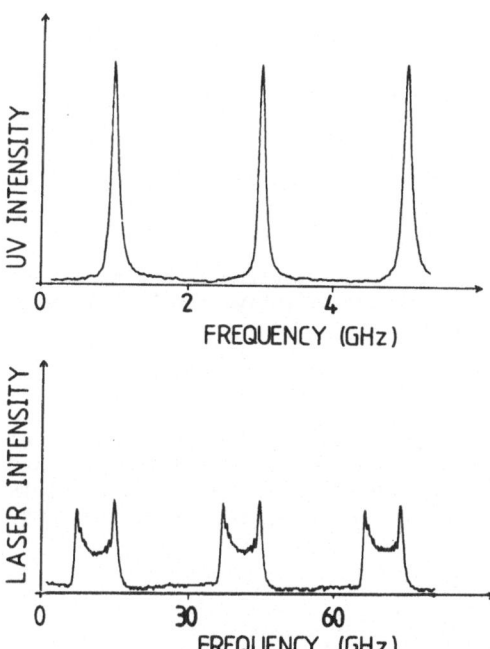

Fig.4 The lower trace shows the FM laser bandwidth of 7.7 GHz as observed with a 30 GHz free spectral range interferometer. The upper trace shows the UV spectrum observed on a 2 GHz free spectral range interferometer.

the resolution of the interferometer. The effect of changing the delay from the $\theta = \pi$ condition is shown in Fig. 5. These curves were taken with the laser operating in the same conditions as those applying to Fig. 4. We have indicated on this figure the measured phase delay $\Delta\theta$ and compared the generated UV spectra with ideal FM spectra to deduce the effective modulation index Γ'. We have found excellent agreement between the predicted and measured modulation index.

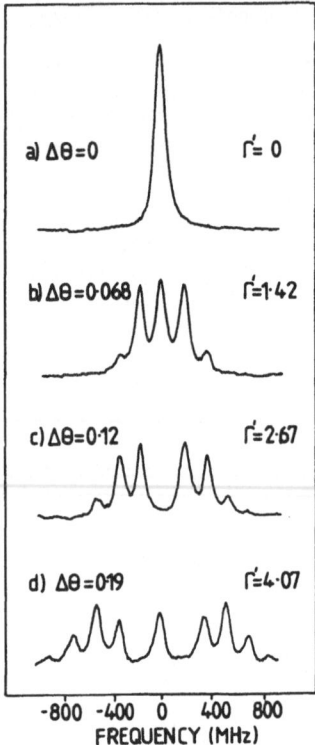

a) $\Delta\theta = 0$ $\quad\quad\quad$ $\Gamma' = 0$

b) $\Delta\theta = 0.068$ $\quad\quad$ $\Gamma' = 1.42$

c) $\Delta\theta = 0.12$ $\quad\quad$ $\Gamma' = 2.67$

d) $\Delta\theta = 0.19$ $\quad\quad$ $\Gamma' = 4.07$

-800 -400 0 400 800
FREQUENCY (MHz)

Fig.5 Sum-frequency spectrum as a function of phase delay $\Delta\theta$. The effective modulation index Γ' is also indicated.

The single frequency UV radiation can be used to perform atomic or molecular spectroscopy. Alternatively, the generated UV could be locked to a stabilised interferometer. Either method would enable the complete FM spectrum to be stabilised to provide a comb of reference frequency over a large bandwidth.

4. DOPPLER-FREE TWO-PHOTON SPECTROSCOPY

Doppler-free two-photon spectroscopy using an FM laser can be thought of in a similar way to single frequency UV generation. However, in this case counter-propagating beams are required and instead of considering sum-frequency generation we consider two-photon excitation.

The experimental arrangement consists of the FM dye laser focussed into a cell containing sodium, at a temperature of about 180°C and tuned to the 3S – 4D transition. This beam is then recollimated and propagates to a mirror after which it returns to the cell [15]. The distance between the cell and mirror is adjusted to be half of the modulation period, corresponding to π phase delay. The two-photon excitation is detected by UV fluorescence from the sodium cell. The dye laser frequency is stabilised by locking to a 2 GHz free spectral range confocal interferometer. This means that the dye laser bandwidth is restricted to at most 2 GHz. The laser carrier frequency is swept by slaving the FM laser to this cavity.

A typical Doppler-free two-photon spectrum is shown in Fig. 6b at π phase delay. Also shown on the same figure is a Doppler-free two-photon spectrum obtained with the same laser system but operated in a single mode. The two spectra are almost identical. Note however, that the Doppler-broadened background is slightly higher in the single frequency case. This is predicted by the Hänsch and Wong theory [10] and is due to the fact that in the FM case the laser bandwidth exceeds the Doppler width and hence the laser power within the Doppler broadened line is smaller than in the single frequency case. The strength and width (8MHz) of the Doppler-free signals are the same in both cases as is expected [10].

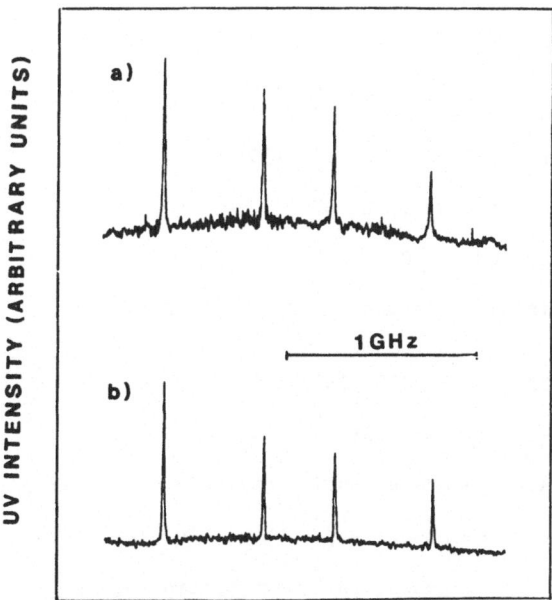

LASER FREQUENCY SCAN

Fig.6a Doppler-free two-photon spectrum of the 3S–4D transition in sodium using single frequency dye laser.

6b The same transition observed using an FM laser of 1.6 GHz bandwidth.

5. CONCLUSIONS

We have developed an FM dye laser capable of producing an FM oscillation over a region in excess of 1THz. We have devised ways in which the spectrum of the FM oscillation can be analysed and compared with a pure FM oscillation. In these techniques of sum frequency UV generation and two-photon Doppler-free spectroscopy, amplitude distortions appear as sidebands and frequency distortions appear as spectral broadening. These methods can also be used to stabilise the FM dye laser to an absolute reference.

As phase modulators are improved we expect to be able to generate even wider bandwidths. This opens up many possibilities for optical metrology and spectroscopy. A further intriguing possibility is that of using dispersion, perhaps in a fibre or grating, to compress this broad spectrum into a train of ultrashort pulses. The bandwidths that we have already achieved suggest that sub-picosecond pulses are possible.

We see no reason why the techniques of frequency modulation that we have developed should not be used in other dyes, and indeed other homogeneously broadened lasers. This will open up possibilities of investigations of atoms throughout the visible region of the spectrum.

REFERENCES

1. J.N. Eckstein, A.I. Ferguson, T.W. Hänsch: Phys. Rev. Lett. $\underline{40}$, 847 (1978).

2. S.E. Harris, R. Targ: Appl. Phys. Lett. $\underline{5}$, 202 (1964).

3. S.E. Harris, Otis McDuff: IEEE J. QE-1, 245 (1965).

4. E.O. Ammann, B.J. McMurtry, M.K. Oshman: IEEE J. QE-$\underline{1}$, 263 (1965).

5. D.J. Kuizenga, A.E. Siegman: IEEE J. QE-$\underline{6}$, 673 (1970).

6. D.J. Kuizenga, A.E. Siegman: IEEE J. QE-$\underline{6}$, 694 (1970).

7. D.J. Kuizenga: IEEE J. QE-$\underline{6}$, 709 (1970)

8. A.I. Ferguson, J.N. Eckstein, T.W. Hänsch: J. Appl. Phys. $\underline{49}$, 5389 (1978).

9. H.P. Weber, E. Mathieu: IEEE J.QE-$\underline{3}$, 376 (1967).

10. T.W. Hänsch, N.C. Wong: Metrologia $\underline{16}$, 101 (1980).

11. D.M. Kane, S.R. Bramwell, A.I. Ferguson: Appl. Phys. B$\underline{39}$, 171 (1986).

12. D.M. Kane, S.R. Bramwell, A.I. Ferguson: Appl. Phys. B$\underline{40}$, 147 (1986).

13. D.M. Kane, S.R. Bramwell, A.I. Ferguson: In Laser Spectroscopy VII ed. by T.W. Hänsch, Y.R. Shen, Springer Ser. Opt. Sci. $\underline{49}$ (Springer Verlag, Berlin 1985) pp.362-365.

14. S.R. Bramwell, A.I. Ferguson, D.M. Kane: Opt. Commun. $\underline{61}$, 87 (1987).

15. S.R. Bramwell, A.I. Ferguson, D.M. Kane: Opt. Commun. (submitted for publication).

Reminiscence of Schawlow at the First Conference on Lasers

D.F. Nelson

AT&T Bell Laboratories, Murray Hill, NJ 07974, USA

Art Schawlow's legendary good naturedness, humor, and showmanship evidenced themselves early in the development of lasers, in fact, at the first session of a conference ever to be concerned with lasers. That conference was the Ann Arbor Conference on Optical Pumping organized by Peter Franken and Richard Sands of the Physics Department of the University of Michigan in June 1959.

I was just completing a post-doctoral appointment at Michigan and had accepted a research position in optical physics at Bell Telephone Laboratories in Murray Hill, New Jersey starting in July of that year. I was interested in pursuing both efficient electroluminescence and the attainment of an "optical maser", as the laser was then generally called. Thus, I attended the Optical Pumping Conference with considerable interest. Little did I realize then that within a little over a year I would be coauthoring a paper on lasers with Art.

Art, having coauthored the now famous Schawlow-Townes paper published in the Physical Review in December 1958, was chairman of Miscellaneous Session I which included a number of papers on proposed optical masers. There were papers by Gordon Gould on pumping processes and optical cavities, J. H. Sanders on possible optical maser action in helium, Ali Javan on attaining a "negative temperature", as a population inversion was then generally called, in neon by resonant collisional excitation with helium, and Irwin Wieder on optical pumping of ruby. The session concluded with remarks by Art Schawlow on the "neighbor lines" in dark ruby arising from chromium ion pairs and their possible use as optical maser transitions, a use that Art was successful in attaining a year and half later.

The paper by Gordon Gould is well remembered by me because of three interesting uses of language, two by Gould and one by Schawlow in a comment. Gould in his paper entitled "The LASER, Light Amplification by Stimulated Emission of Radiation" first introduced the acronym LASER in place of the then current term optical MASER. Many of us would resist that new name for a time, fearing the need to call other devices IRASERs, UVASERs, etc. depending on their wavelength of emission. But the attainment of the laser a year later led to so many dramatically new results that the device demanded a name of its own, not a name derived from the original, but less useful, maser.

The second use of language memorable to me arose when Gordon Gould reported he had applied for $50,000 of federal funding to pursue the laser idea but had been awarded $250,000 (as best I can recall the numbers)! Thus, he referred to funding from Uncle Sam as coming from "Uncle Cornucopia". I have found many occasions in the years since to refer to the largesse of federal funding in many programs as coming from Uncle Cornucopia.

As Gould's paper concluded, Session Chairman Schawlow could not resist - as always - a note of humor. His humor correctly foresaw the laser's main use as an oscillator, not an amplifier, and also expressed the great uncertainty, even improbability, felt by everyone at the time about the possible attainment of a laser. Beginning in mock solemnity and ending in belly-shaking laughter, Art opined that the LASER was likely to be most used as an oscillator and so should be named "Light Oscillation by Stimulated Emission of Radiation", or the "LOSER". Art was certainly right that the LASER is usually a LOSER, but as time has amply shown certainly not a loser!

Part II

Atomic and Molecular Spectroscopy

*A diatomic molecule is
a molecule with one atom
too many*

Laser and Fourier Transform Techniques for the Measurement of Atomic Transition Probabilities

J.E. Lawler

Department of Physics, University of Wisconsin, Madison, WI 53706, USA

ABSTRACT

Radiative lifetimes for many atoms and ions are determined using time-resolved laser-induced fluorescence on slow atomic and ionic beams. These lifetimes are combined with precise branching ratios measured using a Fourier transform spectrometer to determine accurate absolute atomic transition probabilities for the elements in low stages of ionization.

1. INTRODUCTION

A powerful combination of techniques from laser spectroscopy and Fourier transform spectroscopy is playing a central role in determining accurate atomic transition probabilities for the elements in low stages of ionization. This has been one of the most persistent problems of atomic spectroscopy. Classical spectroscopists measured the wavelengths of the major spectral lines of most elements to part per million accuracy. Techniques for measuring Einstein A coefficients for these spectral lines lagged far behind techniques for wavelength measurements. For some elements not a single transition probability was known to within a factor of two before the recent application of laser techniques. The lack of reliable transition probability data has been a serious problem for many fields. The problem has been severe for astronomy, which is in some sense the parent field of spectroscopy. Accurate absolute atomic transition probabilities are essential for determining the abundances of chemical elements in the sun, other stars, and interstellar clouds. These abundance studies provide critical information on the distribution and production of chemical elements throughout the observable universe.

Many techniques have been used to measure atomic transition probabilities including: the Hanle effect, the Hook method, absolute emission measurements on thermal arcs, beam-foil time-of-flight techniques, time-correlated photon counting with pulsed electron beam excitation, and others. Each technique has its own strengths and limitations. None of these older techniques provided the accuracy, convenience, and extremely broad applicability of laser techniques that are now available. Radiative lifetime measurements to ~5% absolute accuracy are now routine using laser techniques. The 5% figure is typical of the systematic plus random uncertainties on many lifetime measurements, and is typical of the level of agreement between measurements by independent groups.[1,2] Accuracy to better than 1% is possible. Measurements on highly refractory species such as Nb, Ta or even Nb^+ or Ta^+ using lasers and hollow cathode atomic/ionic beam sources are now as routine as measurements on the alkalies.[3-7]

The radiative lifetimes do not by themselves determine individual Einstein A coefficients unless one branch dominates decay from the level. The lifetimes provide the essential absolute normalization needed to convert relative intensity measurements or branching ratios into absolute transition probabilities. Fortunately, the vast improvement in radiative lifetime measurements coincided with very important progress on branching ratio measurements. Fourier transform spectrometers, such as the 1.0 meter FTS at the National Solar Observatory on Kitt Peak, are now used to measure branching ratios.[8,9] The 1.0m FTS has extraordinary advantages for spectrophotometry on complex atoms and ions. The 1.0m FTS offers Doppler limited resolution, wavenumber accuracy to one part in 10^8, a very high data collection rate, and has the intrinsic advantage of simultaneous measurement of all spectral elements. An elegant technique for producing a relative intensity calibration of an FTS is available.[9] The potent combination of radiative lifetimes measured using time-resolved laser-induced fluorescence and of branching ratios measured using the 1.0m FTS is rapidly producing an enormous improvement in knowledge of atomic transition probabilities.

This article is a review of recent work with laser and FTS techniques on only part of the periodic table, specifically the transition metals with open 3d, 4d and 5d shells. The transition metals include the Fe group which is quite important in astrophysics, and include other highly refractory metals which have significant technological applications. The very high melting points of many of these metals had previously provided a challenge to experimentalists. The complexity of the spectra and atomic structure makes theoretical calculations of the transition probabilities for these elements very difficult.

2. RADIATIVE LIFETIMES

A low-pressure gas phase sample or an atomic/ionic beam is needed in order to measure radiative lifetimes using time-resolved laser-induced fluorescence. Some of the laser measurements of transition metal lifetimes involved thermal atomic beam sources, including early work on FeI, CoI, and NiI.[10-14] It has become increasingly apparent that discharge (sputter) sources are superior for this work. Hannaford and Lowe at C.S.I.R.O. developed a sputtering cell used for radiative lifetime measurements.[15] Duquette, Salih, and Lawler developed a hollow cathode atomic/ionic beam source.[5,16] Both devices have extremely broad applicability and have been used with substantial success. The hollow cathode beam source has an advantage in that there are no effects due to collisional quenching in a beam environment. Although collisional quenching of longer lived levels is observed at the 0.1 to 5 Torr pressures used in the cell experiments, the effect of collisional quenching can be eliminated by extrapolation to zero pressure. The sputtering cell technique has an advantage in that it limits contamination during work on radioactive elements such as Tc. Figure 1 is a schematic of the sputtering cell experiment.

The hollow cathode atomic/ionic beam experiment which is shown in Fig. 2, will be described in some detail. Many of the ideas which led to the hollow cathode beam source were developed during the author's two year stay at Stanford with the Schawlow-Hänsch group. The hollow cathode atomic/ionic beam source is based on a low pressure, large bore hollow cathode discharge tube. The hollow cathode is used as a beam source by sealing one end of the cathode except for a 1.0 mm diameter opening. The opening is flared outward at 45° to serve as a nozzle for forming an

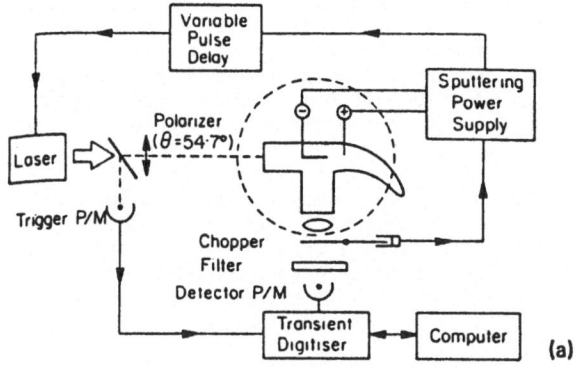

Fig. 1(a) Schematic diagram
of the sputtering cell
experiment developed
by P. Hannaford and
R. M. Lowe at
C.S.I.R.O. in
Melbourne, Australia.
(b) The timing
sequence for
time-resolved
laser-induced
fluorescence
measurement of atomic
and ionic lifetimes.

Timing Sequence (Schematic Time Scale)

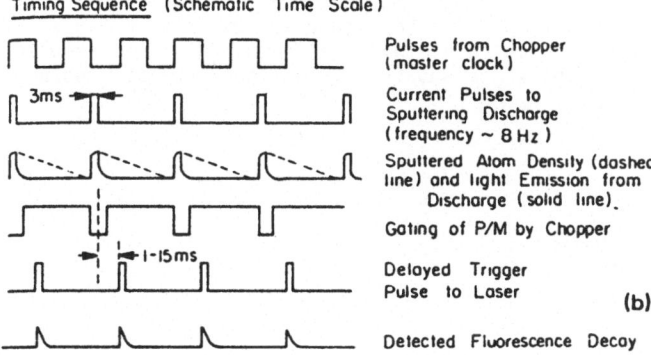

Pulses from Chopper
(master clock)

Current Pulses to
Sputtering Discharge
(frequency ~ 8 Hz)

Sputtered Atom Density (dashed
line) and light Emission from
Discharge (solid line).

Gating of P/M by Chopper

Delayed Trigger
Pulse to Laser

Detected Fluorescence Decay

TOP VIEW

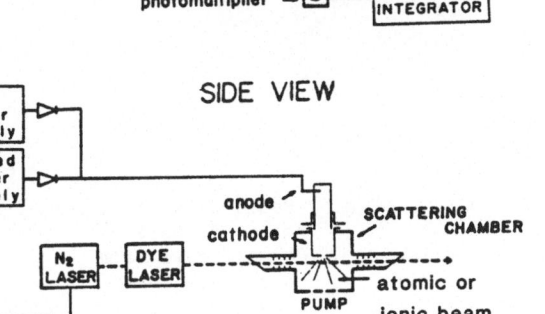

Fig. 2 Schematic diagram of
the hollow cathode
atomic/ionic beam
experiment developed
by D. W. Duquette,
S. Salih and
J. E. Lawler at the
Univ. of Wisconsin.
The experiment is used
for time-resolved
laser-induced
fluorescence
measurements of atom
and ion lifetimes in a
beam environment.

127

uncollimated atomic or ionic beam. The 29 mm inside diameter, 50 mm long stainless steel cathode is lined with a 0.25 mm thick foil of the metal to be studied. The hollow cathode and the scattering chamber are at ground potential. Argon, the carrier gas for the discharge flows continuously into the hollow cathode discharge. A 10 cm diffusion pump is used to evacuate the scattering chamber. The scattering chamber is sealed from the hollow cathode discharge, except for the nozzle, and is maintained at a much lower pressure. The argon pressure in the discharge is usually 0.3 Torr, and the resulting scattering chamber pressure is approximately 10^{-4} Torr of argon. A direct current of 50 mA to 200 mA in the hollow cathode discharge is typically used to produce a beam of neutral atoms. A pulsed power supply which delivers up to 25A for 5 μsec is used to produce an intense burst of slow ($\leq$1eV) ions in the beam. Figure 3 is a scale drawing which shows materials and dimensions for the source.

The source has been used to produce atomic or ionic beams of Mo, Zr, W, Nb, Hf, Rh, Ta, Re, Al, Be, Fe, Co, Ni, Ti, V, Mg, Pt, Ru, and other species. It is believed that the source will work well with all metallic elements. The absolute intensity of the metal atom beam was measured by depositing and weighing a Nb film. The Nb atom beam intensity on axis is 1.8×10^{14} atoms/(sec sr) at 180 mA of discharge current.[3] It is assumed that all Nb atoms stuck to the glass substrate, thus the intensity should be considered a lower limit.

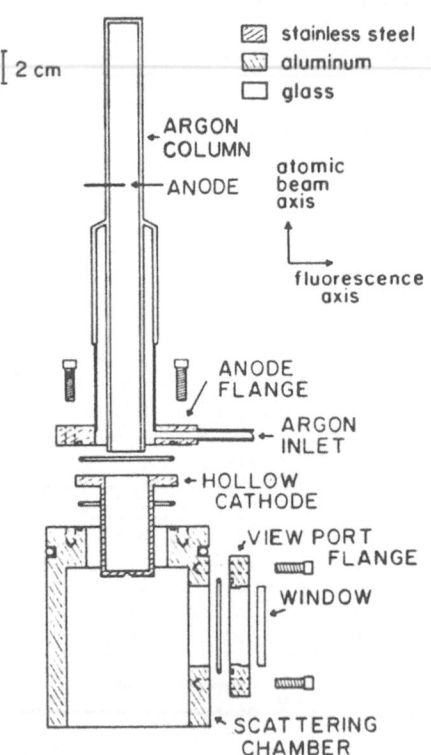

Fig. 3 Scale drawing of the
hollow cathode
atomic/ionic beam
source developed by
D. W. Duquette,
S. Salih, and
J. E. Lawler.

The hollow cathode atomic/ionic beam source and the discharge cell approach share other advantages besides convenience and very broad applicability throughout the periodic table. The free atoms and ions they produce are distributed among many metastable levels. It is routine to study levels which are not directly connected to the ground level, or which are connected to the ground level by VUV transitions. Metastable levels as high as 30,000 cm^{-1} above the ground level of MoI were found to have useful populations when the discharge is pulsed.[17] Metastable levels as high as 14,000 cm^{-1} above the ground level of the ion in NiII were found to have useful populations.[18] Metastable levels in inert gases, including the metastable levels of He at 166,000 cm^{-1} above the ground state, have useful populations. Certain levels in He, which are accessible from the metastable levels, have lifetimes known to better than 1% from theoretical calculations. These levels are very useful for testing the overall performance of the lifetime measuring apparatus.[19] The hollow cathode atomic/ionic beam source is gaining increased acceptance, even among groups which previously preferred thermal sources.[6,7]

The pulsed dye lasers used in the time-resolved laser-induced fluorescence studies are in most cases not too different from the dye laser developed by Hänsch.[20] The intracavity telescopic beam expander introduced by Hänsch has in many cases been replaced by a prismatic beam expander[21], or a grazing incidence grating[22], but much of the dye laser cavity is unchanged. These pulsed dye lasers are now pumped by N$_2$, Nd-YAG, or eximer lasers. The dye lasers typically produce a pulse with a duration of 3 to 8 nsec and a bandwidth of 3 to 8 GHz. The transverse pumping of the dye cell produces an exceeding high ($\sim 10^6$) unsaturated gain. The high unsaturated gain makes the high losses of the cavity tolerable. The 20 cm long cavity used for lifetime experiments at the Univ. of Wisconsin is typical. It has a round trip loss of 99.5%. Only 0.5% of the light survives a round trip because of the 4% reflection of output mirror, the 25% transmission of the double passed prismatic beam expander, and the estimated losses of 50% due to the dye cell and grating. The short, very lossy cavity has an important advantage for lifetime work: it has negligible capacity to store light. Once the pump laser power drops below threshold the dye laser output decreases very abruptly. The abruptness of the dye laser pulse termination is far more important than the pulse duration when measuring short lifetimes. It is not necessary to deconvolute the excitation pulse from the fluorescence data, if the first five nanoseconds of the fluorescence decay curve is discarded. Clean, single exponential decays are observed for lifetimes as short as 3 nsec.

These pulsed dye laser oscillators are often followed by one or more amplifiers in order to produce sufficient power for nonlinear generation of UV light. Frequency doubling crystals of KDP and KB5 provide tunability to 217 nm in the UV. Dye lasers pumped by high power Nd:YAG or eximer lasers will provide tunability into the VUV. Raman shifting of the dye laser frequency in H$_2$ provides continuous tunability to 138 nm.[23,24] The shorter wavelengths will result in excitation of very short lived levels. Fortunately each nonlinear step, whether frequency summing or doubling in a crystal, Raman shifting in H$_2$, or four wave sum frequency mixing in metal vapor, will further sharpen the temporal distribution of the dye laser pulse.

The simplicity and very broad tunability of the pulsed dye lasers is an enormous advantage. Spatially resolved laser-induced fluorescence on fast ionic beams is a good technique[25], but it has not been used as widely as time-resolved laser-induced fluorescence on slow atomic and ionic beams.

Spatially-resolved laser-induced fluorescence requires an accelerator and is limited to wavelength ranges where c.w. dye lasers are available. The requirement for a c.w. laser is not fundamental, but is dictated by signal-to-noise considerations.

Several different approaches are used to filter, detect, and log fluorescence signals in time-resolved laser-induced fluorescence experiments.[1,3,10] The use of a grating monochromator to filter the fluorescence is not necessary because of the selective laser excitation. A grating monochromator offers certain advantages: it can provide additional confidence that the laser excitation line is not blended, has been correctly classified, and is correctly identified in the experiment. The substantial loss of fluorescence signal intensity due to the grating monochromator is not a disadvantage if time-correlated photon counting is used in detecting and logging the fluorescence. Typically ~0.1 photon per laser pulse is detected in time-correlated photon counting experiments. The low counting rate is essential to prevent severe distortion of the fluorescence decay curve.

It is desirable to collect as much fluorescence as possible when using a boxcar integrator or transient digitizer to log the fluorescence signals. The very fast f/1 collection system shown in Fig. 2 is used with dye or interference filters.[3] Minimal spectral filtering provided by dye or interference filters can sometimes be used to block all laser light scattered from windows etc., if the branching ratios are favorable. It remains necessary to detect fluorescence at the laser wavelength in many cases, thus it is important to minimize the amount of scattered laser light. The use of very high quality Brewster windows and extensive light baffling is essential to minimize scattered laser light. Background light from the discharge in cell experiments and from hollow cathode beam source in beam experiments is not a problem. The discharge is pulsed in the cell experiments and the laser is fired after an appropriate delay as indicated in Fig. 1.[1] The several millisecond delay is chosen such that light emission from the discharge has died away, but sputtered metal atoms are still present in the cell. Background light from the hollow cathode atomic beam source shown in Fig. 2 is often 10^4 times weaker than the peak fluorescence signals.[3] This high signal-to-background ratio is achieved without spectral filtering.

The use of a boxcar integrator provides a substantial (10^2 or 10^3) advantage over time-correlated photon counting in data collection rate.[3] A transient digitizer, though, provides a 10^4 to 10^5 advantage over time-correlated photon counting in data collection rate.[1] Thousands of photons can be logged per laser shot using a transient digitizer. It is often possible to collect an excellent decay curve in seconds using a transient digitizer. The Tektronix 7912AD digitizer provides mV sensitivity, an analogue bandwidth of 0.5 GHz, and a sampling rate up to 100 GHz.

The primary advantage of a fast boxcar or a time-correlated photon counting system is that they have the potential to measure subnanosecond lifetimes. This advantage has not yet been exploited in work on transition metals, but it should be important for VUV studies.

Many good photomultipliers are available which are well suited for time-resolved laser-induced fluorescence studies. Several groups use the 1P28A.[15,16] It is inexpensive and rugged. It provides high gain (~10^6), linearity, and freedom from ringing when used in a properly designed base.

The base should be designed for very low inductance, and it should include bypass capacitors and small damping resistors.[26] The 1P28A provides excellent linearity to peak anode currents of 10 mA. Although single photon spikes are 2 nsec wide, they have very steeply rising and falling edges. The 1P28A has adequate electronic bandwidth for lifetimes as short as 2 or 3 nsec. Faster photomultipliers will be necessary in the VUV. Several faster, new types of photomultipliers are available.[27]

The large signals achieved in time-resolved laser-induced fluorescence experiments are important in reducing statistical errors, but the greatest strengths of the experiments are their nearly complete freedom from systematic errors. The selective excitation provided by tunable dye lasers eliminates the radiative cascading problem which plagued beam-foil time-of-flight experiments. Minimal spectral filtering of the fluorescence is occasionally necessary to block cascade fluorescence from lower levels in laser-induced fluorescence experiments. Collisional quenching is not observed in experiments involving laser excitation of atoms or ions in a beam. Systematic studies to test for collisional quenching of long-lived levels are routinely performed by varying the beam intensity, but no collision quenching has been observed. Collisional quenching of long-lived levels is observed at the 0.1 to 5 Torr pressures used in the cell experiments, but it is not difficult to extrapolate the observed lifetime to zero pressure.[1] The corrections are usually comparable to the 3 to 8% uncertainties on the lifetimes. Radiation trapping on strong resonance lines of metals is not a problem. Although routine tests are performed by varying the beam intensity when studying strong resonance lines, radiation trapping on resonance lines of metal atoms has not been observed in beam experiments using a hollow cathode source. Radiation trapping can also be easily avoided in the cell experiments. Distortion of fluorescence decay curves by Zeeman quantum beats is avoided by making zero field (<20 mG) measurements on short lifetimes (<300 nsec), and high field (30 G) measurements on long lifetimes.[3] Any possible distortion of the fluorescence decay curve due to zero-field hyperfine quantum beats can be avoided by using "magic" angle polarization.[1]

Time-resolved laser-induced fluorescence is used routinely on lifetimes from a few nanoseconds to a few microseconds. New detection systems and new lasers are making it possible to measure lifetimes substantially less than a nanosecond.[28] The accuracy of time-resolved laser-induced fluorescence measurements of lifetimes over a microsecond has been limited by error due to atoms leaving the observation region before radiating. Recent experiments demonstrate that this effect can be largely eliminated by using time-of-flight selection in the atomic beam experiments.[29] Lifetimes greater than a microsecond are now being measured to accuracies of 5%. Time-of-flight selection is achieved by pulsing the hollow cathode atomic beam source. The laser excitation occurs some distance (~1 cm) from the source nozzle. The delay between the current pulse to the hollow cathode discharge source and the laser pulse is used to select atoms of a particular velocity. Error due to atoms leaving the observing region before radiating is greatly reduced by extrapolating to infinite delay (zero atomic velocity). Some recent measurements on the $4s^2 4p\ ^2P_{1/2}$ level of ScI are shown in Fig. 4.

The various precautions and systematic studies are important in establishing the accuracy of the lifetime measurements. The high level of agreement routinely achieved by independent groups using time-resolved laser-induced fluorescence gives a great deal of credibility to the 3 to 8% accuracy claimed for the lifetime measurements. Three independent groups published lifetime measurements in NbI during 1982.[3,30,31] All three

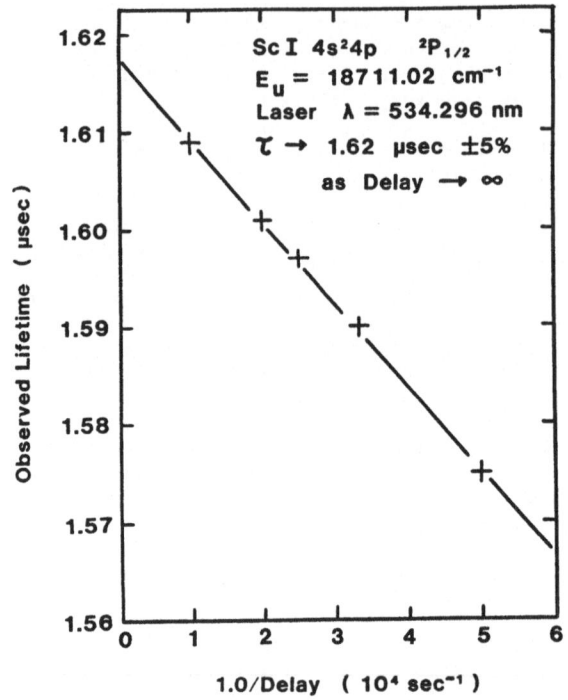

Fig. 4 Lifetime measurements on a ScI velocity selected beam using time-resolved laser-induced fluorescence. The horizontal axis which is labeled 1.0/Delay in units of 10^4 sec^{-1} may also be read as atomic velocity in units of 10^4 cm/sec. Each data point is an average lifetime determined from thousands of fluorescence decay curves recorded using a transient digitizer. The ±5% total uncertainty includes systematic and random effects.

groups used time-resolved laser-induced fluorescence on Nb atomic beams. To the best of this author's knowledge the groups were unaware of each others work prior to publication. Duquette and Lawler measured lifetimes for 50 NbI levels.[3] Rudolph and Helbig measured lifetimes for 6 NbI levels.[30] Kwiatkowski, Zimmerman, Biemont, and Grevesse measured lifetimes for 11 NbI levels.[31] The superior performance of the hollow cathode atomic beam source used by Duquette and Lawler is evident from the larger number of measurements they made. All lifetimes measured by more than one of the groups are compared in Table I. The excellent agreement indicated by the comparison in Table I is now considered routine for lifetime measurements made using time-resolved laser-induced fluorescence.

Table II is a list of some recent research papers reporting transition metal radiative lifetimes measured using laser-induced fluorescence. Accuracy of 3 to 8% is achieved in most of the measurements reported in Table II. Measurements accurate to better than 1% are possible with extraordinary care using time-resolved laser-induced fluorescence.

3. BRANCHING RATIOS

A radiative lifetime by itself determines an Einstein A only if a single transition dominates the decay of the level. Branching ratios for many levels in metal atoms and some levels in ions can be derived from the Corliss and Bozmann (CB) monograph.[80] The extensive CB monograph is from absolute emission measurements on intense arcs. Absolute transition

Table I. A comparison of NbI radiative lifetimes measured using time-resolved laser-induced fluorescence. Uncertainties of approximately 5% were claimed for all measurements.

	Level	Energy (cm^{-1})	Lifetime (nsec)		
			Ref[3]	Ref[30]	Ref[31]
$4d^3 5s(a^3F)5p$	$z^4F_{5/2}$	23574	44.5		42.7
	$z^4F_{7/2}$	24015	38.6		37.3
	$z^4F_{9/2}$	24507	35.9		33.2
$4d^4(a^5D)5p$	$y^6F_{1/2}$	23985	8.5		8.0
	$y^6F_{3/2}$	24165	8.6		8.3
	$y^6F_{7/2}$	24770	8.5		8.1
	$y^6F_{9/2}$	25200	8.2	8.6	7.9
	$y^6F_{11/2}$	25680	7.8	7.5	
$4d^3 5s(a^5P)5p$	$y^6D_{7/2}$	26832	25.8		25.3
	$y^6D_{9/2}$	27420	15.1	14.7	
$4d^4(a^5D)5p$	$x^6D_{1/2}$	26552	8.0		8.3
	$x^6D_{3/2}$	26713	7.3	7.1	
	$x^6D_{5/2}$	26983	7.0	7.3	6.7
	$x^6D_{7/2}$	27427	7.7	7.9	7.9

Table II. Recent radiative lifetime measurements on 3d, 4d, and 5d transition metals using laser-induced fluorescence.

Atom or Ion	Number of Levels	Reference	Atom or Ion	Number of Levels	Reference
ScI	13	[32]	CrI	3	[37]
	59	[29]		1	[38]
				28	[39]
ScII	8	[25]			
	15	[29]	MnI	2	[40]
TiI	17	[33]	MnII	5	[41]
TiII	18	[34]	FeI	10	[10]
				1	[11]
VI	9	[33]		6	[12]
	12	[35]		25	[14]
	39	[36]	FeII	13	[42]

Table II (continued)

Atom or Ion	Number of Levels	Reference	Atom or Ion	Number of Levels	Reference
CoI	8	[12]	ZrII	1	[15]
	10	[43]		20	[51]
CoII	14	[19]		5	[1]
NiI	8	[13]	NbI	50	[3]
	11	[44]		6	[30]
NiII	12	[18]		11	[31]
CuI	4	[45]	NbII	7	[5]
	2	[1]		27	[54]
AgI	2	[1]	MoI	6	[16]
	1	[61]		9	[55]
CdI	14	[62]		6	[30]
	16	[63]		56	[56]
LaI	4	[64]		14	[17]
LaII	7	[65]	MoII	15	[57]
HfI	25	[66]	RuI	12	[58]
				50	[59]
TaI	35	[4]	RhI	13	[31]
				22	[4]
TaII	6	[6]	PdI	9	[60]
	10	[7]	WII	3	[6]
WI	15	[67]	ReI	9	[70]
	13	[68]			
	39	[69]	OsI	6	[71]
ZnI	11	[46]	IrI	25	[72]
	5	[47]	PtI	15	[73]
YI	9	[48]	AuI	2	[74]
	10	[49]	HgI	3	[75]
	34	[50]		1	[76]
YII	14	[50]		1	[77]
				1	[78]
ZrI	11	[15]	HgII	1	[79]
	34	[51]			
	11	[52]			
	15	[53]			
	36	[1]			

probabilities from this source are known to have many large errors, but branching ratios derived from it are reasonably accurate for branches greater than 0.1.[50] Errors in CB absolute transition probabilities due to lack of local thermodynamic equilibrium, and to inaccurate knowledge of the electron temperature and metal atom density, do not affect the accuracy of branching ratios derived from the CB monograph. Intensity and wavelength-dependent errors in the photographic calibration do affect the branching ratios. The CB monograph is nevertheless a useful source of branching ratios for stronger (≥ 0.1) branches.

Remeasurement of the branching ratios is desirable because of the enormous advantage of photoelectric detectors over the photographic system used by CB. Some branching ratios have been remeasured using grating monochromators, but it has become increasingly clear that Fourier transform spectrometers (FTS) are very well suited to this work. The 1.0m FTS at the National Solar Observatory is probably the best instrument in the world for spectrophotometry on complex atoms and ions.[8] High spectral resolution and good absolute wave number accuracy are important to reduce line blending and to insure correct line identification. The 1.0m FTS provides Doppler limited resolution and absolute wave number accuracy to one part in 10^8. A high data collection rate is also desirable for spectrophotometry on complex atoms or ions. The 1.0m FTS records a million point interferogram covering a spectral region from the edge of the VUV to the IR in approximately one hour. The spectra are produced in a convenient computer format for analysis, because the interferograms are recorded and transformed digitally. The 1.0m FTS has a five decade dynamic range with very good linearity. Fourier transform spectrometers also have an intrinsic advantage over grating monochromators for spectrophotometry. A grating monochromator which sequentially scans each spectral element will map any small drift in source intensity into a branching ratio error. The interferogram recorded by a FTS represents simultaneous measurements on all spectral elements. A FTS is therefore less sensitive to small drifts in source intensity.

Conventional hollow cathode discharge lamps are often used in spectrophotometry with the 1.0m FTS. These lamps are simple, very reliable, and broadly applicable. The lamps usually contain a few Torr of Ne or Ar. The metal emission lines are strong and are broadened primarily by the Doppler effect. The translational temperatures are often only slightly above room temperature. The low collision rate in the lamps implies severe departures from local thermodynamic equilibrium, but this is not a concern in branching ratio measurements. It is important to verify that the lamps are optically thin. A test for radiation trapping is performed by recording spectra at several discharge currents, and verifying that the branching ratios are unchanged. Radiation trapping is sometimes observed. Accurate branching ratios are then determined by extrapolating to zero discharge current.

The relative intensity calibration as a function of wave number is the primary concern when using the 1.0m FTS for spectrophotometry. Whaling and collaborators, who were the first to use the 1.0 FTS for spectrophotometry in the visible and UV, devised an elegant method for determining the relative efficiency versus wave number of the FTS.[9] Overlapping sets of Ar branching ratios are used to construct the efficiency versus wave number curve for the FTS. The selected ArI and ArII branching ratios are for transitions between excited levels which are not prone to radiation trapping. The selected branching ratios cover the region from 4300 cm^{-1} to 34000 cm^{-1}, and are reasonably accurate ($\sim 4\%$).[9,81] The use of ArI and

ArII branching ratios to calibrate the FTS has important advantages. Argon is often the carrier gas in the hollow cathode discharge. Thus a calibration spectrum is recorded with each metal spectrum. Furthermore the calibration automatically includes any effects due to variations in the transmission of the source window or due to variations in the reflectance of the back of the hollow cathode. The 1.0m FTS has recently been used to measure branching ratios in FeI [82,83], CoI [84], YI and YII [50], MoI [17,56], CoII [19], VI [36], CrI [85], RhI [86], NbI [87], HfI [87], Ta [69], and WI [69].

4. OTHER DEVELOPMENTS

It is essential to include all major branches when combining radiative lifetimes and emission branching ratios to determine transition probabilities for individual spectral lines. Spectra from the 1.0m FTS are usually analyzed to determine branching ratios using an interactive computer. All known energy levels are entered in the computer and the spectra are searched for every possible transition between known energy levels. The tables of Atomic Energy Levels by C. E. Moore are often used in the search.[88]

Although the 1.0m FTS is well suited to spectrophotometry in the IR, the possibility of strong IR branches beyond the long wavelength limit of the spectra is occasionally a cause for concern. If the level of interest has a relatively short lifetime (~10 nsec) and strong visible or UV branches, then one can be quite confident that the visible and UV branches are dominant. An IR branch from such a level would need an extraordinarily large oscillator strength to have a significant branching ratio.

Short-lived levels with strong visible or UV branches serve as reference levels in an experiment by Duquette, Den Hartog and Lawler.[87] This "missing branch experiment" is used to screen emission branching ratios for strong unidentified IR branches. The magnitudes of the laser-induced fluorescence signals produced by driving two transitions sharing a common lower level are compared in this experiment. One transition is to the reference level and the second is to the level to be tested for missing IR branches. If the test level has strong unidentified IR branches with a total branching ratio R, then the fluorescence from the test level will be suppressed by a factor of $(1-R)^2$. The technique is sensitive to the square of the missing branches because the failure to include IR branches in predicting the visible or UV fluorescence affects both laser excitation and subsequent fluorescence. The use of a common lower level eliminates any need to determine an atomic density. The spectral fluence produced by each dye laser as it is scanned across the spectral lines must be measured, and the relative efficiency versus wavelength of the fluorescence collection system must be known. The missing branch experiment is performed on atoms in a beam using much of the same apparatus used in the lifetime experiments. The beam environment eliminates any problems from collisional quenching of the fluorescence. The hollow cathode atomic beam source makes the missing branch experiment very broadly applicable. It can be applied to any metallic element, and many metastable levels have sufficient populations to serve as lower levels for laser excitation using the hollow cathode beam source. The missing branch experiment has been applied to long-lived levels of Hf, Nb, W, and Ta.[69,87] Only the decays of the z^5F levels of HfI were found to be dominated by IR branches. The strong IR branches were subsequently identified and measured on FTS spectra.

The missing branch experiment described here is used to verify the completeness of emission branching ratio. Relative magnitudes of laser-induced fluorescence signals from transitions sharing a common lower level could also be used in a fashion analogous to the powerful "bowtie" method of Cardon et. al.[89] The bowtie method combines emission oscillator strengths and relative absorption measurements in a least square adjustment. The use of relative laser-induced fluorescence signals instead of relative absorption signals will make the bowtie method far more broadly applicable.

5. APPLICATIONS IN ASTRONOMY

Although accurate atomic transition probabilities are needed in many fields, the need is most acute in astronomy. The wealth of new data on atomic transition probabilities from laser spectroscopy and Fourier transform spectroscopy is already having an impact in astronomy. The new transition probability data have resulted in greatly improved solar abundance determinations for La [65], Zr [51], Co [84], Pd [60], Re [70], Y [50], Rh [4,31], Mo [90], Ru [58], Os [71], V [36], and Nb [54].

These very accurate spectroscopically determined solar abundances are in excellent agreement with elemental abundances determined from chondritic meteorites. The discrepancy between a Nb solar abundance determined using NbI lines and the Nb abundance in chrondritic meteorites has been resolved.[31,54] Line misidentification and/or blending in the solar spectrum was the source of the problem. The NbI lifetimes and oscillator strengths are accurate. Recent work on NbII transitions has resolved the apparent disagreement between the spectroscopic solar abundance and that determined from chrondritic meteorites.[54] The high level of agreement between spectroscopically determined solar abundances and abundances from chrondritic meteorites is not accidental. The meteorites are believed to be a representative sample of matter in the solar system. The elemental abundances should agree for most non-volatile elements.

The precision solar abundance studies are important for many reasons. In some sense the solar abundances work paves the way for precision spectroscopic studies on other stars with elemental abundances much different than the sun. An exciting example is recent work on χ Cygni and o Ceti by Dominy and Wallerstein.[91] Certain red giants with enhanced heavy element abundances contain the unstable element Tc. The longest lived Tc isotope has a half-life of only 2×10^5 years which is short on stellar time scales. The enhanced heavy element abundances and the existence of Tc is believed to be due to neutron capture during shell flashes. Spectroscopic elemental abundance studies of the solar and stellar atmospheres, and of interstellar clouds,will ultimately produce a more quantitative understanding of the production and distribution of chemical elements.

6. SUMMARY

A combination of developments in laser spectroscopy and in Fourier transform spectroscopy is leading to enormous progress in determining atomic transition probabilities for the elements in low stages of ionization. Time-resolved laser-induced fluorescence is widely used to determine radiative lifetimes with small (~5%) systematic and random uncertainties. The development of sputter techniques such as the hollow cathode atomic/ionic beam source makes it possible to apply laser-induced fluorescence to neutral atoms and ions of all metallic elements. The application of powerful Fourier transform spectrometers in the visible and UV is providing branching ratios which complement the lifetime measurements. It is hoped that in a few years it will be possible to assemble a compendium of atomic transition probabilities, as extensive as the CB monograph, but with a vastly improved accuracy of 5 to 10%. The recent progress in measuring atomic transition probabilities is resulting in precision elemental abundance determinations in the sun, other stars, and in interstellar clouds. The abundance studies are essential for developing an understanding of the production and distribution of elements in the observable universe.

ACKNOWLEDGMENTS

The preparation of this manuscript was supported by NSF Grant AST85-20413. I wish to acknowledge the profound influence of Art Schawlow on my life and work.

LITERATURE

1. P. Hannaford, R. M. Lowe: Opt. Engineering 22, 532 (1983)
2. J. Richter: Physica Scripta T8, 70 (1984)
3. D. W. Duquette, J. E. Lawler: Phys. Rev. A 26, 330 (1982)
4. S. Salih, D. W. Duquette, J. E. Lawler: Phys. Rev. 27, 1193 (1983)
5. S. Salih, J. E. Lawler: Phys. Rev. 28, 3653 (1983)
6. M. Kwiatkowski, F. Naumann, K. Werner, P. Zimmermann: Phys. Lett. A103, 49 (1984)
7. W. Schade, V. Helbig: Phys. Lett. A115, 39 (1986)
8. J. W. Brault: J. Opt. Soc. Am. 66, 1081 (1976)
9. D. L. Adams, W. Whaling: J. Opt. Soc. Am. 71, 1036 (1981)
10. H. Figger, K. Siomos, H. Walther: Z. Physik 270, 371 (1974)
11. K. Siomos, H. Figger, H. Walther: Z. Physik A272, 355 (1975)
12. H. Figger, J. Heldt, K. Siomos, H. Walther: Astron. and Astrophys. 43, 389 (1975)
13. J. Heldt, H. Figger, K. Siomos, H. Walther, Astron. and Astrophys. 39, 371 (1975)
14. J. Marek, J. Richter, H. J. Stahnke: Physica Scripta 19, 325 (1979)
15. P. Hannaford, R. M. Lowe: J. Phys. B14, L5 (1981)
16. D. W. Duquette, S. Salih, J. E. Lawler: Phys. Lett. A83, 214 (1981)
17. W. Whaling, P. Chevako, J. E. Lawler: J. Quant. Spectrosc. Radiat. Transfer 36, 491 (1986)
18. J. E. Lawler, S. Salih: Phys. Rev. A (in press).
19. S. Salih, J. E. Lawler, W. Whaling: Phys. Rev. A31, 744 (1985)
20. T. W. Hänsch: Appl. Optics 11, 895 (1972)
21. G. K. Klauminzer: U. S. Patent 4, 127, 828, Nov. 28 (1978)
22. M. G. Littman, H. J. Metcalf: Appl. Optics 17, 2224 (1978)
23. H. S. Schomburg, H. F. Dobele, B. Ruckle: Appl. Phys. B30, 131 (1983)
24. K. G. H. Baldwin, J. P. Marangos, D. D. Burgess, M. C. Gower: Optics Comm. 52, 351 (1985)
25. O. Vogel, L. Ward, A. Arnesen, R. Hallin, C. Nordling, A. Wännstrom: Physica Scripta 31, 166 (1985)
26. J. M. Harris, F. E. Lytle, T. C. McCain: Analytical Chem. 48, 2095 (1976)
27. B. Leskovar, Laser Focus/Electro-Optics 20, 73 (1984)
28. I. Yamazaki, N. Tamai, H. Kume, H. Tsuchiya, K. Oba: Rev. Sci. Instrum. 56, 1187 (1985)
29. E. A. Den Hartog, G. Marsden, J. E. Lawler, J. T. Dakin. V. Roberts: Bull. Am. Phys. Soc. (in press) Presented at the 39th G. E. C. (1986)
30. J. Rudolph, V. Helbig: Phys. Lett. A89, 339 (1982)

31. M. Kwiatkowski, P. Zimmermann, E. Biemont, N. Grevesse: Astron. and Astrophys. 112, 337 (1982)
32. J. Rudolph, V. Helbig: J. Phys. B15, L1 (1982)
33. J. Rudolph, V. Helbig: J. Phys. B15, L599 (1982)
34. M. Kwiatkowski, K. Werner, P. Zimmerman: Phys. Rev. A31, 2695 (1985)
35. A. Doerr, M. Kock, M. Kwiatkowski, K. Werner, P. Zimmerman: J. Quant. Spectrosc. Radiat. Transfer 33, 55 (1985)
36. W. Whaling, P. Hannaford, R. M. Lowe, E. Biemont, N. Grevesse: Astron. Astrophys. 153, 109 (1985)
37. R . M. Measures, N. Drewell, H. S. Kwong: Phys. Rev. A16, 1093 (1977)
38. H. S. Kwong, R. M. Measures: Appl. Optics 19, 1025 (1980)
39. M. Kwiatkowski, G. Micali, K. Werner, P. Zimmermann: Astron. Astrophys. 103, 108 (1981)
40. H. D. Kronfeldt, J. R. Kropp, A. Subaric, R. Winkler: Z. Physik A322, 349 (1985)
41. M. Kwiatkowski, G. Micali, K. Werner, P. Zimmermann: J. Phys. B15, 4357 (1982)
42. P. Hannaford, R. M. Lowe: J. Phys. B16, L43 (1983)
43. J. Marek, K. Vogt: Z. Physik A280, 235 (1977)
44. U. Becker, H. Kerkhoff, M. Schmidt, P. Zimmermann: J. Quant. Spectrosc. Radiat. Transfer 25, 339 (1981)
45. H. Kerkhoff, G. Micali, K. Werner, A. Wolf, P. Zimmermann, Z. Physik-A300 , 115 (1981)
46. H. Kerkhoff, M. Schmidt, P. Zimmermann: Z. Physik A298, 249 (1980)
47. M. Chantepie, J. L. Cojan, J. Landais, B. Laniepce, A. Moudden, M. Aymar: Opt. Comm. 51, 391 (1984)
48. P. Hannaford, R. M. Lowe: J. Phys. B15, 65 (1982)
49. J. Rudolph, V. Helbig: J. Phys. B15, L1 (1982)
50. P. Hannaford, R. M. Lowe, N. Grevesse, E. Biemont, W. Whaling: Astrophys. J. 261, 736 (1982)
51. E. Biemont, N. Grevesse, P. Hannaford, R. M. Lowe: Astrophys. J. 248, 867 (1981)
52. D. W. Duquette, S. Salih, J. E. Lawler: Phys. Rev. A25, 3382 (1982)
53. J. Rudolph, V. Helbig: Z. Physik A306, 93 (1982)
54. P. Hannaford, R. M. Lowe, E. Biemont, N. Grevesse: Astron. Astrophys. 143, 447 (1985)
55. M. Kwiatkowski, G. Micali, K. Werner, P. Zimmermann: Phys. Lett. A85, 273 (1981)
56. W. Whaling, P. Hannaford, R. M. Lowe, E. Biemont, N. Grevesse: J. Quant. Spectrosc. Radiat. Transfer 32, 69 (1984)
57. P. Hannaford, R. M. Lowe: J. Phys. B16, 4539 (1983)
58. E. Biemont, N. Grevesse, M. Kwiatkowski, P. Zimmermann: Astron. Astrophys. 131, 364 (1984)
59. S. Salih, J. E. Lawler: J. Opt. Soc. Am. B2, 422 (1985)
60. E. Biemont, N. Grevesse, M. Kwiatkowski, P. Zimmermann: Astron. Astrophys. 108, 127 (1982)
61. K. P. Selter, H. J. Kunze, Astrophys. J. 221, 713 (1978)
62. H. Kerkhoff, M. Schmidt, U. Teppner, P. Zimmermann, J. Phys. B13, 3969 (1980)
63. M. Chantepie, J. L. Cojan, J. Landais, B. Laniepce, A. Moudden, M. Aymar: Optics Comm. 46, 93 (1983)
64. B. R. Bulos, A. J. Glassman, R. Gupta, G. W. Moe: J. Opt. Soc. Am. 68, 842 (1978)
65. A. Arneson, A. Bengtsson, R. Hallin, J. Lindskog, C. Nordling, T. Noreland: Physica Scripta 16, 31 (1977)
66. D. W. Duquette, S. Salih, J. E. Lawler: Phys. Rev. A26, 2623 (1982)
67. D. W. Duquette, S. Salih, J. E. Lawler, Phys. Rev. A24, 2847 (1981)

68. M. Kwiatkowski, G. Micali, K. Werner, M. Schmidt, P. Zimmerman: Z. Physik A$\underline{304}$, 197 (1982)
69. E. A. Den Hartog, D. W. Duquette, J. E. Lawler: J. Opt. Soc. Am. B $\underline{4}$, 48 (1987)
70. D. W. Duquette, S. Salih, J. E. Lawler: J. Phys. B$\underline{15}$, L897 (1982)
71. M. Kwiatkowski, P. Zimmermann, E. Biemont, N. Grevesse, Astron. Astrophys. $\underline{135}$, 59 (1984)
72. D. S. Gough, P. Hannaford, R. M. Lowe: J. Phys. B$\underline{16}$, 785 (1983)
73. D. S. Gough, P. Hannaford, R. M. Lowe, J. Phys. B$\underline{15}$, L431 (1982)
74. P. Hannaford, P. L. Larkins, R. M. Lowe: J. Phys. B$\underline{14}$, 2321 (1981)
75. E. N. Borisov, A. L. Osherovich, V. N. Yakovlev, Opt. Spectrosc. $\underline{47}$, 109 (1979)
76. E. N. Borisov, A. L. Osherovich: Opt. Spectrosc. $\underline{50}$, 346 (1981)
77. J. A. Halstead, R. R. Reeves: J. Quant. Spectrosc. Radiat. Transfer $\underline{28}$, 289 (1982)
78. P. van de Weijer, R. M. M. Cremers: J. Appl. Phys. $\underline{57}$, 672 (1985)
79. J. C. Bergquist, D. J. Wineland, W. M. Itano, H. Hemmati, H. U. Daniel, G. Leuchs: Phys. Rev. Lett. $\underline{55}$, 1567 (1985)
80. C. H. Corliss, W. R. Bozman: Experimental Transitional Probabilities for Spectral Lines of Seventy Elements, U. S. Natl. Bur. of Stand. Monograph $\underline{53}$, (U. S. G. P. O. Washington, DC 1962)
81. K. Danzmann, M. Kock: J. Opt. Soc. Am. $\underline{72}$, 1556 (1982)
82. D. L. Adams, W. Whaling: J. Quant. Spectrosc. Radiat. Transfer $\underline{25}$, 233 (1981)
83. M. Kock, S. Kroll, S. Schnehage: Physica Scripta T$\underline{8}$, 84 (1984)
84. B. L. Cardon, P. L. Smith, J. M. Scalo, L. Testerman, W. Whaling: Astrophys. J. $\underline{260}$, 395 (1982)
85. G. P. Tozzi, A. J. Brunner, M. C. E. Huber: Mon. Not. R. Ast. Soc. $\underline{217}$, 423 (1985)
86. D. W. Duquette, J. E. Lawler: J. Opt. Soc. Am. B$\underline{2}$, 1948 (1985)
87. D. W. Duquette, E. A. Den Hartog, J. E. Lawler: J. Quant. Spectrosc. Radiat. Transfer $\underline{35}$, 281 (1986)
88. C. E. Moore, Atomic Energy Levels, U. S. Natl. Bur. Stand. Natl. Stand. Ref. Data Ser. $\underline{35}$, (U. S. G. P. O. Washington DC 1971)
89. B. L. Cardon, P. L. Smith, W. Whaling: Phys. Rev. A$\underline{20}$, 2411 (1979)
90. E. Biemont, N. Grevesse, P. Hannaford, R. M. Lowe, W. Whaling: Astrophys. J. $\underline{275}$, 889 (1983)
91. J. F. Dominy, G. Wallerstein: Astrophys. J. (in press) (1987)

Atomic Engineering of Highly Excited Atoms

M.H. Nayfeh

Department of Physics, University of Illinois at Urbana-Champaign,
1110 W. Green Street, Urbana, IL 61801, USA

1   Introduction

Recent astronomical observations in the direction of the object Cassiopeia A have detected interstellar radio emissions at the lowest frequencies observed to date, in the band 7 - 16 MHz [1]. These probably originate from atoms of enormous size, about a million times larger than a normal hydrogen atom and larger than the size of a bacterium, which would be the largest atoms yet encountered in any environment. The idea of gigantic atoms is fascinating in itself. However, its scientific appeal is not due just to its *Guinness Book of Records* aspect. In fact work in progress in laboratories is not competing with this record, but is rather aimed at the deliberate manipulation of atoms. This concept has long been sought but it has only become possible using highly excited (large) atoms. Although the production and detection of such large atoms involve some of the most advanced techniques of laser and atomic physics, their study represents in some sense a dramatic return to the earliest modern ideas of atomic structure: the model of Niels Bohr, the centenary of whose birth has just been celebrated [2].

2   Bohr's Atomic Theory

In 1913 Niels Bohr developed the first quantum theory of hydrogen [2]. It retained major elements of a classical theory in that it represented the hydrogen atom in terms of an electron in orbit about a proton, just like a miniature model of a planet orbiting the sun. However, it introduced a new assumption which was not based on any principle of classical mechanics: Nature allows only those orbits for which the angular momentum is an integral multiple of Planck's constant h. For a circular orbit of speed v and radius r, this condition is expressed as

$mvr = nh$, where m is the mass of the electron and n takes the integer values 1, 2, 3, ... for the different allowed orbits. Classical mechanics requires that the centripetal force which holds the electron in orbit be provided by the inverse-square-law electrical force between electron and proton: $(mv^2)/r = q^2/(4\pi\epsilon_o)r^2$ (using MKS electrical units). These two relations together yield values for the radii, speeds, periods $(T = 2\pi r/v)$, and binding energies E of the allowed orbits, which are summarized in Table 1.

The main prediction of the Bohr theory concerns the wavelengths of light that will be emitted or absorbed when the electron makes a transition from one orbit to another. The predicted values of wavelength agree very well with those actually observed: to within a few parts per million, which was at the time virtually the limit of experimental accuracy. On this basis alone the Bohr model can be judged a spectacular success. However, there is no obvious way to extend this model to the other atoms in the periodic table. For example, the neon atom has 10 electrons, and in its normal state all of these are contained within a volume comparable to that of the most tightly bound orbit of hydrogen (n=1). Despite many ingenious attempts, it was never found possible to describe many-electron atoms in terms of classical mechanics with a subsidary quantization assumption. About a decade after Bohr's theory, quantum mechanics was developed by Schrödinger, Heisenberg, and others.

Table 1: Quantum mechanical results for hydrogen

Quantity	Dependence on n	Value for $n = 1$	and for $n = 100$
Orbital radius, r	n^2	5.3×10^{-11} m	5.3×10^{-7} m
Speed, v	n^{-1}	2.2×10^6 m/s	2.2×10^4 m/s
Period of Revolution, T	n^3	1.5×10^{-16} m/s	1.5×10^{-10} m/s
Binding energy, E	n^{-2}	13.6 eV	1.36×10^{-3} eV
Radiative Lifetime, τ	$n^{4.5}$	—	~ 40m/s
Cross section for excitation from the ground state σ	n^{-3}	—	$\sim 10^{-26}$ m^2

The new theory confirmed many of the predictions of Bohr's theory, put them on a more solid basis, and solved all of its shortcomings. One of these is the inability of Bohr's theory to give a method for calculating excitation cross sections σ, lifetimes τ, and the intensities of the spectral lines, the former two of which are summarized in Table 1. Electronic orbits or states in the new theory are described not only by Bohr's original quantum number n, but also by three additional quantum numbers: the orbital angular momentum quantum number l which gives the angular momentum of the electron about the nucleus; the azimuthal quantum number m_l, which gives the direction of l (projection of l along a specified direction); and m_s, the projection of the electron's spin angular momentum along a specified direction. For each n, l can take the values 0, 1, 2, ... up to $n - 1$, and m_l takes the values $- l$, $-l + 1$, ...$l - 1$, and l, and m_s takes two values: $+1/2$, and $-1/2$. According to our present understanding, the Bohr energy levels of hydrogen are split into fine structure and hyperfine structure, which are due respectively to interaction of electron and proton spins with the magnetic fields associated with their orbits. The fine and hyperfine splittings are 10^{-5} and 10^{-7} of the energy difference between Bohr levels in hydrogen. Much larger fine and hyperfine splittings are present in complex atoms as a result of the interaction of the active electron with the other electrons. In some cases the "fine" structure splittings are larger than the energies given by the Bohr model. Bohr's theory has thus come to be regarded as an important, but merely transitional, state between classical and quantum mechanics.

3 Emergence of Interest in Highly Excited States

In the past ten years there has been a great revival of interest in the Bohr picture and its consequences [3,4]. Surprisingly, the initial stimulus for this renewal came not from studies of atomic hydrogen, but rather from those of complex atoms – the very species that contributed to the downfall of the model! How can this be?

The simple technical answer is that any atom becomes hydrogenic when it is in a highly excited state. From Table 1 we see that the radius of a stable Bohr orbit increases as the square of n (now called the principal quantum number). As n increases, the electron is placed ever further from the proton; and if the electron is sufficiently far out it will be unable to discern the difference between a proton and any other object with one unit of positive electric charge (this is analogous to saying that the orbit of a planet about the sun depends only on the total mass of the sun, and not how that mass is distributed within the sun). In particular, the proton can be replaced by a complex atom with one electron removed. Thus the real question to be answered is how and why highly excited states have become a topic of interest. We believe it is originally due to three key developments in different areas of physics.

The first of these developments that brought about this field took place in 1965 following the remarkable astrophysical discovery of radio recombination lines by Höglund and Metzger [5]. These lines are due to transitions between adjacent highly excited states (Rydberg states) of atomic hydrogen, helium, and carbon that are populated primarily by radiative recombination, in which a free electron emits a photon and is captured into a high-n orbit about an ion. The rates at which such recombination processes occur are very low compared to those of ordinary (three-body) chemical reactions, in addition to the fact that it is not selective, and so it is very difficult to study or to utilize radiative recombination in terrestrial laboratories. (However, another recombination process — dielectronic recombination — has been the focus of much recent experimental work[6], and it too takes place via highly excited states - see article in [4].)

Since electron-ion recombination is a weak and nonselective process, and hence cannot be used to produce highly excited states in the laboratory, alternative means involving excitation of the atomic ground state must be employed. Two major problems then arise: first, the cross section for excitation decreases rapidly with n (see Table 1), so that a very high spatial density of excitation is required; second, the difference in energy between successive Rydberg states also decreases with n, so that a highly monochromatic source of excitation is required for selective population of a given state. Both of these problems were solved by another major development, in an area completely distinct from radio astronomy: the invention of the narrow-band tunable dye laser by Hänsch [8], which advanced laser technology initiated by Schawlow and Townes [7] by a major step. Although this laser design has since found applications in diverse problems in physics, chemistry and biology, the initial motivation for its development was the study of the fine structure of optical transtions of hydrogen - a structure that was not predicted by Bohr's theory, and whose accurate measurement is necessary for the determination of the cornerstone of the fundamental constants, the Rydberg constant, which is the binding energy of the ground state of atomic hydrogen [9].

Finally, during the 1960s evidence was accumulating for the validity of the quark model of elementary particles. The great predictive success of this model was shadowed by a fact that was difficult to explain: no isolated quark had ever been observed. These particles are postulated to have distinctive electrical properties (i.e. a charge of -e the charge of the electron), and would give rise to characteristic spectra if they were bound to an atomic nucleus. It became apparent that the new laser technology offered the possibility of detecting a single atom of a desired species in a sample containing up to 1020 other atoms and molecules, and so it provided a technique for searching for very rare elementary particles. This was the primary stimulus behind the effort [10] to achieve single atom detection by selective excitation and ionization, a process which necessarily involved the production of excited states. It should be noted that the further development of single-atom detection techniques has not yet resulted in any major discovery in particle physics, although work in this area continues, but it has had a significant impact on analytical chemistry and materials science [11].

4   The Preparation of Large Atoms

With this initial motivation for Rydberg state excitation, derived from interest in astrophysics and in the possibilites for single-atom detection schemes, basic spectroscopy of Rydberg states soon blossomed into a major subfield of atomic physics.

The first generation of experiments involved complex atoms because their excitation is within the reach of the new visible tunable dye lasers. This is certainly true for most alkalis and some rare earth atoms whose ionization potentials are less than 5 to 6 eV and whose low lying excited states typically lie at about 2 eV. Double or triple photon excitation by one or two lasers at visible wavelengths of 2 - 2.5 eV is sufficient to excite high Rydberg states of these atoms. Because of the narrow band and high intensity available from these lasers, excitation efficiencies of near 10 % can be achieved.

Hydrogen, on the other hand, has an ionization potential of 13.6 eV, and its first low lying excited state occurs at 10.2 eV, thus making it out of reach of the dye lasers. Thus, hydrogen, the simplest of all atoms, continues to be the most popular atom to atomic theorists, while sodium became the most popular atom in laser spectroscopy experiments. Since sodium has one electron outside a closed shell, it does in many respects behave as a true one-electron atom.

Nevertheless, highly excited states of the hydrogen atom possess unique properties, most notably the ways in which they respond to external electric and magnetic fields. Thus there has been constant interest in applying tunable laser technology to the excitation of hydrogen, and in the past few years several approaches have been successful. We review some of these in order to indicate the diversity of experimental effort on this problem.

One method, developed largely at Yale University, utilizes a process in which a beam of protons of a few keV kinetic energy is shot through a gaseous target [12]. The protons pick up an electron from the gas (charge exchange) to form hydrogen atoms. In this charge exchange process, the atomic hydrogen beam exciting the target contains atoms in a large number of excited states, with relative populations determined by the various charge exchange cross sections and the state lifetimes. The basic process therefore suffers from a lack of selectivity. It is possible, however, to employ an elaborate scheme of electric field quenching of many of the excited states and further excitation of a remaining state with an infrared laser to produce a well-defined population of a given highly excited state.

Another method that bypasses the use of tunable dye lasers relies on excitation with a fixed frequency laser beam and an appropriate utilization of the Doppler effect. This approach has been developed at Los Alamos Laboratory [13]. A large particle accelerator, originally designed to generate mesons for nuclear and particle physics, produces a beam of H^{-1} ions (this is a hydrogen atom with an extra electron attached to it) travelling at 84% of the speed of light; these are passed through a thin foil, resulting in the stripping of the extra electron to yield a relativistic beam of atomic hydrogen. This beam is crossed with a laser beam of fixed wavelength. By varying the angle at which the two beams cross, the energy of the photons in the rest frame of the atoms can be adjusted into resonance with transitions between the ground state and various high-lying states. However, due to the energy spread of available atomic hydrogen beams, the resolution attainable with this technique is fairly limited, and to date only states with $n < 14$ have been selectively excited.

The complexity and unavailablity of accelerators has kept highly excited hydrogen out of reach of small laboratory experiments. Recent advances, however, in nonlinear optical processes in certain types of crystals and gases have allowed conversion of tunable visible dye laser radiation into tunable ultraviolet radiation with reasonable efficiency. These methods have been applied to excitation of high-n states of atomic hydrogen at the University of Bielefeld [14], in West Germany, and at our laboratory at the University of Illinois [15]. In this first scheme, hydrogen is excited to the 2p state by absorption of a single photon from a 1215 Å pulsed tunable beam followed by an absorption of a single photon from a 3660 Å pulsed tunable beam that excites a high n state. In our scheme (shown in Fig. 1) two photons from a 2430 Å tunable beam excite the 2s state from the ground state, and subsequent absorption of a single photon from a 3660 Å tunable beam excites a high n state. Although single-photon excitation with 1215 Å is more efficient than the two photon excitation with 2430 Å, it is in some respects more convenient to utilize the two-photon process. This is due to the fact that it is much easier and more efficient to produce the 2430 Å radiation than the 1215 Å radiation and it does not suffer from resonance trapping (emission and reabsorption of radiation) which prevents a well-defined geometry of excitation.

In our scheme the exciting radiation (2430 Å) is capable of directly photoionizing the $n = 2$ states, which constitutes a loss of the efficiency

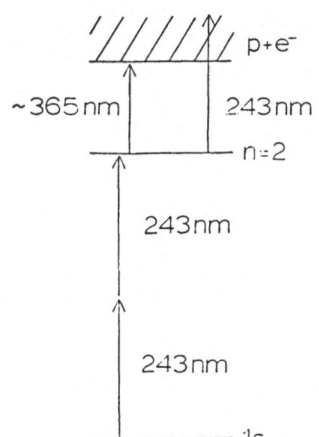

Fig. 1. Scheme for exciting a high n state.

of the excitation of the highly excited states. Such competition is usually taken into consideration in the design of the experiment, and excitation efficiency of 10^{-7} has been achieved from the ground state with a loss of 10^{-8} to direct photoionization, under atomic beam conditions. A simplified experimental setup is shown in Fig. 2. Atomic hydrogen produced by a discharge is pumped out of the discharge, collimated into an atomic beam, and enters a region between two electrodes that supply a dc electric field. The atomic beam is crossed by the two collinear laser beams in the field region. The produced charge is pushed by the electric field and travels about 1m before it gets detected by a single ion detector. The time of flight provides mass analysis of the detected charge. By this means we are able to excite states with n up to about 60, as shown in Fig. 3. The loss of resolution beyond $n = 60$ is due to the 1 cm^{-1} bandwidth of the 3660 Å pulsed laser [15,17].

When continuous wave lasers (CW) rather than pulsed lasers are used it is possible to selectively excite states of higher n because one can attain much narrower bandwidth. Laser technology has just been extended to allow CW excitation of hydrogen [18]. The highest n reached to date is that of $n = 280$ in barium taken at the Free University of Berlin [16] using CW lasers. The excitation of a hydrogen n state was monitored by ionizing it by an external pulsed electric field of a few milliseconds which is turned on after the excitation has taken place in

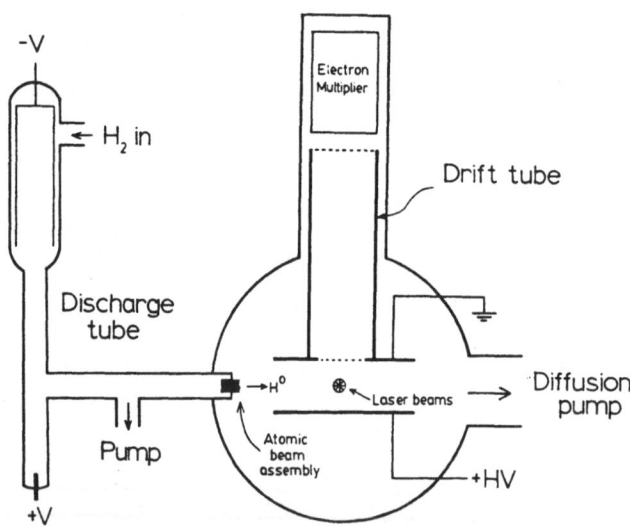

Fig. 2. Experimental Apparatus.

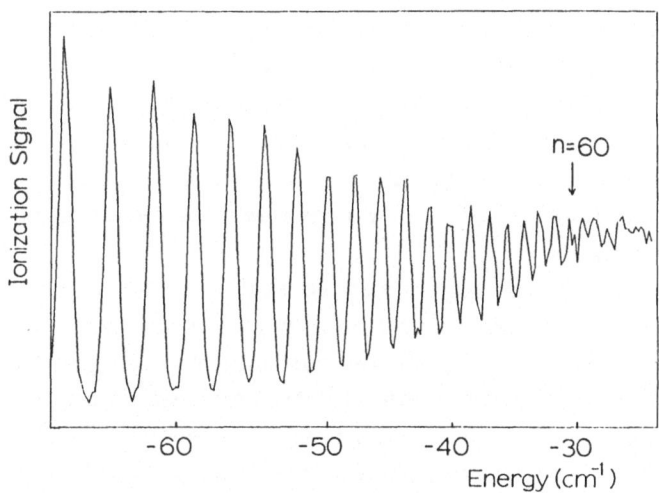

Fig. 3. Rydberg Spectrum of hydrogen.

order not to interfere with the excitation process itself. This method is very sensitive, selective (by appropriate choice of field amplitude one can ionize a specific $n > n_o$ without affecting the $n < n_o - 1$ state, and so on), and is actually at the heart of strong field effects that will be discussed in the latter sections. The barium Rydberg states on the other hand were monitored by colliding them with a beam of other atoms, thus utilizing another property of excited states: their sensitivity to collisional interactions. However this technique does not have the n selectivity of the field ionization method.

5 Sensitive Probes

We have seen that highly excited states are gigantic in size (geometrical cross section πr^2 is very large), thus making them very fragile, i.e. the highly excited electron can be easily removed by environmental effects such as collision with atoms of the same kind or with foreign atoms and molecules, ions or electrons if present, walls of the container, etc. Moreover highly excited atoms are also very sensitive to external electric or electromagnetic fields. As such, these atoms promise to be the most sensitive detectors of weak external fields whether they are of collisional or of electromagnetic origin. The collsional sensitivity has been illustrated most dramatically by the recent astrophysical observation alluded to above [1]. The observation of the 16 MHz emission point to the existence of very large atoms in the interstellar gas clouds. These are large enough such that collisional effects by particles in the medium are becoming observable in spite of the extremely low density (1 atom/cc) via the broadening they cause to the radio lines characteristic of their transitions. This has provided new diagnostic methods for the density and temperature of the clouds whose study provides information about the process of star formation.

One promising application of highly excited states is in the development of very sensitive detectors of low-energy radiation, especially microwaves. It is relatively easy to build instruments operating in the visible wavelength range which are capable of detecting a single photon. This is because a visible photon has enough energy (a few eV) to eject an electron from a solid, and the free electron can be accelerated to arbitrarily high energy by an electric field. Thus unlimited amplification can be achieved if the photon can produce a free electron. The energies of microwave photons lie in the range 10^{-3} – 10^{-4} eV, which is far too small to liberate an electron from ordinary matter. However, this range coincides with the energies of transitions between experimentally accessible Rydberg states: for instance, the transition between $n = 30$ and $n = 31$ requires 10^{-3} eV. As noted above, we can use a pulsed electric field to monitor the excitation of n to $n + 1$. Thus it is often possible in principle to adjust the electric field strength so that the state n is long-lived while state $n + 1$ is ionized fairly quickly. A group of Rydberg atoms prepared in the state n could then serve to detect photons of the appropriate energy. Unfortunately, it has not yet been possible to implement this scheme in a practical fashion, due mostly to the difficulty of maintaining a large population of highly excited states.

6   Atomic Engineering

As the strength of the external field increases, so will the atomic response. It is interesting to consider sufficiently high fields such that the electronic interactions with the external and the Coulomb fields are comparable, that is when none of them dominates. In this case the interaction is highly nonlinear in the external field amplitude. However, it is almost impossible to create fields in the laboratory which are strong enough to disrupt atoms in their normal states: for instance, the electric field on an electron in the ground state of the hydrogen atom has a strength of 5×10^9 V/cm. A similar value is obtained for the valence electrons in all atoms, and it is about 1000 times greater than fields which can be maintained steadily in the laboratory. On the other hand, for the $n = 30$ state in hydrogen, the Coulomb field of the nucleus can be overcome by an external electric field of only 5 kV/cm. Thus entry of atomic physics into the strong-field regime has been accomplished by dealing with highly excited states, rather than by generating enormous laboratory fields. The same statement also applies to the interaction of atoms with external magnetic fields.

These points are clearly illustrated by examining the potential energy of the electron of a hydrogen atom placed in an external electric field $\mathbf{F}$ (along the z axis), and simultaneously in a magnetic field $\mathbf{B}$.

$$V = -\frac{e^2}{r} - e\mathbf{F} \cdot \mathbf{r} + \frac{1}{8}\frac{e^2}{c^2}(\mathbf{r} \times \mathbf{B})^2 + \frac{e}{2c}\mathbf{l} \cdot \mathbf{B},$$

where e is the elementary unit of charge and $\mathbf{l}$ is the orbital angular momentum. The Coulomb term dominates at small r (normal atomic size), whereas at very large r (highly excited atoms), the external fields dominate. At some intermediate distances the three potentials become comparable and none of them dominate. It is in this intermediate regime that the concept of atomic engineering has been realized.

6.1 Electric Fields

To explain the concept of atomic engineering we consider the case where hydrogen is immersed in a static electric field only. We utilize Fig. 4a which shows the potential of a hydrogen atom for a cut along the z axis in the presence of an external electric field along the z axis. Several interesting features of the combined Coulomb and Stark potential should be noted. First, it is evident that the potential is not spherically symmetric; the electricfield leads to a lowering of the ionization potential of the atom and creation of a potential barrier in the $z > 0$ half space. Thus the state of energy higher than the top of the barrier E_c can classically escape, that is ionize; in other words the motion for this state in this half space is *unbounded*. On the other hand in the $z < 0$ half space the electric field leads to a rise in the potential towards infinity; consequently the motion of the electron is *bounded* for all energies including positive energies in this half spac, something that does not occur in the isolated atom case.

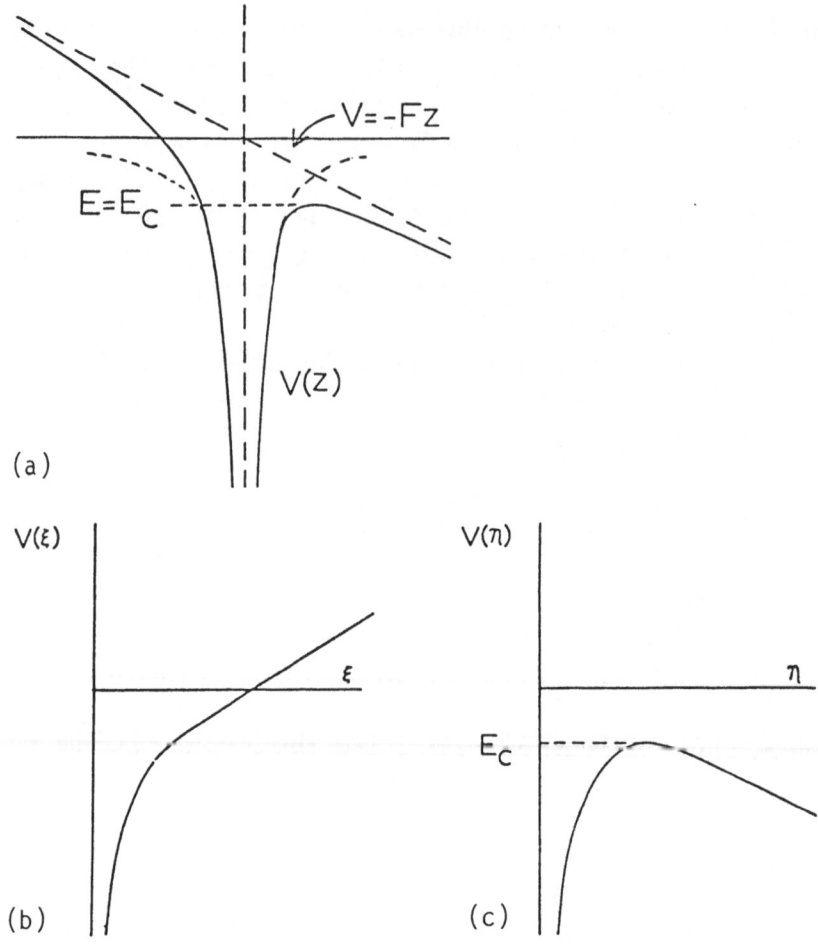

Fig. 4. Potential of hydrogen in a strong external electric field.

Thus the atom in the presence of the field will have quantized energy levels for all energies including positive energies that will *spontaneously* ionize for energies above E_c. The positions of these levels are governed by the binding well in the $z < 0$ region, while their lifetime against ionization is governed by the degree of coupling to the free motion in the $z > 0$ region.

Although the potentials in the $z = 0$ plane are very useful in bringing out some features of the interaction, they are not very useful for quantitative calculation. This is because the nonspherical symmetry of the potential makes the interaction non-separable: that is, it cannot be separated into three independent one dimensional motions in spherical coordinates. The interaction, however, is separable in parabolic coordinates $\xi = r + z$, $\eta = r - z$, and ϕ the azimuthal angle, with quantum numbers n_1, n_2, and m respectively. The effective potentials for the ξ and η motions shown in Figs. 4 b,c in fact, have good resemblance to that of the z cut in the $z < 0$ and $z > 0$ regions respectively, and hence govern the energy of the system (location of the energy) and the ionization lifetimes of these levels, respectively. The quantum number m_l is common to both parabolic and spherical descriptions, and the principle quantum number $n = n_1 + n_2 + |m_l| + 1$. The spherical l and parabolic n_1, n_2 quantum numbers do not have a one-to-one correspondence: a state with definite values of n_1 and n_2 is composed of many different values of l.

One important property of the atom that comes out of this procedure is the fact that only a fraction of the nuclear charge $Z_1 < 1$ drives the ξ motion and hence dictates the energy of the system, while the rest of the charge $Z_2 = 1 - Z_1$ drives the free η motion and hence dictates its lifetime. Thus the presence of an external electric field provides us with a situation where the nuclear charge that drives the bounded motion can be varied, in a near continuous fashion. Considering the fact that the physical and chemical identity of isolated atoms is defined by the nuclear charge, then it is clear that we have at our hand a means for creating new "types" of atoms.

We will now discuss the preparation and nature of the new types of atoms by discussing their spectroscopic properties such as ionization lifetimes, charge distributions (or excitation dipole moments), and branching ratios (or excitation strengths). To do so we will consider the positive and negative energy regimes separately, starting with the former.

Let us assume that atomic hydrogen is immersed in laser radiation of energy just larger than 13.6 eV, the ionization potential of hydrogen, and whose polarization is along the external dc electric field in which the atom is immersed. Because of this choice of polarization, the electron gets an initial kick along the dc field, and the energy of the system is raised by 13.6 eV, thus rising to zero total energy. The electron can now execute bound motion even for this positive energy. The motion of the electron is nearly a one-dimensional motion with the orbit resembling a cigar whose axis is along the external field, the nucleus being located inside it near its lower tip (Fig. 5a). This specifically tailored atom lives on the order of 5×10^{-13} s (giving very broad widths), and the electron

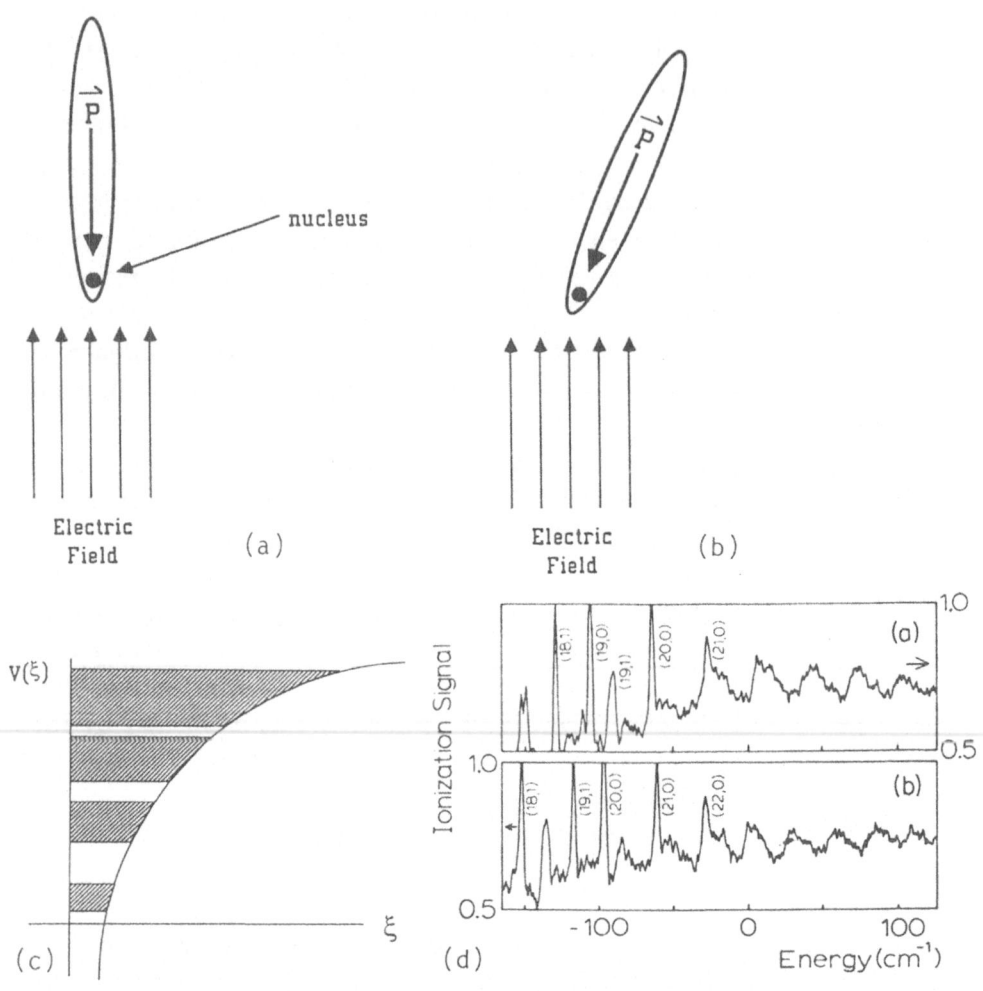

Fig. 5(a). Schematic orbit of a giant dipole aligned along the field.
Fig. 5(b). Schematic orbit of a giant dipole at an angle to the field.
Fig. 5(c). Schematic energy levels above $E = 0$.
Fig. 5(d). Spectrum of hydrogen in a field of 5 kV/cm (top) and 3.5 kV/cm (bottom).

executes on the average about 5 rounds before it breaks away from the proton on its own, and it is found to spend most of its time away from the nucleus, near the upper tip of the cigar. If the electron were initially kicked perpendicular to the field (laser polarization perpendicular to the external field), the cigar would have been created at an angle with the field. (See Fig. 5b.)

The extraordinary thing about this cigar atom is that such a "separated charge" distribution gives a dipole moment **P** which points *opposite* to the external field. Moreover, the dipole is very large since the separation of the charge (length of the cigar) is about 1600 Å, hence giving dipole moments that are 3000 times larger than those of normal atoms. For this reason we call these atoms "giant dipole" atoms. However, in general, one cannot exclusively prepare these types of atoms without preparing the highly excited normal atom since, first of all, the excitation has to start from the ground state of the normal atom which is only weakly affected by electric fields and secondly both Coulomb and Stark fields will have to compete. Therefore, after the excitation process we always have a superposition of these two types with the branching ratio depending on a number of parameters including the total energy of the system, field strength, and the properties of the exciting laser radiation [19-23]. For example, the "visibility" of the giant dipoles, which is a measure of how much they rise above the accumulated smooth continuum, tends to be very small (4% at 5 kV/cm). This visibility gets worse at higher energy because these states get closer to each other in energy as shown in Fig. 5c. Those states were first seen in complex atoms such as rubidium, sodium, barium, krypton, and yttrium during 1978-1983 [25-29], but were found to have strengths that are smaller than is predicted for hydrogen. Theories that included the effect of core electrons explained the reduction of the strength [21]. The first observation of the giant dipoles in hydrogen was made in 1984 in our laboratory at the University of Illinois [30]. (See Fig. 5d.) Similar observation was also achieved at the University of Bielefeld [14].

Considering the shortness of their lifetimes, and the low efficiency of excitation, it is clear that experimentation with these "new atoms" will not be easy unless these two properties are enhanced. Recently we have been able to improve the efficiency [31] and to produce giant dipoles that live much longer than 10^{-12} s [17] in atomic hydrogen. We will discuss the efficiency first. The scheme we devised for this purpose relies on a process we call multistage shaping or charge shape tuning of the charge of the atom. In one-photon excitation from the ground state one effectively starts from a spherically symmetric charge distribution (zero dipole moment), and tries to mold it by a single operation into a giant dipole whose charge is highly focussed along the field. On the other hand in multistage shaping one uses one photon to create from a ground state a not too large dipole of charge distribution that is focussed along the field at an intermediate state followed by another photon absorption from this intermediate state that produces larger dipole whose charge is even more focussed along and so on till one excites the giant dipole in a highly focussed distribution along the field.

The ability to create moderately large focussed dipoles as intermediates is the key to the success of the multistage shaping operation [17,32]. This is explained in Fig. 6 for a two-stage process using as an intermediate $n = 2$ of hydrogen. Because the level splittings in $n = 2$ of hydrogen are small enough (0.3 cm^{-1}) such that an electric field imposed on the atom which is larger than 5 kV/cm will be able to mix all of these sublevels and hence their charge distributions (each has a zero dipole) to produce distinct dipole distributions needed for the shaping process.

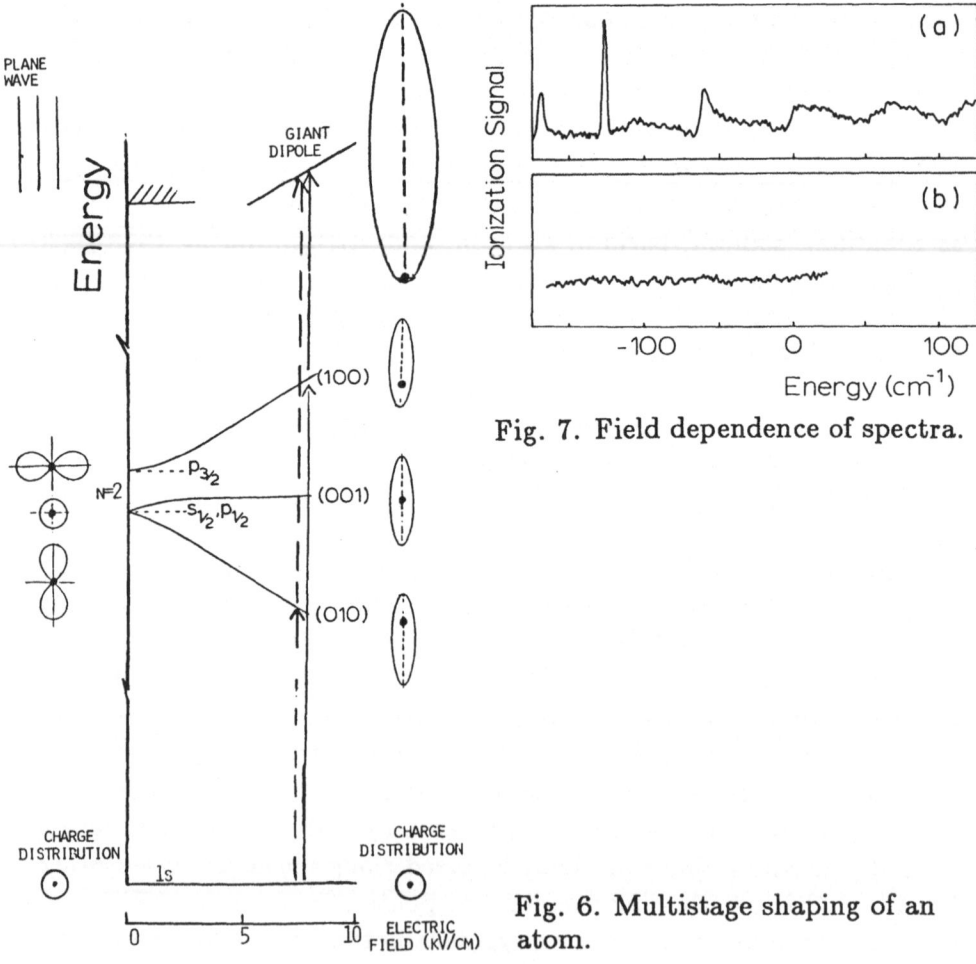

Fig. 7. Field dependence of spectra.

Fig. 6. Multistage shaping of an atom.

Our calculations show that by utilizing the up-field extended dipole of $n = 2$ as an intermediate, the efficiency can be increased from 10 to 30 %, whereas by utilizing the down-field extended dipole the efficiency is reduced to 1 %. These were confirmed in our hydrogen experiment as shown in Fig. 7. Our further calculations using higher n states whose charge can be focussed along the field more easily, as shown in Fig. 8, and hence can be matched or tuned more closely to the charge of the giant dipoles, showed dramatic effects on the efficiencies [33]. The use of, for example, $n = 9$ as the intermediate step in the process, rejects almost completely the excitation of the spectrum of the normal atom in favor of the one-dimensional atom. Results for $n = 1$, $n = 2$, and $n = 9$ are shown in Fig. 9, along with a schematic of the focussing effect on the intermediates.

The enhanced efficiency is very nice, but it is found that it is practically not possible to increase the lifetime of these giant dipoles in this positive energy region by too much. Such inability is related to the fact

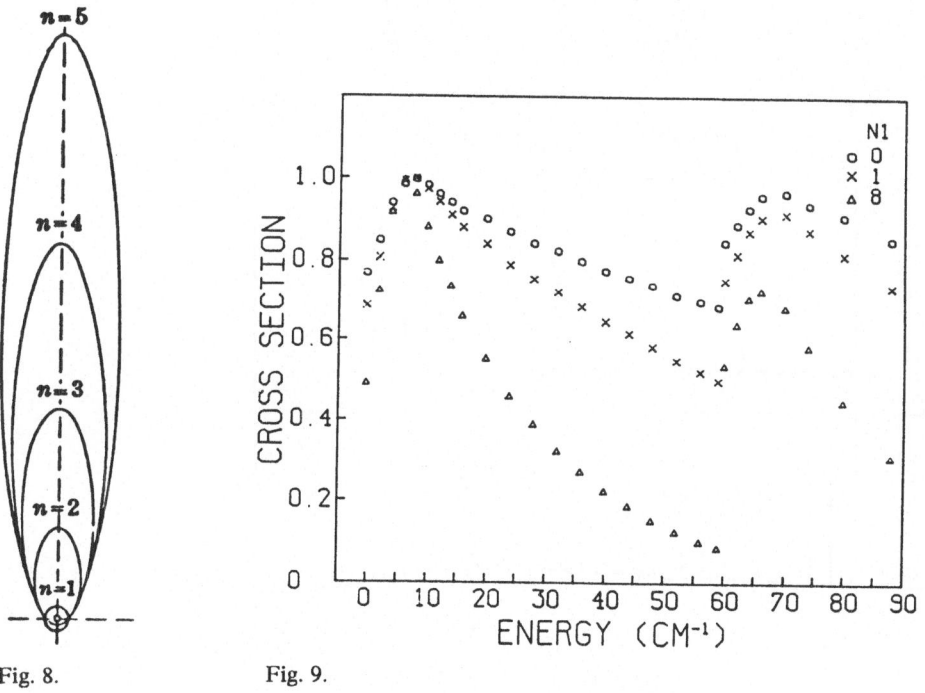

Fig. 8.   Fig. 9.

Fig. 8. Intermediate state dependence of final wavefunction.
Fig. 9. Cross sections for different intermediate states.

that the bound motion of all these giant dipoles in this region are driven by nearly the same charge, most of the nuclear charge $Z_1 \sim 1$, which also dictates very similar orbits where the nucleus is located at the lower tip of the cigar. However, it is found that such enhanced efficiency can be extended to the negative energy region where it is also possible to

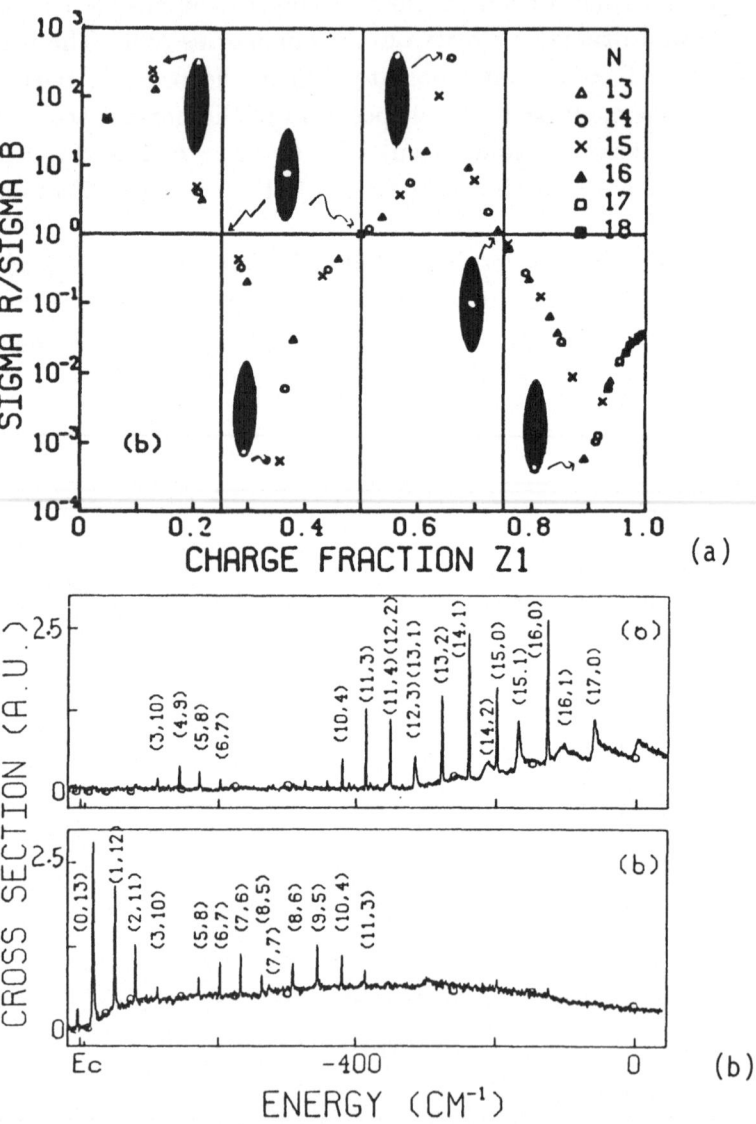

Fig. 10(a). Position of nucleus as a function of charge fraction.
Fig. 10(b). The giant dipole spectrum.

produce giant dipoles that live quite long. Therefore we will now discuss such promising negative energy regions between $E = 0$ and $E = E_c$.

In this region the giant dipole atoms take on different properties than the one in the positive energy region. Firstly, the fraction of the charge that drives the bound motion can be varied from 0 to 1 by varying the energy of the system, and consequently the position of the nucleus inside the cigar can also be controlled. In Fig. 10a we plotted an indicator of the location of the nucleus inside the cigar as a function of Z_1, along with sketches of some possible orbits. This indicator is related to the ratio of the area of the orbit below the nucleus to that above the nucleus. The figure shows a remarkable property: for the fractional charges $Z_1 = 1/4$, $1/2$, and $3/4$ the nucleus is located at the center of the cigar (the atom has zero dipole moment). These fractional charges thus constitute lines across which the direction of the giant dipole reverses. In the first and third quarters the dipole is along the imposed field whereas in the other two quarters it is opposite to it. Given the size of the orbit (which can be calculated), one can use the above indicator to determine the magnitude of the giant dipole. Moreover, we have recipes to cook up giant dipoles of given Z_1 values. These features and others have been recently confirmed by our experiments. The giant dipole spectrum is shown in Fig. 10b.

Examination of the spectrum indeed shows a variety of widths (lifetimes) that range from quite short to quite long. In fact there are giant dipoles that do not show up in our spectrum because they live longer than the time of measurement, which is 100 ns, or because they radiatively decay before they ionize and hence do not get detected. Again there are systematics to the ionization lifetime as a function of Z_1 and hence as a function of energy that makes the selection of a giant dipole of given specification possible.

6.2 Magnetic Fields

Atomic engineering can also be achieved using external magnetic fields or simultanous electric and magnetic fields. The effect of a magnetic field on a highly excited atom is very different from that of an electric field. As can be seen from Fig. 11a, a magnetic field is described by a potential which increases in all directions away from the nucleus. Thus it does not polarize the atom in a given direction, but rather confines the motion of the electron in the plane perpendicular to the field thus

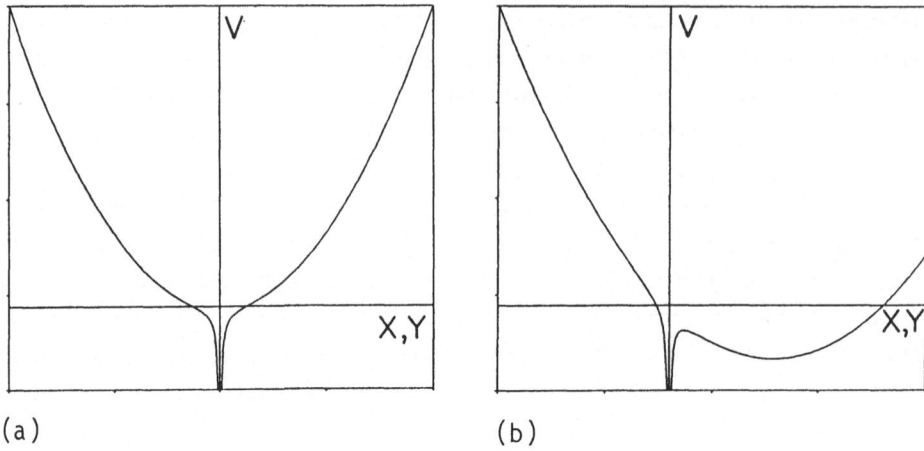

Fig. 11. Hydrogen potential in a magnetic field (a) and in crossed electric and magnetic fields (b).

giving a near two-dimensional atom. This confinement makes it possible to produce localized states with energies in excess of the ionization energy. Such states were first seen [34] in photoabsorption by barium in 1969. They are called "quasi-Landau resonances" because of an analogy between their theoretical description and Landau's theory of a free electron in a magnetic field. The quasi-Landau resonances were observed in hydrogen just two years ago [14,35]. Recently, new experiments with higher resolution have suggested the existance of many long lived states with energies above $E = 0$. These findings have stimulated further theoretical efforts. Isolated classical orbit calculations show good agreement with the experimental results, however, the problem is not yet fully understood.

As indicated by our previous discussion of the dynamics of hydrogen in an electric field, the key to understanding is the identification of a coordinate system in which the motion along each coordinate can be treated independently. For hydrogen in an electric field, the parabolic coordinates ξ, η, and ϕ provide such a system; in the case of a magnetic field, on the other hand, it is almost certain no such coordinate system exists. Many such nonseparable systems are encountered in other branches of physics and no general method has been developed for their theoretical description. At present some initial progress on the theoretical understanding of highly excited hydrogen in a magnetic field has been achieved with the aid of large scale numerical calculations [36].

6.3 Simultaneous Electric and Magnetic Fields

The simultaneous application of electric and magnetic fields to a Rydberg atom represents the ultimate in atomic engineering. It is an area which at present is not as well explored as that of pure electric or magnetic field studies, but it has yielded some interesting results. First of all, one can investigate in greater detail the dynamical symmetry of the hydrogen atom, which is responsible for the l-degeneracy of states in zero field and the stability of the blue giant dipole states. This dynamical symmetry, which can be represented in terms of the four-dimensional rotation group O(4), yields a remarkably orderly spectrum for the hydrogen atom in combined electric and magnetic fields of moderate strength: remarkable in the sense that its simplicity cannot be attributed to any geometrical symmetry of the field configuration. This effect has recently been observed in high angular momentum states of alkali atoms [37].

The regime in which both electric and magnetic fields are strong compared to the atomic field, as shown in Fig. 11b, has not yet been studied closely. However, there are theoretical predictions of a new type of motion in a highly excited atom in perpendicular electric and magnetic fields. In this case the electron experiences an electric force that tends to stretch it in a cigar shape, and a magnetic force that tends to squeeze it in the perpendicular plane. The balance of the two forces results in forcing the electron to orbit around two centers: the nucleus and the saddle point [38]. Experiments to search for such states are in progress.

6.4 Circular Atoms

There is a case of atomic engineering of a Rydberg state n which actually utilizes weak dc electric field but with additional help from absorption of $n - 1$ microwave photons [39]. With each photon imparting to the atom, in an ordered fashion, a unit of angular momentum, the atom ends up in the state of the highest possible angular momentum, $l = n - 1$. The high angular momentum prevents the electron from coming close to the nucleus, and in fact the electron executes a circular orbit around the nucleus which is in fact the closest approximation to the original Bohr model of the atom. Such states are the counterpart to the giant dipole atoms; they have zero electric dipole moments but quite large magnetic dipole moment. The production of such states in sodium atoms has been achieved by researchers at the Massachusettes Institute of Technology (MIT).

6.5 Atoms Between Large Plates: Squeezed States

Finally a property of large atoms that makes another type of atomic engineering possible is the fact that radiation emitted or absorbed between the adjacent highly excited levels is in the microwave region of the electromagnetic spectrum. When such a highly excited atom is injected into a microwave cavity or waveguide of a size that is comparable to the atom's characteristic transition wavelength λ_o, then a number of interesting effects may occur [40,41]. For example, if the cavity size is such that the wave spontaneously emitted by the atom cannot propagate in the guide because λ_o is larger than the so called cutoff wavelength of the guide λ_c, then the wave will not be emitted, thus preventing the spontaneous emission of the atom and consequently lengthening its lifetime. Alternatively this process can be looked at as an engineering process of the properties of the generated electromagnetic radiation. Electromagnetic radiation can propagate in free space with any wavelength and direction, but the introduction of a guide restricts the propagation to only certain wavelengths and directions. Such propagation effects are well known in electrical engineering, but it is only recently that their implications for the generation of radiation from a quantum system such as highly excited atoms placed inside cavities have been realized. One exciting possibility is the engineering of what is called "squeezed states" of electromagnetic waves. In a normal wave, the levels of noise associated with its amplitude and its phase are balanced; however, a squeezed state is a state where some of the noise in either one of them is channeled to the other. Since some detectors can be designed to utilize either the phase or the amplitude separately, then such process promises extreme improvement in detection sensitivity [42].

6.6 Complex Atoms

It is relevant to examine the possibility of atomic engineering in complex atoms. In complex atoms, the highly excited electron interacts with the rest of the electrons in the system (electronic core) in addition to its interaction with the Coulomb field of the nucleus and with the external field. This additional interaction is short range, i.e. it is only important for distances on the order or less than the size of the ground state atom. If the highly excited electron stayed away from the electronic core, the effect of this interaction would be negligible and as a result the excited atoms would be nearly hydrogenic. In atomic engineering, however, the orbit of the highly excited electron does penetrate the core, thus making the system not quite hydrogenic. The effect of such penetration has the following implications. First of all it mixes and smears the Coulomb-Stark interaction to the degree of preventing selective excitation of the giant dipoles, resulting in the diminishing of their dipole moment. Also it causes mixing among states of short and long lifetimes, resulting in shortening of the lifetimes of the long-lived ones. Another serious effect is the prevention of the multistage shaping concept that we use for the enhancement of excitation strength. This is because the interaction with the electronic core destroys the near degeneracy in l, that is, it causes very big splittings in the l substates of low lying states, thus rendering the external field ineffective in partially molding their charge distributions into distinct dipole distributions, operations that are the backbone of the concept of the enhancement.

7   One Dimensional Atoms in Microwave Fields

There has been a much interest in the question of existence of chaotic behavior in quantum mechanical systems whose classical analogs are known to be nonintegrable and exhibit chaotic behavior [50]. The interest stems from the fact that the allowance for quantum effects in the systems gives rise to substantial anomalies in the manifestation of stochasticity even when the initial population of the system is quasiclassical. The periodically kicked classical and quantum rotors have been analyzed extensively by many researchers [51-53]. In some studies it was found that the quantum correlation functions attenuate much more slowly than in the classical limit, ultimately reducing the rate of quantum diffusion. In another study [53] it was noted that one cannot tell with certainty whether a quantum state of the rotor describes a chaotic or a regular state beyond the quantum resolution $\sqrt{\hbar}$. Other studies of quantum systems, however, have shown that under certain conditions and for evolution over finite times, quantum dynamics can be brought closer to the classical stochastic approach [50].

Experiments to study chaotic behavior in the response of atoms and molecules to radiation fields are underway in a number of laboratories including ours [54-57]. One experiment deals with the collisionless dissociation of polyatomic molecules interacting with coherent IR laser

163

radiation. Another system of interest is that of highly excited atomic hydrogen ionized by a microwave field [55-57].

Because an analysis of the hydrogen system involves many degrees of freedom (the electron moves in a 6-dimensional phase space), the detailed numerical and analytical investigation of this process is time consuming, and at this date remains incomplete [58]. A system that is amenable to theoretical analysis is the one-dimensional hydrogen atom (involving two dimensions in phase space) that we have been experimenting with recently [59]. These atoms are prepared by laser excitation of hydrogen in the presence of strong dc electric fields. They offer not only the nearest approximation to the ideal one-dimensional problem, but they offer systems of low quantum numbers as well as high ones so that a wide range of quantum numbers can be studied.

We have theoretically analyzed stochasticity in highly excited atomic hydrogen in the presence of a microwave field and a dc electric field using a classical one-dimensional model similar to that of an electron system over a helium surface. We have determined in detail the effect of the dc field on the threshold of global stochasticity and on the number of states trapped in the nonlinear resonances [60]. Our results indicate that the number of trapped states may not always increase, and may decrease with increasing dc field contrary to the case of a system of surface electrons.

8 Detection by Electric Field Ionization

The concept of detection of neutral atoms by electric field ionization is a consequence of atomic engineering, although it has been known since the mid 1970s [43]. In fact this technique is one among many that have been developed during the past decade for the detection of ever smaller concentrations of neutral atoms, including the ultimate detection of single atoms [10].

Traditional optical spectroscopy has been widely used to identify and characterize various substances through the measurement and interpretation of the spectra arising from the absorption or emission of radiant energy by samples under investigation. In techniques measuring absoprtion, the test material is illuminated by a continuous light source, such as an incandescent lamp, and the transmission of light is monitored. The spectral lines of the sample appear as dark lines in the transmitted light. Although this method can be very selective, its sensitivity is only moderate ($10^9/cm^3$).

In traditional emission methods, samples are subjected to flames or electrical discharges to excite the electrons of the sample to higher orbits, and the radiant energy (fluorescence) they emit during relaxation to the ground state is monitored. This method is a more sensitive measurement than the absorption method; however, it is less selective because many excited states can be populated from all the species present, with the result that the spectral lines of the predominant materials swamp the weaker emission of the dilute constituents.

Spectroscopic technique was improved in both sensitivity and selectivity with lasers. Atoms are excited from the ground state by a resonant intense laser beam, with no ions produced and with many fewer competing resonances, and the emission light can be observed directly, making the method simultaneously very sensitive and very selective. It was possible with this laser-based method to measure sodium densities down to $100/cm^3$, an improvement over the sensitivities of existing analytical methods of eight orders of magnitude. Further studies using the resonance-fluorescence method resulted in measurements of $3 \times 10^3 /cm^3$ sodium atoms and $10^3/cm^3$ uranium atoms.

There is a scheme that matches the selectivity of resonance fluorescence and surpasses its sensitivity which was developed by Hurst, Nayfeh and Young (then at Oak Ridge National Laboratory) [10]. Atoms are stimulated to a sequential series of excited states by causing them to absorb a photon of frequency ν_1, then a photon of frequency ν_2, and so on. The number of allowable absorptions is arbitrary depending on the atom being investigated. In all cases, however, the final step involves the photoionization of the atom: the absorption of a photon results in breaking the atom into an ion pair – an electron and a positive ion, or the application of a pulsed dc electric field that also results in the ionization of the atom. The produced charge – either the electron, the ion, or both – is detected in the method, which is called resonance ionization spectroscopy (RIS). Resonance ionization differs from resonance fluorescence in a very basic way. With ionization, the detection scheme involves the measurement of massive particles–electrons or ions that can be easily controlled through their diffusion times (this time can be on the order of a few microseconds) — whereas with fluorescence, measurements are made of photons, which are difficult to control. The ability to control electrons makes elimination of wall events possible, resulting in considerable reduction of background emission interference. In the resonance-fluorescence scheme, the detector must be focused at less than 4π solid angle to avoid the main unscattered beam; in the ionization scheme, 100% collection efficiency is achieved. In addition, the measurement of small numbers of electrons or ions is much less complicated than the measurement of small numbers of photons — another advantage of the resonance ionization method over the fluorescence method. In fact, with this improved sensitivity even detection of single atoms has been achieved. Later efforts and variations on the resonance fluorescence have also pushed its sensitivity to the single atom limit.

The concept of electric field ionization of excited atoms can be easily explained using Fig. 4. One can see that the presence of the electric field lowers the ionization potential of the atom in half of the space by an amount given by $E = 2\sqrt{F}$.

Thus all excited states of the atom whose energies occur higher than the new potential spontaneously ionize with different rates. For states lying below this threshold (the top of the barrier), the ionization rates are very slow for the atom to be detected as a pair of charges. In atomic engineering one excites the atom from the ground state in the presence of the electric field, however, when the field is used as a detector only, the electric field is switched on after the excitation has taken place. What is very attractive about this method is that it has some energy selectivity. If we have an ensemble of atoms in which a number of excited states are populated, then by increasing the strength of the pulsed electric field continuously from zero, we can map out the population of all of these excited states.

Although in principle one can use field ionization or photoionization in the final step of the detection, for low lying excited states (2-3 eV binding) photoionization is more appropriate since the electric field needed is too large, while the required radiation is visible radiation that is readily available. On the other hand, when one is dealing with highly excited states one finds it is more convenient to use field ionization, since photoionization with visible laser is not efficient because the binding energy (a few meV) is much too small relative to the photon energy. Moreover, the required electric field is no more than a few kV/cm which can be generated easily and cheaply. In most cases, ionization efficiencies of near unity can be achieved.

9   Large Atoms in Physical Environments

One question which arises immediately is the extent to which a highly excited state can survive in a typical gaseous medium. For sufficiently large n, the electron's orbit will enclose a number of atoms of the gas, and it might be thought that these would disrupt the stability of the Rydberg state. However, as was first pointed out by Fermi [44], when many neutral gas atoms intervene between the Rydberg electron and the ion, they can be represented by a continuous dielectric medium. This effectively reduces the electric charge on the ion, but it does not qualitatively alter the basic Bohr picture of the dynamics. Thus it is possible for a Rydberg atom to enclose tens of thousands of neutral atoms of a background gas and still remain stable - a feat which, in a famous passage, Breene [45] has compared to the ability of a nanny goat to pick its kid from a large herd. What limits that stability of a Rydberg state is the effect of hard collisions between the slowly moving, nearly free electron and individual atoms of the gas: the distance the atom travels between collisions must be greater than the de Broglie wavelength of the electron. Analysis of this effect has made it possible to deduce from Rydberg spectra the cross sections for electron scattering by atoms and molecules at very low energies, which are extremely difficult to measure directly.

In the presence of strong external fields, giant dipoles in complex atoms can be produced with picosecond lifetimes. Giant dipoles in hydrogen on the other hand can be produced with lifetimes on the order of nanoseconds and even microseconds [46-48]. How do collisions affect a giant dipole or a disk-like atom? How does the lifetime depend on the direction of the relative velocity with respect to the quantization axis of the giant atom? What is the role of the geometrical cross section of the atom? Our recent work has shown that large distortions of the electronic charge distribution, in the case of giant dipoles, make these atoms extremely active and render the geometrical cross section (area of the orbit) irrelevant as an indicator of the collisional activity [46]. We have recently investigated the simultaneous interactions of highly excited atoms with external dc electric fields and depolarizing collisional interactions with electrons, ions, or neutral atoms [60]. We find for the first time that the electric field enhances by many orders of magnitude the depolarization cross section. The interaction is so long range that the excitation duration (pulse width) governs the time over which the interaction takes place. Although this long range dipole lives less than 10^{-11} seconds and the excitation pulses are less than 10 ns, the interaction is strong enough to cause appreciable depolarization even at number densities of ions as low as $10^8/\text{cm}^3$. In other words, the electric field renders the geometrical cross section irrelevant as an indicator of the collisional activity [60,45]. An alternative indicator of the collisional activity is the distance from the nucleus to the classical turning point.

Another aspect to consider is whether the "size" of Rydberg atoms could ever be perceived in the familiar mechanical sense. Atoms of $n = 280$ have been prepared where the orbital radius is of a few micrometers. This is a distance scale just at the limits of common optical microscopy, and one on which machining of objects can be carried out, e.g. by electroforming techniques. A recent experiment [49] at the Ecole Normale Supérieure in Paris has taken a step towards direct comparison of the size of a Rydberg atom with that of a classical object. A gold foil was perforated with an array of rectangular slits several micrometers wide. A beam of Rydberg atoms was directed towards the slit, and the fraction of atoms passing through was measured as a function of principal quantum number. The results were found to be consistent with a simple "hard sphere" model, in which a Rydberg atom passes through a slit if its "edge" doesn't hit the side. The apparent size of the Rydberg atoms as determined from these results does indeed vary as n^2, though with a constant of proportionality larger than that given by the Bohr model. Of course, the specific mechanism by which the Rydberg beam is attenuated must be understood in terms of electromagnetic interaction between the atoms and the foil. Nevertheless this experiment does provide some sort of classical measurement of atomic dimensions, as well as a demonstration of an effect of a macroscopic object upon a highly excited atom.

10   Acknowledgements

I would like to acknowledge the input of Dr. Charles Clark of NBS Gaithersburg, MD, to this article. Some parts of it were prepared by both of us for different purposes. Finally, I would like register deep gratitude to my graduate student, Tom Sherlock, for his patience and technical expertise in typesetting this paper.

References

[1] K. R. Anantharamaiah, W. C. Erickson, and V.Radhakrishnan: Nature 315, 647 (1985); W. D. Watson, ibid, p. 630 - 631.

[2] Physics Today 38, 23-72, (1985).

[3] R. F. Stebbings: Science 193, 537 (1976); T. F. Gallagher: In *Advances in Atomic and Molecular Physics*, 14, ed. by B. Bederson, (Academic, New York 1978) p. 365; D. Kleppner: *Progress in Atomic Spectroscopy*, ed. by W. Hanle and H. Kleinpopper, (Plenum, New York 1979) p. 713.

[4] See articles in *Atomic Excitation and Recombination in External Fields*, ed. by M. H. Nayfeh and C. W. Clark (Gordon and Breach, New York 1985).

[5] B. Höglund and P. G. Metzger: Science 150, 339 (1965).

[6] J. B. A. Mitchell, C. T. Ng, J. L. Foraud, D. P. Levac, R. E. Mitchell, A. Seu, D. B. Miko, and J. Wm. McGowan: Phys. Rev. Lett. 50, 335 (1983); D. S. Belic, G. H. Dunn, T. J. Morgan, D. W. Mueller, and C. Timmer: ibid 50, 339 (1983); P. F. Dittner, S. Datz, P. D. Miller, C. D. Heath, P. H. Stelson, C. Bottcher, W. B. Dress, G. D. Alton, and N. Neskovic: ibid 51, 31 (1983).

[7] A. L. Schawlow and C. Townes: Phys. Rev. A 112, 1940 (1958).

[8] T. W. Hänsch: Appl. Optics 11, 895 (1972).

[9] T. W. Hänsch, M. H. Nayfeh, S. A. Lee, S. M. Curry, and I. S. Shahin: Phys. Rev. Lett. 32, 1336 (1974); B. P. Kibble, W. R. C. Rowley, R. E. Shawyer and G. S. Series: Proc. Phys. Soc. of London 6, 1079 (1973).

[10] M. H. Nayfeh: Am. Sci. 67, 204 (1979); G. S. Hurst, M. G. Payne, S. D. Kramer, and J. P. Young: Rev. Mod. Phys. 51, 767 (1979).

[11] See articles in *Resonance Ionization 1984*, ed. by G. S. Hurst, and M. G. Payne (The Institute of Physics, Bristol 1984).

[12] J. E. Bayfield, L. D. Gardner, and P. M. Koch: Phys. Rev. Lett. 39, 76 (1977);
P. M. Koch: Phys. Rev. Lett. 41, 99 (1978).

[13] H. C. Bryant : Phys. Rev. A 27, 2889, 2912 (1983);
W. W. Smith, C. Harvey, J. E. Stewart, H. C. Bryant, K. B. Butterfield, D. A. Clark, J. B. Donahue, P. A. M. Gram, D. Macarthur, G. Comtet, T. Bergman, in *Atomic Excitation and Recombination in External Fields*, ed. by M. H. Nayfeh, and C. Clark (Gordon and Breach, New York 1985).

[14] K. H. Welge and H. Rottke: In *Laser Techniques in the Extreme Ultraviolet-OSA*, Boulder, Colorado, 1984, ed. by S. E. Harris and T. B. Lucatorto, AIP Conf. Proc. No. 119 (AIP, New York 1984), pp. 213-219;
H. Rottke and K. H. Welge: Phys. Rev. A 33, 301 (1986).

[15] W. L. Glab and M. H. Nayfeh: Opt. Lett. 8, 30 (1983);
W. Glab: Ph.D Thesis, University of Illinois, 1984 (unpublished).

[16] H. Rinneberg, J. Neukammer, G. Jonsson, H. Hieronymous, A. Konig, and K. Vietzke: Phys. Rev. Lett. 55, 382 (1985).

[17] M. H. Nayfeh, K. Ng, and D. Yao: In *Atomic Excitation and Recombination in External Fields*, ed. by M. H. Nayfeh and C. W. Clark, (Gordon and Breach, New York 1985).

[18] T. W. Hänsch: In *Proceedings of the International Laser Science Conference*, ed. by W.C. Stwalley and M. Lapp (Am. Inst. of Phys. 1986).

[19] E. Luc-Koenig and A. Bachelier: Phys. Rev. Lett. 43, 921 (1979).

[20] A. R. P. Rau and K. T. Lu: Phys. Rev. A 21, 1057 (1980).

[21] D. A. Harmin: Phys. Rev. A 24, 2491 (1981); Phys. Rev. Lett. 49, 128 (1982); Phys. Rev. A 26, 2656 (1982).

[22] U. Fano: Phys. Rev. A 24, 619 (1981).

[23] W. D. Kondratovich and V. N. Ostrovsky: Zh. Eksp. Teor. Fiz. $\underline{4}$, 1256 (1982).

[24] C. W. Clark, K. T. Lu, and A. F. Starce: In *Progress in Atomic Spectroscopy C* ed. by H. Beyer and H. Kleinpoppen (Plenum, New York 1984) p. 247

[25] R. R. Freeman, N. P. Economou, G. C. Bjorklund, and K. T. Lu: Phys. Rev. Lett. $\underline{41}$, 1463 (1978).

[26] T. S. Luk, L. DiMauro, T. Bergeman, and H. Metcalf: Phys. Rev. Lett. $\underline{47}$, 83 (1981).

[27] S. Feneuille, S. Liberman, E. Luc-Koenig, J. Pinard, and A. Taleb: Phys. Rev. A $\underline{25}$, 2853 (1982).

[28] W. Sandner, K. A. Safinya, and T. F. Gallagher: Phys. Rev. A $\underline{23}$, 2448 (1981).

[29] W. Glab, G. B. Hillard, and M. H. Nayfeh: Phys. Rev. A $\underline{28}$, 3682 (1983).

[30] W. L. Glab and M. H. Nayfeh, Phys. Rev. A $\underline{31}$, 530 (1985).

[31] W. L. Glab, K. Ng, D. Yao, and M. H. Nayfeh: Phys. Rev. A $\underline{31}$, 3677 (1985).

[32] M. H. Nayfeh, K. Ng and D. Yao In *Laser Spectroscopy VII*, T. Hänsch and R. Shen, eds. Springer Ser. Opt. Sci., Vol. 49 (Springer, Berlin Heidelberg 1985) p. 71
K. Ng, D. Yao and M. H. Nayfeh, Phys. Rev. A (april) (1987).

[33] Y. P. Ying and M. H. Nayfeh: Phys. Rev. A $\underline{35}$, (1985).

[34] W. R. S. Garton and F. S. Tomkins: Astrophys. J. $\underline{158}$, 839 (1969).

[35] A. Holle and K. Welge: In *Laser Spectroscopy VII*, ed. by T. Hänsch and R. Shen, Springer Ser. in Opt. Sci., Vol. 49 (Springer, Berlin, Heidelberg 1985) p. 81

[36] C. W. Clark and K. T. Taylor: Nature $\underline{292}$, 437 (1981); J. Phys. B $\underline{15}$, 1175 (1982);
G. Wunner: Phys. Rev. A (in press 1986);
P. O'Mahony and K. T. Taylor: J. Phys. B (in press 1986).

[37] F. Peuent, D. Delande, F. Biraben, and J. C. Gay: Opt. Comm. 49, 184 (1984).

[38] C. W. Clark, E. Korevaar, and M. G. Littman: Phys. Rev. Lett. 54, 320 (1985).

[39] R. G. Hulet and D. Kleppner: Phys. Rev. Lett. 51, 1430 (1983).

[40] P. Dobiasch, G. Rempe and H. Walther: In *Laser Spectroscopy*, ed. by T. Hänsch and R. Shen, Springer Ser. Opt. Sci., Vol. 49 (Springer, Berlin, Heidelberg 1985) pg. 62

[41] S. Haroche, C. Fabre, P. Goy, M. Gross, J. M. Raimond, A. Heidmann and S. Reynaud: In *Laser Spectroscopy VII*, ed. by T. Hänsch and R. Shen, Springer Ser. Opt. Sci., Vol. 49 (Springer, Berlin, Heidelberg 1985) pg. 62

[42] M. D. Levenson and R. M. Shelby: In *Laser Spectroscopy VII*, ed. by T. Hänsch and R. Shen, Springer Ser. Opt. Sci., Vol. 49 (Springer, Berlin, Heidelberg 1985) pg. 250

[43] T. W. Ducas, M. G. Littman, R. R. Freeman and D. Kleppner: Phys. Rev. Lett. 35, 36 (1975);
V. S. Letokhov, In *Tunable Lasers and Applications*, ed. by A. Mooradian, T. Jaeger, and P. Stokseth, Springer Ser. Opt. Sci., Vol. 3 (Springer, Berlin, Heidelberg 1976) p. 122

[44] F. Fermi: Nuovo Cimento 11, 157 (1934).

[45] R. G. Breene, Jr. : *The Shift and Shape of Spectral Lines*, (Pergamon, Oxford 1961), p. 301.

[46] T. Yoshizawa and M. Matsuzawa: J. Phys. B17, L 485 (1984);
R. L. Becker and A. D. MacKellar: Bull. Am. Phys. Soc. 29 (1984).

[47] M. H. Nayfeh, G. B. Hillard, and W. Glab: Phys. Rev. A 32, (1985).

[48] K. T. Lu: In *Atomic Excitation and Recombination in External Fields*, ed. by M. H. Nayfeh, C. W. Clark, (Gordon and Breach, New York 1985) p.507

[49] C. Fabre, M. Gross, J. M. Raimond and S. Haroche: J. Phys. B16, L 671 (1983).

[50] G. Casati, B. V. Chirikov, F. M. Israelev, J. Ford: In *Stochastic Behavior in Classical and Quantum Systems*, ed. by G. Casati and J. Ford, **Lecture Notes in Physics**, Vol. 93, (Springer, New York 1979).

[51] D. R. Grempel, S. Fishman, and R. E. Prange: Phys. Rev. Lett. 49, 833 (1982).

[52] M. V. Berry: Physica (Amsterdam) 10D, 369 (1984).

[53] S. J. Chang and K. J. Shi: Phys. Rev. Lett. 55, 269 (1985).

[54] E. V. Shuryak: Zh. Eksp. Teor. Fiz. 71, 2939 (1975); Sov. Phys. JETP 44, 1070 (1976).
P. I. Belobrov, G. P. Berman, G. M. Zaslavski, and A. P. Slivinskii: ibid. 76, 1960 (1979); 49, 993 (1979).

[55] J. E. Bayfield and L. A. Pinnaduwage: Phys. Rev. Lett. 54, 313 (1985).

[56] P. M. Koch: J. Phys. (Paris), Colloq. 43, C2-187 (1982).

[57] D. Humm and M. N. Nayfeh: to be published.

[58] R. V. Jensen: Phys. Rev. A 30, 386 (1984); in *Chaotic Behavior in Quantum Systems*, ed. by G. Casati, (Plenum, London 1985).

[59] M. H. Nayfeh, D. Yao, Y. Ying, D. Humm, K. Ng and T. Sherlock: In *Advances in Laser Science - 1*, AIP Conference Proceedings 146, 370 (1986);
See also [32a]

[60] M. H. Nayfeh and D. Humm in *Proceedings of the International Workshop on Photons and Continuum States*, ed. by N. Rahman, (Cortona, Italy 1986), (Springer, Berlin, Heidelberg 1987)

Study of Small but Complete Molecules - Na$_2$

Hui-Rong Xia and Zu-Geng Wang

Department of Physics, East China Normal University,
Shanghai 200062, People's Republic of China

1. INTRODUCTION

Several years ago we felt somewhat puzzled to hear that "A diatomic
molecule is a molecule with one atom too many." This was said by Professor
Arthur L. Schawlow during one lecture in a series on High Resolution Laser
Spectroscopy at East China Normal University in Shanghai. We have no doubt
of this now. In fact, a diatomic molecule as simple as a neutral sodium
dimer, with only two outer-orbital electrons was not well understood.
Schawlow's comment:

> "To find something new, you never need to know
> everything about a subject. You only need to
> recognize one thing that is unknown."

This impressed us the most! Therefore, which "unknown" would be a good one
for Mrs. Xia to study when she began research in Schawlow's group during
April of 1980?

At that time, KENNETH C. HARVEY [1] had observed two-photon lines in Na$_2$,
and J. P. WOERDMAN [2] had reported the two-photon line at 16601.88 cm^{-1} in
Na$_2$ and had identified the rotational quantum number, but the upper
electronic state was not identified. Success in determining the energy
level constants of the unknown high-lying gerade states in Na$_2$ was achieved
by Professor Schawlow and his talented students through the development of
two-step polarization labeling spectroscopy. As many as 20 new states were
found.[3] One was the level identified by WOERDMAN, however it was
uncertain as to which one it was.

Professor Schawlow proposed an experiment for Mrs. Xia with his graduate
student Gerard P. Morgan (Fig. 1), and it made a good beginning for our
study on sodium dimers. Our paper summarized the research on this subject
during the time we were Visiting Scholars at Stanford in 1980 and 1982,
respectively, to the present time at East China Normal University. We
combined various nonlinear laser spectroscopic techniques for the study of
Na$_2$. This was done on near-resonant enhanced two-photon transitions to
populate high-lying singlet (earlier) and triplet states (recently) and on
various mechanisms of optically pumped stimulated emission of radiation
distributed from infrared (earlier) to violet or ultraviolet regions
(recently), see Fig. 2.

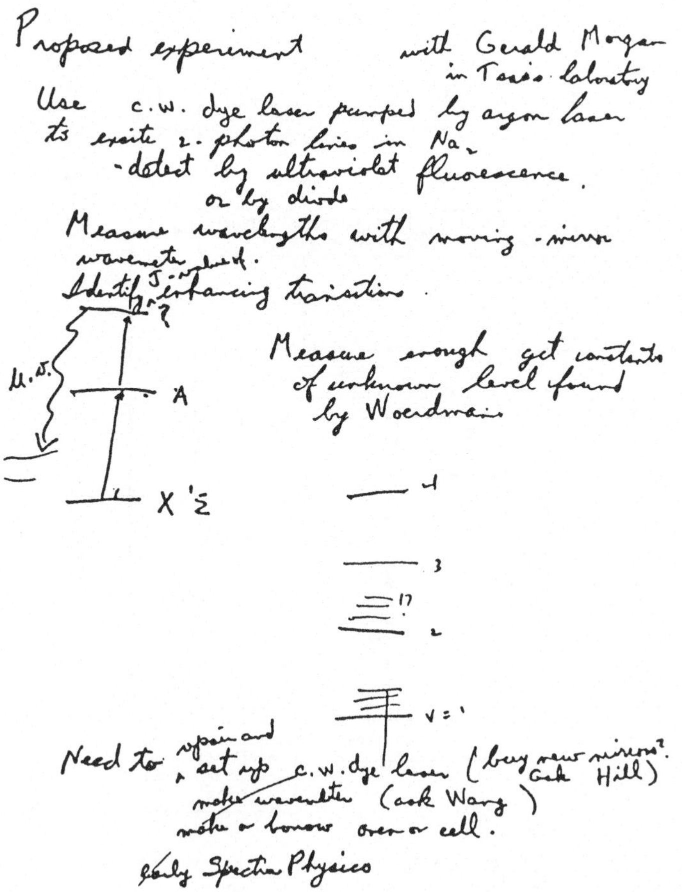

Proposed experiment   with Gerald Morgan
in Tsao's laboratory

Use c.w. dye laser pumped by argon laser
to excite 2-photon lines in Na₂
- detect by ultraviolet fluorescence.
  or by diode
Measure wavelengths with moving-mirror
wavemeter ....
Identify enhancing transition.

Measure enough get constants
of unknown level found
by Woerdman

Need to ... set up c.w. dye laser (buy new mirrors?
        make wavemeter (ask Wang)   ask Hill)
        make or borrow oven or cell.
Early Spectra Physics

Fig. 1. A proposed experiment by Professor Schawlow for Xia in June, 1980

2. COMPREHENSIVE IDENTIFICATION OF EQUAL-FREQUENCY MOLECULAR TWO-PHOTON TRANSITIONS

It was surprising to us initially, that with careful scanning of the single mode cw dye laser, covering 1000 cm^{-1} to around 6000A, we observed only 79 two-photon lines with very different intervals in Na$_2$. The total number of observed two-photon transitions was much smaller than that which could occur between the numerous rovibrational levels of the ground state and excited states. The conventional one-photon molecular spectral structure was distorted beyond recognition. A concrete analysis of the transition rates for a branch of lines of a particular two-photon band revealed the truth. In fact, each observed transition was mainly enhanced by a near coincidence of the laser frequency with a suitable one-photon transition. Only those transitions with intermediate levels less than about 1 cm^{-1} from two-photon resonances were strong enough to be measured. It

175

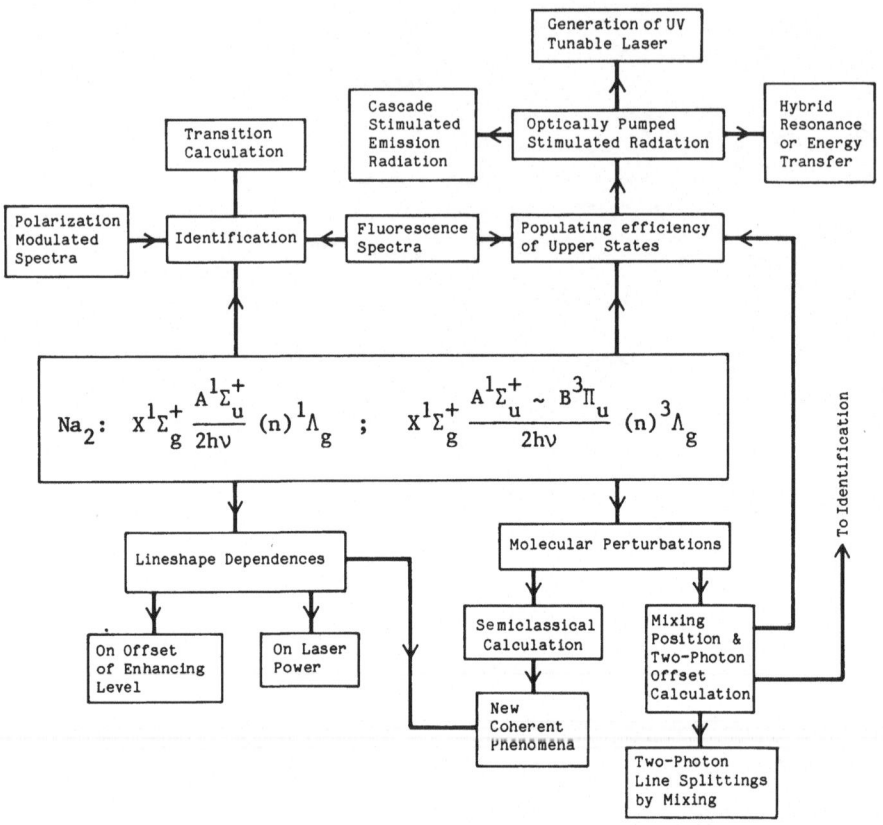

Fig. 2. The general scheme of study for Na$_2$ from the time at Stanford University to the present time at East China Normal University

stimulated our curiosity to calculate the two-photon absorption frequencies and their respective enhancing levels, as described in Ref. 4. We then plotted curves of optical frequency as a function of rotational energy for the one- and two-photon transitions, seeking the rotational quantum number at which the two curves cross. This level provides the strongest enhancement of the two-photon transition. With this figure we then could identify the two-photon transitions and analyze the dramatically altered band structures if the constants of the related states were known. Or we could fit the constants for an unknown high-lying state to the observed two-photon lines.

Indeed there are two types of excitation for two-photon transitions with equal frequencies, as illustrated in Fig. 3. The first is shown in Fig. 3b, this occurred between the singlet states in Na$_2$. The observed two-photon lines were calculated and 57 of 79 lines were recognized as having frequencies very close to those of transitions between the ground state $X^1\Sigma_g^+$ and either of two excited g-parity states, $^1\Sigma_g^+$ or $^1\Pi_g$, at 33000 cm^{-1}, observed by SCHAWLOW, et al.[3] The identifications were associated with

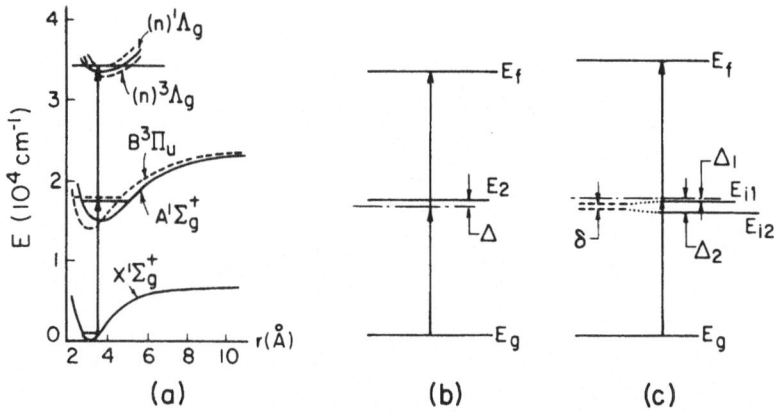

Fig. 3. Schemes of the near-resonant molecular two-photon transitions in
Na$_2$: (a) respective electronic states,
(b) two-photon transition between singlet states,
(c) a mixing-level enhancing two-photon transition

two complementary experiments: taking the fluorescence spectrogram around
the exciting wavelength and recording cw Doppler-free two-step excitation
spectra, for part of the lines.[4]

Recently, the above-mentioned methods have been used to study the second
type of molecular two-photon transitions, efficiently populating the excited
triplet states as shown in Fig. 3c. We observed dozens of lines between
6200 and 6800A distinguished from the mentioned spectral structures. It
was interesting to note that the spectral density for the second type of
transition was in some wavelength regions much greater than that for the
first type of transition. In fact, the ground state in Na$_2$ is singlet, so
this kind of two-photon transition is allowed and strong enough to be
noticed only if there was a near resonant enhancing level to be singlet-
triplet mixed. Due to their wavefunction sharing, there are two branches
of one-photon transitions enhancing one branch of the two-photon band, thus
corresponding to two different regions providing the strong enhancement of
the two-photon transitions (see framed in Fig. 4b). Furthermore, a new
coherent effect, predicted by the semiclassical calculation for the four-
level-system two-photon processes may make the spectral density even
heavier.[5] We have undertaken to determine the energy constants of the
triplet states.

While working on triplet states we met new problems. Where the levels
mix, there are level shifts of the zero-approximation energy values. The
suggested polarization modulated two-photon spectroscopy was demonstrated
to be helpful and to distinguish molecular two-photon branches of lines in
the whole scanning region.[6] However, for examining the specifications of
the upper states, the complementary experiment of recording fluorescence
spectra was extended over a much wider range from UV (3000A) to near
infrared (8000A), as shown in Fig. 5.
The comparison of the fluorescence spectra for different excitation
approaches, including three energy transfer processes in addition to
molecular two-photon excitation transition, helps us to determine the
population efficiencies for the high-lying states.[7]

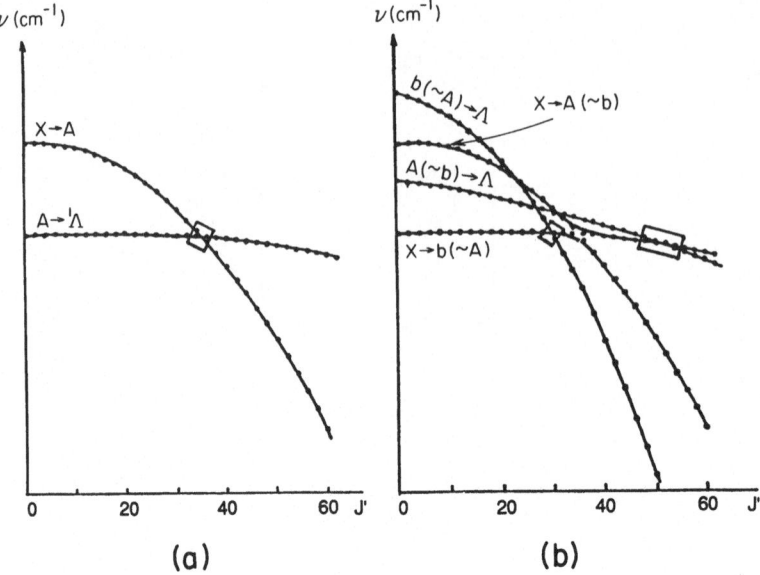

Fig. 4. Frequency of light for one- and two-photon transitions:
(a) for the first approach shown in Fig. 3a, $X:X^1\Sigma_g^+$; $A=A^1\Sigma_u^+$; $^1\Lambda:(n)^1\Lambda_g$,
(b) for the second approach shown in Fig. 3b. Dots represent unperturbed frequencies; Frames indicate near-resonant enhancing positions for equal-frequency two-photon transitions; $X:X^1\Sigma_g^+$; $^3\Lambda:(n)^3\Lambda_g$; $A(\sim b)$ or $b(\sim A)$: $A^1\Sigma_u^+ \sim b^3\Pi_u$ mixing

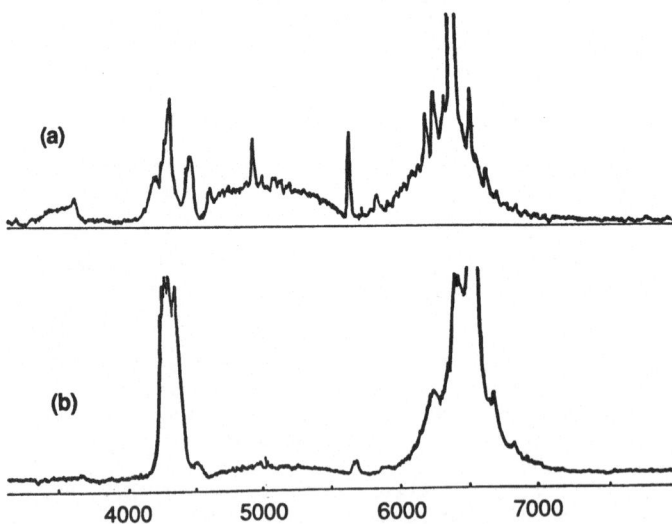

Fig. 5. Fluorescence spectra of two-photon absorption, corresponding to the upper states:
(a) singlet-triplet coupled level,
(b) quite pure triplet state

178

3. LINESHAPE DEPENDENCES AND SPECTRAL STRUCTURES
OF MOLECULAR TWO-PHOTON TRANSITIONS

Every step of our studies was encouraged by Professor Schawlow. His frequent questions always began with "Anything new?... ." During the measurements of the Doppler-free two-photon lines we found that the strongest lines eventually occurred with wide linewidths. When Professor Schawlow heard this, he suggested that they may be two-photon transitions and we should check this by changing the laser power. Thus it was confirmed that the wide linewidth was the Doppler-broadened pedestal formed by absorbing two photons from one beam. Compared to the Doppler-free peak above it, the Doppler-broadened background increased so fast for some of the lines that the high ratio of DF/DB was as small as 1/4! (Fig. 6a). To find the lineshape dependences, Mrs. Xia and Mr. Yan observed one- and two-photon transitions simultaneously to determine the offset of the enhancing level by measuring the frequency interval between the Lamb dip of the one-photon transition center and the DF peak of the related two-photon transition.[8] The measured offset ranged from about 3 GHz down to as little as 34 MHz. The Doppler-free peak became weaker as the offset decreased.

The source of the violently reduced peak ratio was proved to be due to the real population of the intermediate level with an offset smaller than the Doppler width. The effect was aggravated by increasing laser power. The background can be eliminated by intermodulation or polarization modulation spectroscopic approaches.[9]

Recently we have observed more complicated spectral structures for the two-photon transitions linked with triplet states. As mentioned above, we observed, for example, over 10 lines within a few wavenumbers with different lineshapes, except one peak line similar to the appearance as shown in Fig. 6a. Also, there were symmetric multi-peak lines, shown in Fig. 6b, and different kinds of asymmetric shapes. The lineshapes help us to recognize the mixing positions and the structures help us to determine

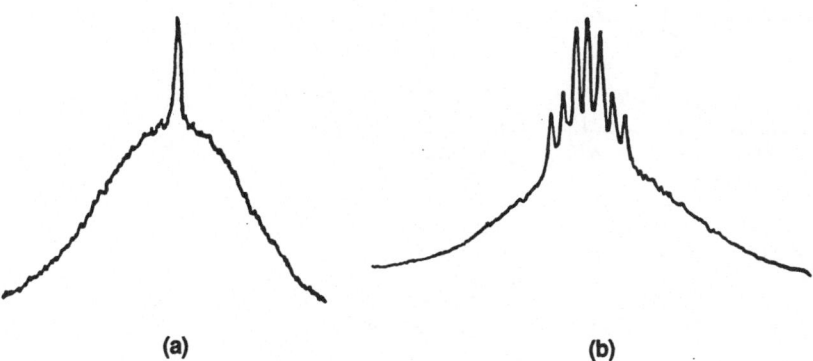

(a)   (b)

Fig. 6.   Molecular two-photon line shapes:
   (a) a typical trace for the near-resonant two-photon transitions
       between singlet states (Ref. 8),
   (b) one kind of the line structure for the two-photon transitions
       enhanced by the singlet-triplet mixing levels up to high-lying
       triplet states.

the total number spin of the molecule, as well as the total angular
momentum of the electrons projected on the internuclear axis and the total
angular momentum of the molecules inclusive of the nuclear spin.[10]

4. GENERATION OF THE OPTICALLY PUMPED STIMULATED EMISSION OF RADIATION BASED ON MOLECULAR ELECTRONIC TRANSITIONS IN Na_2

Generating stimulated emission of radiation by optical pumping is an
important part of the studies concerning Na_2, as pointed out by Professor
Schawlow. At that time, we demonstrated stimulated emission of radiation
in the infrared region near 0.91 μm with vibrational and rotational quantum
numbers identified, and observed violet diffuse band stimulated emission at
4300A (Fig. 7c).[11] Stimulated radiation in the UV region was not observed,
although we did try to avoid the parity limitation of homonuclear molecules
by adding potassium to the heat-pipe oven to form NaK molecules to increase
energy transfer efficiency.

In addition, we have performed a number of stimulated emission processes
between the excited bound states in Na_2 providing stimulated emission
spectral lines distributed at different wavelength regions. We have
observed the generation of UV signals by different mechanisms, including (1)
molecular two-photon pumping,[12] (2) atomic two-photon pumping with energy
transfer to Na_2,[13] and (3) molecular and atomic hybrid resonance
processes.[14] The stimulated emission spectra were different, as shown
in Fig. 7.

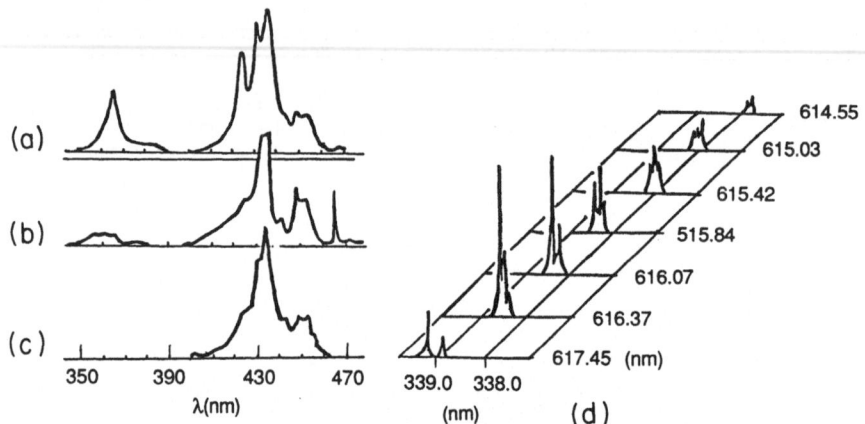

Fig. 7. Optically pumped UV stimulated tunable radiation:
   (a) by molecular two-photon pumping (Ref. 12),
   (b) by energy transfer from atoms to molecules (Ref. 13),
   (c) without UV stimulated radiation by UV one-photon pumping,
   (d) by molecular and atomic hybrid resonance pumping

5. CONCLUSION

Our time working with Professor Schawlow at Stanford University was
very joyful and fruitful, and even today there still remains a great
indelible influence on our research. We will always remember the words

he wrote for us during his second visit to East China Normal University
in 1984:

There are a lot of simple and beautiful
things left for us to find.

Arthur L. Schawlow
October 30, 1984

Professor Schawlow has shown us that **"OPTICS IS LIGHT WORK,"** and we
believe this to be true!

6. REFERENCES

1. K. C. Harvey, Thesis: M.L. Report No. 2442,
   Stanford University (1975).
2. J. P. Woerdman, Chem. Phys. Lett., $\underline{43}$, 279 (1976).
3. N. W. Carlson, A. J. Taylor, K. M. Jones, and A. L. Schawlow,
   Phys. Rev. A$\underline{24}$, 822 (1981).
4. G. P. Morgan, H.-R. Xia, and A. L. Schawlow,
   J. Opt. Soc. Am. $\underline{72}$, 315 (1982).
5. H.-R. Xia, J.-W. Xu, and I.-S. Cheng, to be published.
6. J.-G. Cai, H.-R. Xia, and I.-S. Cheng,
   ACTA Optica Sinica $\underline{6}$, 212 (1986).

7. H.-R. Xia, J.-W. Xu, J.-G. Cai, and I.-S. Cheng, in Proceedings of the National Conference on Laser Spectroscopy (China, 1986), p. 16-19.
8. H.-R. Xia, G.-Y. Yan, and A. L. Schawlow, Opt. Comm. 39, 153 (1981).
9. G.-Y. Yan and H.-R. Xia, Scientia Sinica, 28 505 (1985).
10. H.-R. Xia, L.-S. Ma, J.-W. Xu, and I.-S. Cheng, to be published.
11. Z.-G. Wang, Y.-J. Wang, G. P. Morgan, and A. L. Schawlow, Opt. Comm. 48, 398 (1984).
12. Z.-G. Wang, K.-C. Zhang, X.-L. Tan, and I.-S. Cheng, Acta Optica Sinica, 6, 1081 (1986).
13. Z.-G. Wang, L.-S. Ma, H.-R. Xia, K.-C. Zhang, and I.-S. Cheng, Opt. Comm. 58, 315 (1986).
14. Z.-G. Wang, X.-L. Tan, K.-C. Zhang, and I.-S. Cheng, to be published.

Laser-Driven Ionization and Photoabsorption Spectroscopy of Atomic Ions

W.T. Hill III[1,a] *and C.L. Cromer*[2]

[1]University of Maryland, IPST, College Park, MD 20742, USA
[2]National Bureau of Standards, Gaithersburg, MD 20899, USA

Abstract

The application of laser-driven ionization techniques to photoabsorption spectroscopy of atomic ions will be discussed. A summary of the experimental results which confirm that a collisional mechanism is responsible for the nearly complete ionization following laser irradiation is given along with a bibliography of the photoabsorption measurements reported to date. The importance of these investigations, demonstrated in two studies involving the Ba nuclear sequence (i.e., Ba, Ba^+, and Ba^{++}) and the Xe isoelectronic sequence (i.e., Xe, Cs^+ and Ba^{++}), will be discussed. Finally, a tabulation of quantum defect parameters for Xe, Cs^+ and Ba^{++} based on a re-analysis of the Xe-sequence spectra via a shifted R-matrix quantum defect approach, will be presented.

1. Introduction

Uniform, nearly complete ionization of an atomic vapor following resonant laser irradiation has proven to be an extremely useful way to prepare atomic ions for photoabsorption studies. To date, this resonant laser-driven ionization (RLDI) technique has been successful in producing quantifiable densities of Li^+ [1], Na^+ [2,3], Cs^+ [4], Sr^+ [5], Ca^+ [5,6], Ba^+ [5,7-13], Ba^{++} [12-15] and Mn^+ [16] ions. The technique empowers one with an ability to strip away a prespecified number of electrons from a given nucleus and, conversely, to select the strength of the Coulomb field (i.e., nuclear charge) for a specific valence configuration. Thus, systematic changes along sequences of ions, which have either the same nuclear charge or valence structure, can be investigated. When coupled with recent advances in theoretical and empirical approaches to analyze complicated spectra, RLDI provides the atomic physicist with a unique opportunity to study the details of many electron effects, such as autoionization, heretofore not possible.

The focus of this article is the application of laser-driven ionization techniques to study atomic physics problems. To that end, the paper is divided into three parts. To familiarize the reader with RLDI, the first section will present a brief review of: (1) the mechanism responsible for the uniform ionization; (2) the key experimental results which validate the belief in the mechanism; and (3) the conditions and parameters necessary to achieve nearly complete ionization. The second section will be devoted to photoabsorption studies in the vacuum ultraviolet (VUV) region of the spectrum which have been made with the aid of RLDI. The discussion will center primarily on two important experiments involving Cs^+, Ba^+ and Ba^{++} ions but this section will include a bibliography of all the VUV

183

measurements employing RLDI reported to date. The paper will conclude with a short discussion of future experiments that laser-driven ionization techniques could make possible.

2. Resonant Laser-Driven Ionization Mechanism

The first observation of nearly 100% ionization of a Na atomic vapor following resonant excitation by a pulsed dye laser was reported by LUCATORTO and MCILRATH [2] more than ten years ago. Several processes contribute to the ionization, but the universality of the phenomenon, independent of atomic species, corroborates the belief that the primary mechanism is based on superelastic collisions between free electrons and laser-excited atoms followed by collisional ionization by the free electrons, as first suggested by MEASURES [17]. It has also been demonstrated that doubly charged ionic vapors can be produced, with nearly the same efficiency, by resonantly exciting the singly charged ions [12-15]. In principle, the technique can be applied repeatedly to achieve any desired stage of ionization provided that an appropriate laser and resonant transition can be found. (At the writing of this article there have been no reports of ionization beyond the second stage.)

Experimental observations show that RLDI is capable of producing complete ionization of a 10 - 20 cm column of an atomic vapor on a microsecond time scale when: (1) the atomic densities fall in the range of 10^{13} - 10^{16} cm^{-3}; (2) the laser energy is $\geq$ 0.15 J; and (3) the laser pulse length is of the order of 500 ns. (These laser requirements are met by flashlamp pumped dye lasers.)

The mechanism can be summarized as follows. The energy necessary to drive the process is extracted from the pulsed laser field via resonant absorption. Upon irradiation, an excited state population roughly equal to the ground state population is created while, at the same time, a few free electrons are produced. The electrons are generated through a variety of processes, such as multiphoton ionization, associative ionization, or laser-assisted Penning ionization [3,7]. These processes, whose rates vary with atomic species, are generally too inefficient to be solely responsible for the total ionization observed; however, they do provide a source of initial electrons to seed the primary ionization mechanism [3, 7].

The primary mechanism can be qualitatively understood by considering the tendency for the electrons to be in equilibrium with the atomic vapor. In the case of a Na vapor, for example, with the laser saturating the $3\ ^2S_{1/2} \rightarrow 3\ ^2P_{1/2}$ transition (i.e., creating nearly equal population in the two states) the "apparent temperature" of the vapor (treating Na as a two-level system) will be almost infinite. Thus, the electrons will undergo superelastic collisions with the laser-excited atoms (a process in which the electrons gain energy while de-exciting the $3\ ^2P$ atoms) in an attempt to bring the electron-atom system into equilibrium [3]. In order for equilibrium to exist between the electrons and the atoms, however, collisional processes must dominate radiative processes. As a consequence, equilibrium will generally exist between the electrons and the closely spaced high-lying levels (which have long radiative lifetimes) but not with the lower levels [11,18]. Nevertheless, the electrons will gain energy through superelastic collisions and heat up the entire electron-vapor system--the energy gained by the electron will be collisionally transmitted back to the vapor resulting in the high-lying levels becoming populated. A fraction of the atoms in high-lying states will be photo-ionized during the laser pulse which adds to the pool of free electrons while the remainder of the atoms will be ionized by the free electrons. The percentage of ionization will depend on the temperature of the electron-vapor system through the Saha-Boltzmann equation [19].

Complete ionization can only be achieved if the laser pulse is long enough to maintain the excited population so that the electron-vapor system can gain sufficient energy, otherwise only partial ionization will occur. Under long pulse excitation the mechanism is self-terminating because once the neutral population is depleted the energy link to the laser field is broken and, as a consequence, the vapor comes to equilibrium at the next higher stage of ionization.

A few investigators have suggested that the ionization is principally due to radiative effects [9,20]. Some of these suggestions have been quite controversial [21]. Although radiative processes are very important in some systems and contribute to some degree to the production of seed electrons in other systems, experimental results do not support a generic radiative process. On the other hand, there have been several important experiments which provide convincing evidence in support of a collisional mechanism. A few key results are summarized in Table 1.

Table 1. Evidence for Collisional Ionization Mechanism

Experimental Result	Reference
1. Direct observation of superelastic collisions between electrons and excited Ba and Na atoms.	22, 23
2. Ion yields depend on the laser pulse length; when pulse length is too short ($\leq$ 50 ns) electron-atomic vapor does not have time to heat up.	24
3. Ion yields depend on the atomic density; when densities are too low ($\leq 10^{13}$ cm^{-3}) collision rates are too slow.	5, 25
4. The onset of ionization lags behind the beginning of the laser pulse and the peak ion densities occur near the end of the pulse or when the pulse has ended (see Fig. 1).	1, 5-9
5. When several atomic vapors are contained in the same cell and only one is resonantly excited all species show appreciable ionization.	5
6. Ionization can be quenched by momentum-changing elastic collisions with He atoms which cool the electrons.	26

It should be emphasized that complete ionization is reached with nearly the same laser fluence ($\sim$ 5 J/cm^2) for all atomic systems even though the number of photons necessary to ionize the laser-excited atoms varies from one in Cs [4] to three in Ba$^+$ [7]. At the same time, the intermediate near-resonance structure, enhancing multiphoton ionization out of the excited states is vastly different for the different atomic systems. Typically, 100% ionization is reached when the oscillator strength (f) of the transition to which the laser is tuned is of the order of 1. In contrast, even for Ca [6] and Mn [16] in which f is of the order of 10^{-5}, 50% ionization or better was still achieved. These differences in the atomic structure and the transition strengths would produce a strong intensity dependence in the ionization if radiative processes were primarily responsible. One also observes that the ionization time development is of the same order of magnitude for all systems [1,5-9,11]. Figure 1 shows this for Li (f ~ 1) and Ca (f ~ 10^{-5}, intercombination line). In summary, the experimental results overwhelmingly support a primary mechanism based on collisional processes and not on multiphoton ionization processes.

The ionization process has also been studied numerically in an attempt to prove or disprove the collisional or radiative mechanisms [3,7,9,11,17, 27-29]. The results of these model calculations have been mixed. This is largely due to the complexity of the calculation, which requires an accurate knowledge of level populations, radiative rates and collision cross sections. Unfortunately, many of these parameters are not well known, making the numerical results less decisive than the experimental results.

Figure 1 shows that in addition to ions, appreciable densities of excited neutrals are present in the vapor at early times. Less transient excited neutrals are also a by-product of quenching of RLDI by over pressuring the vapor with He gas. (See entry 6 in Table 1 and [22].) Thus, RLDI can be used to prepare excited neutrals for photoabsorption as well.

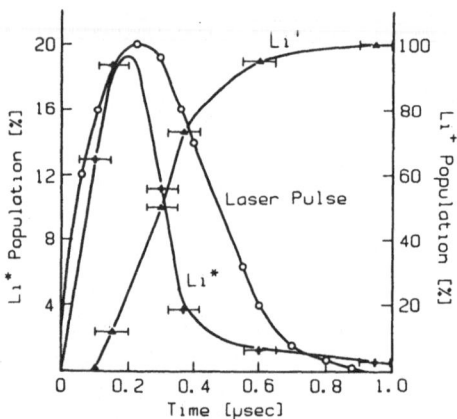
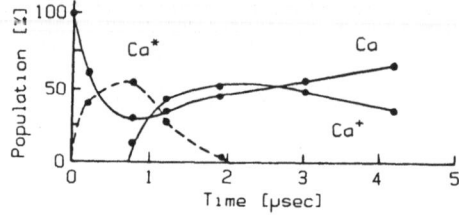

Fig. 1. Evolution of excited neutral and ion populations for Li (from [1]) and Ca (from [6])

Finally, it is interesting to note that it has been recently shown that appreciable ionization occurs on a similar time scale after non-resonant excitation of atomic vapors in the presence of large Ar atmospheres [30]. The mechanism responsible here is based on the laser-fragmentation of metal clusters catalyzed by the Ar atoms. (Cluster formation in He atmospheres is much less efficient.) Cluster fragmentation generates many excited state atoms in addition to ions, so that free electrons can gain energy as they do after resonant excitation and thus create significant ionization, albeit less than 100%.

3. Photoabsorption Spectroscopy of Atomic Ions and Excited Neutral Atoms

Attention will now be turned toward photoabsorption experiments. The studies summarized in Table 2 were obtained with an apparatus similar to that schematically shown in Fig. 2. When doubly charged ions are desired, a second flashlamp pumped dye laser must be added. The studies include photoabsorption from both inner and valence shell electrons and from ions as well as excited states of neutral atoms.

The background radiation in these experiments was generated by either a BRV electrical spark [32] or a laser- produced plasma light source [33] both of which emit radiation in the VUV in the 5 - 200 nm region. These light sources provide time resolution on the order of 10 - 50 ns. The vapors were contained in oven/heatpipe cells and the photoabsorption spectrum was dispersed by a grazing-incidence spectrograph with either photographic film [7] or a photoelectric detector [34] placed in its image plane. Since this radiation does not propagate through the air, the optical paths connecting the light source, the absorption region and the detector were enclosed and evacuated. For the two experiments discussed below, the dye lasers were tuned to 459.3 nm in Cs^+, 553.5 nm in Ba^+ while a second laser tuned to the 493.4 nm resonance line of Ba^+ was used to generate Ba^{++}. More details about the experiments can be found in the references (see [4,12-15]).

Table 2. Photoabsorption Spectra of Ions and Excited Neutrals

System	Lower State Configuration		Electron Shell(s) Excited	Reference
Li^*	$(1s^2 2p)$	^2P	1s	1
Li^+	$(1s^2)$	^1S	1s	1
Na^*	$(2p^6 3p)$	^2P	2p	31
Na^+	$(2p^6)$	^1S	2p	2
Cs^+	$(5p^6)$	^1S	5p	4
Ca^*	$(3p^6 4s 4p)$	^3P	3p	6
Ca^+	$(3p^6 4s)$	^2S	3p	6
Mn^+	$(3p^6 3d^5 4s)$	^7S	3p	16
Ba^+	$(4d^{10} 5p^6 6s)$	^2S	4d	12,13
	$(4d^{10} 5p^6 5d)$	^2D	4d	12,13
Ba^{++}	$(4d^{10} 5p^6)$	^1S	4d, 5p	12-19

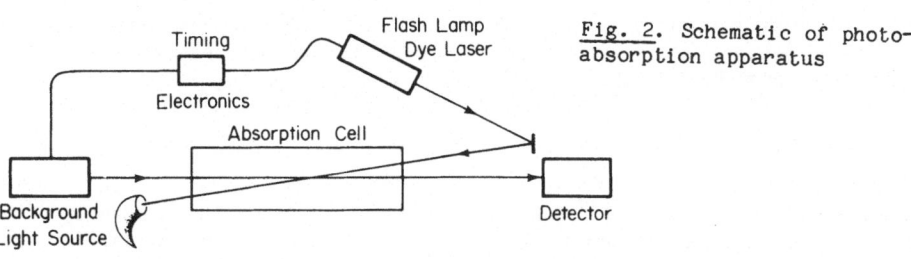

Fig. 2. Schematic of photoabsorption apparatus

3.1. Ba, Ba+, Ba++ Sequence

Figure 3a shows the photoabsorption spectra between 14.6 and 8.3 nm (85 - 150 eV) in Ba, Ba^+ and Ba^{++} involving inner shell $4d^{10} \rightarrow 4d^9$ nf,εf transitions (taken from [12]). In Ba and Ba^+ the spectrum consists of a broad resonance feature (i.e., excitation of an εf continuum state) while in Ba^{++} there are several sharp resonances (i.e., excitation of nf autoionizing states) prior to the onset of the broad feature. The main structural difference in the $4d \rightarrow f$ photoabsorption spectra between Ba (and Ba^+) and Ba^{++} is due to the reduction in the screening associated with the removal of the 6s valence electrons. As a consequence, the f potential, which has a double well shape in Ba and Ba^+, is flattened out in Ba^{++}. In response, there is more overlap between the 4d wavefunction and the nf wavefunctions (autoionizing states) because the nf wavefunctions are contracted closer to the nucleus in Ba^{++}.

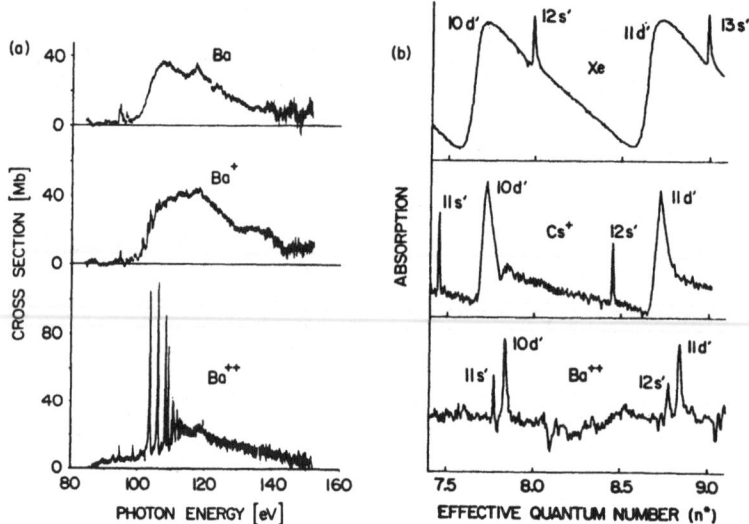

Fig. 3. Photoabsorption spectra of: (a) $4d^{10} \rightarrow 4d^9$ nf, εf for Ba, Ba^+ and Ba^{++} (from [12]) and (b) $5p^6 \rightarrow 5p^5 \, ^2P_{1/2}$ $ns_{1/2}$, $nd_{3/2}$ for Xe (from [35]), Cs^+ (from [4]) and Ba^{++} (from [14])

The significance of this observation extends beyond gas-phase atomic physics when the Ba and Ba^{++} spectra are compared with $4d \rightarrow f$ photoabsorption spectra of divalent Ba compounds. In particular, Fig. 14 of [13] shows that the spectrum of $BaBr_2$ is similar to that of atomic Ba while that of BaF_2 begins to resemble Ba^{++}. This progression suggests that the Ba 6s electrons start to become de-localized as the electron affinity increases from Br to F.

3.2. Xe, Cs+, Ba++ Sequence

Figure 3b shows several densitometer traces of autoionizing Rydberg resonances of Xe, Cs^+ and Ba^{++} involving transitions from the ground $5p^6 \, ^1S_0$ state to states designated as $5p^5 \, ^2P_{1/2}$ $ns_{1/2}$, $nd_{3/2}$. In the figure, the

$ns_{1/2}$ and $nd_{3/2}$ resonances are given the labels ns' and nd', respectively. These states lie above the $^2P_{3/2}$ ionization limit and autoionize into the concomitant ϵs and ϵd continua. The s and d continua are the above threshold continuation of the $5p^5\ ^2P_{3/2}\ ns_{1/2}$, $nd_{3/2}$, $nd_{5/2}$ Rydberg series which converge to the $^2P_{3/2}$ ionization limit. These five interacting Rydberg series and their associated continua comprise the five channels of excitation observed in the Xe-like ions.

The spectra in Fig. 3b are plotted vs. the effective quantum number n^* which is defined through:

$$E = I \frac{Z_{eff}^2 R}{(n^*)^2} \tag{1}$$

with E being the energy of the transition (in cm^{-1}), I the ionization limit to which the autoionizing resonances are converging (i.e., $^2P_{1/2}$), Z_{eff} the effective charge on the nucleus and R the Rydberg constant. On an n^* scale, the ns' and nd' resonances appear with a period of one. Displaying the spectra in this way eliminates wavelength dependence and allows the resonances with the same (ns' or nd') classification to be compared on an equal footing.

Figure 3b shows that, as the Coulomb field strength is increased (i.e., in going from Xe to Ba^{++}), the resonances respond in the following manner: (1) the positions of the s' and d' resonances are shifted to higher n^* and (2) the widths of the d' resonances (which are measures of the probabilities for autoionization) are reduced. This behavior is associated with the change in the effective potential and the tendency of the system to approach the hydrogenic limit as Z_{eff} increases [14].

This striking demonstration of the influence Z_{eff} has on the autoionizing widths also shows the intimate connection between autoionization widths and the strength of perturbations in the bound portion of the spectrum. This can be seen by comparing Figs. 3b and 4. Figure 4 displays the energy positions of the bound states (i.e., all Rydberg states below the $^2P_{3/2}$ ionization limit) for Xe, Cs^+, and Ba^{++} in a Lu-Fano diagram [37]. The energy positions are taken from [36,4,15] respectively for Xe, Cs^+ and Ba^{++}. Each experimental point is plotted with the abscissa being the effective quantum number, $\nu_{1/2}$ ($= n^*$ from (1)), relative to the upper $^2P_{1/2}$ ionization limit and the ordinate being the effective quantum number modulo 1, $\nu_{3/2}$\{mod 1\}, relative to the lower $^2P_{3/2}$ ionization limit. (An equation similar to (1) with I given by the $^2P_{3/2}$ threshold defines $\nu_{3/2}$.) The solid curves connecting the points are derived from the parameters given in Table 3 which were determined from a numerical fit to the experimental data using a shifted R-matrix formulation of the Multichannel Quantum Defect Theory (MQDT) equations [38,39]. Theoretical MQDT parameters obtained from a relativistic random phase approximation calculation by CHENG [40] were used as starting values for the fit. Some of the parameters could not be determined from the experimental data and the theoretical values were retained; these values are indicated by an asterisk (*) in Table 3.

The Lu-Fano plot is composed of three curves running across the unit cell for the three series converging to the lower limit and two curves running up the cell for the series converging to the upper limit; these curves are repeated for each unit cell. The parameters denoted by δ_i in

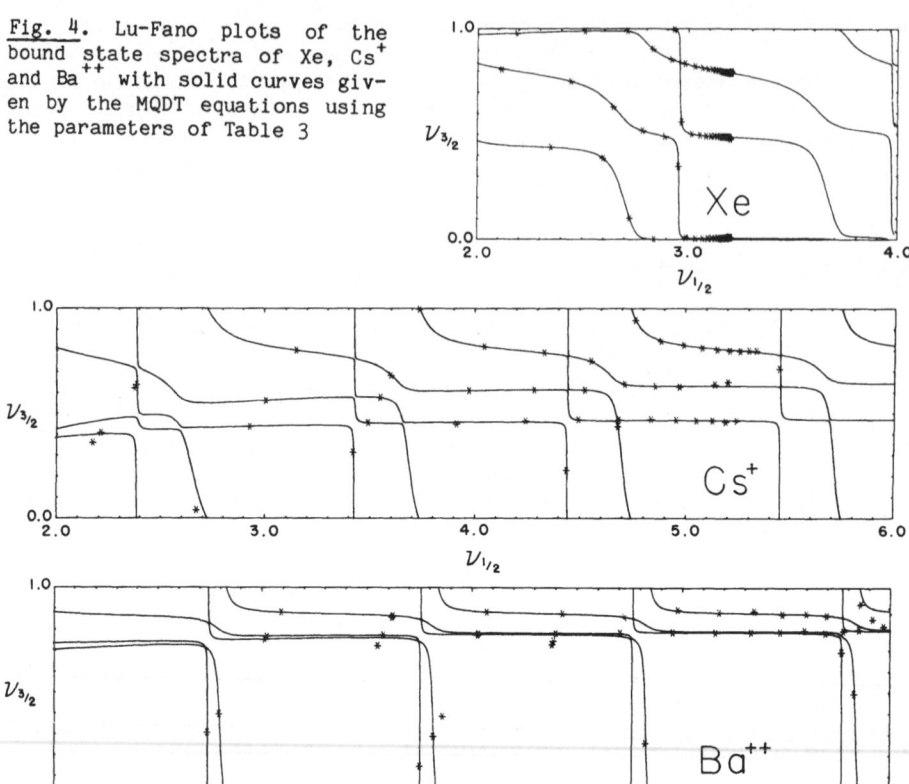

Fig. 4. Lu-Fano plots of the bound state spectra of Xe, Cs$^+$ and Ba^{++} with solid curves given by the MQDT equations using the parameters of Table 3

Table 3 correspond to the quantum defect of each channel in the absence of interchannel perturbations. If there were no perturbations, the curves running across the cell would be straight and horizontal and would pass through the points $(1 - \delta_1)$, $(1 - \delta_2)$, and $(1 - \delta_3)$ while the other two curves would be straight and vertical and pass through the points $(1 - \delta_4)$ and $(1 - \delta_5)$. Because of interchannel interaction, the curves are not straight and avoid crossing each other. The size of the avoided crossings is given by the R_{ij} parameters in Table 3 which are a measure of the strength of the perturbations among the bound resonances below the $^2P_{3/2}$ threshold. For example, R_{24} for Xe describes the avoided crossing in the center of the Xe Lu-Fano plot. (The separation of the parameters into those responsible for intrachannel (δ_i) and interchannel (R_{ij}) interactions is the primary reason for using the shifted R-matrix formulation of MQDT to fit the experimental levels. In addition, the fitting procedure using this parameterization is more transparent because each parameter is associated with a particular, localized section of the Lu-Fano curve [38, 39].) One can immediately see, as pointed out by HILL et al. [14], that (1) the large avoided crossings are linked to large autoionization widths above the $^2P_{3/2}$ threshold (compare Figs. 3b and 4) and (2) the perturbations are reduced as Z_{eff} increases (see R_{24} and R_{25} values in Table 3).

Table 3. Shifted R-matrix MQDT parameters for Xe, Cs^+ and Ba^{++}

Parameters[†]	Xe	Cs^+	Ba^{++}
δ_1^0	0.9938(20)	0.5289(25)	0.2028(51)
δ_2^0	0.5131(16)	0.3617(26)	0.2053(55)
δ_3^0	0.2042(19)	0.1940(31)	0.1201(51)
δ_4^0	1.0374(17)	0.5487(26)	0.2292(33)
δ_5^0	0.3260(19)	0.2917(37)	0.1635(36)
R_{14}^0	0.0442(100)	0.0537(100)	0.0435*
R_{24}^0	-0.0729(47)	-0.0072*	0.0056*
R_{34}^0	-0.0113(296)	0.0085*	-0.0018*
R_{15}^0	0.0635(59)	0.0122*	0.0367*
R_{25}^0	0.2397(35)	0.1004(95)	0.1045*
R_{35}^0	-0.4209(35)	-0.2814(68)	-0.1255(292)
δ_1^1	-0.5324(201)	-0.2029(74)	-0.1722(169)
δ_2^1	-0.8990(112)	-0.4980(82)	-0.2301(186)
δ_3^1	-0.0759(169)	-0.0467(114)	-0.0357(195)
δ_4^1	-0.4592*	-0.2285(106)	-0.1379(85)
δ_5^1	1.1516*	-0.2939(220)	-0.1688(94)
R_{14}^1	-0.0430*	-0.0418*	-0.0270*
R_{24}^1	0.6324*	0.0799*	0.1982*
R_{34}^1	-0.2913*	-0.0720*	-0.0260*
R_{15}^1	-0.2516*	0.0546*	-0.2712*
R_{25}^1	-1.1251*	-0.1262*	0.0955*
R_{35}^1	-0.6291*	0.6828*	0.2144*

*These values could not be determined from experimental data [36,4,15] and are equal to theoretical values [40] which were used as starting values for the reduction.

†The channels:

$$5p^5 \; {}^2P_{3/2} \; ns_{1/2}, \; nd_{3/2}, \; nd_{5/2}; \; 5p^5 \; {}^2P_{1/2} \; ns_{1/2}, \; nd_{3/2},$$

are numbered consecutively. The energy dependence in the parameters is defined by:

$$\delta_i = \delta_i^0 + \varepsilon\delta_i^1 \text{ and } R_{ij} = R_{ij}^0 + \varepsilon R_{ij}^1,$$

with ε being the energy in atomic units. Values in parentheses indicate the uncertainty in the last digits.

4. Future Experiments

There are several new spectroscopic experiments yet to be performed in conjunction with RLDI; a few examples will now be given. As the flashlamp pumped dye laser technology improves, allowing more intense radiation to be generated in the blue and near UV regions of the spectrum, higher stages of ionization can be achieved. The rare earth elements would provide good candidates for creating the third state of ionization, for example. Furthermore, the cluster fragmentation ionization process (in Ar atmospheres [30]) can be employed to prepare moderate ion densities of samples which do not have suitable transitions for RLDI to be used. At the same time, the fragmentation process probably generates ionized clusters which could be the subject of investigations. The ability to select the percentage of ionization by adding an appropriate amount of He buffer gas [26] will permit one to investigate partially ionized vapors. For instance, line broadening of neutral atomic states in the presence of their own ions might be studied. These mixed vapors might also produce the right environment to create molecular ions. In addition to the spectroscopic studies, pure or partially ionized vapors might be employed as new nonlinear optical media for frequency conversion. Although the suggestions cited are related to atomic physics, laser-driven ionization also has application to plasma physics problems [17,27].

Acknowledgments

The authors would like to thank T. B. Lucatorto, T. J. McIlrath and K. Yoshino for kindly permitting their data to be reproduced; K. T. Cheng for making available the unpublished results of his RRPA calculation for Xe, Cs^+ and Ba^{++}; T. B. Lucatorto for helpful discussions; and B. P. Turner, D. J. Davis and M. G. Spell for technical assistance in preparing this manuscript. This work is sponsored in part by the National Science Foundation under grant PHY-84-51284, the Research Corporation and the National Bureau of Standards.

---

[a]National Science Foundation Presidential Young Investigator

References

1. T. J. McIlrath, T. B. Lucatorto: Phys. Rev. Lett. 38, 1390 (1977).
2.. T. B. Lucatorto, T. J. McIlrath: Phys. Rev. Lett. 37, 428 (1976).
3. B. Carré, F. Roussel, P. Breger, G. Spiess: J. Phys. B 14, 4289 (1981).
4. T. J. McIlrath, J. Sugar, V. Kaufman, D. Cooper, W. T. Hill, III: J. Opt. Soc. Am. B3, 398 (1986).
5. C. H. Skinner: J. Phys. B 13, 55 (1980).
6. B. F. Sonntag, C. L. Cromer, J. M. Bridgeis, T. J. McIlrath, T. B. Lucatorto: In Proc. of the Third Topical Meeting of Short Wavelength, Coherent Radiation, Monterey, CA (March, 1986).
7. T. B. Lucatorto, T. J. McIlrath: Appl. Opt. 19, 3948 (1980).
8. H. A. Bachor, M. Koch: J. Phys. B 13, L369 (1980).
9. H. A. Bachor, M. Kock: J. Phys. B 14, 2793 (1981).
10. R Künnemeyer, M. Koch: J. Phys. B 16, L607 (1983).
11. L. Jahreiss, M. C. E. Huber: Phys. Rev. A28, 3382 (1983).
12. T. B. Lucatorto, T. J. McIlrath, J. Sugar, S. M. Younger: Phys. Rev. Lett. 47, 1124 (1981).

13. T. B. Lucatorto, T. J. McIlrath, W. T. Hill, III, C. W. Clark: In Inter. Conf. X-Ray and Atomic Inner-Shell Physics, ed. by B. Crassman, AIP Conf. Proc. 94, 584, (1982).
14. W. T. Hill, III, K. T. Cheng, W. R. Johnson, T. B. Lucatorto, T. J. McIlrath, J. Sugar: Phys. Rev. Lett. 49, 1631 (1982).
15. W. T. Hill, III, T. B. Lucatorto, J. Sugar, K. T. Cheng: to be submitted to Phys. Rev. A. (1987).
16. J. W. Cooper, C. W. Clark, C. L. Cromer, T. B. Lucatorto, B. F. Sonntag, F. S. Tomkins: to be submitted to Phys. Rev. A Rapid Communications (1987).
17. R. M. Measures: J. Quant. Spectrosc. Radiat. Transfer 10, 107 (1970).
18. M. Mitchner, C. H. Kruger: In Partially Ionized Gases (John Wiley and Sons, NY, 1972).
19. K. R. Lang: In Astrophysical Formulae, (Springer-Verlag, NY, 1974), p. 244.
20. J. M. Salter: J. Phys. B12, L763 (1979).
21. C. H. Skinner: J. Phys. B13, L637 (1980); T. J. McIlrath, T. B. Lucatorto, J. Phys. B 13, L641 (1980).
22. D. F. Register, S. Trajmar, G. Csanak, S. W. Jensen, M. A. Fineman, R. T. Poe: Phys. Rev. A28, 151 (1983); J. M. Bizau, B. Carré, P. Dhez, D. L. Ederer, P. Gerard, J. C. Keller, P. Koch, J. C. LeGouët, J. L. Picqué, F. Roussel, G. Spiess, F. Wuilleumier: In Laser Spectroscopy VI, Proc. 6th Intern. Conf. Interlaken, Springer Series in Optical Sciences, ed. by H. P. Weber, W. Lüthy (Springer-Verlag, NY, 1983).
23. I. V. Hertel, W. Stell, Adv. At. Mol. Phys. 13, 113 (1977); J. L. LeGouët, J. L. Picqué, F. Wuilleumier, J. M. Bizau, P. Dhez, P. Koch, D. L. Ederer: Phys. Rev. Lett. 48, 600 (1982).
24. T. Stacewicz: Opt. Commun. 35, 239 (1980); C. Bréchignae, P. H. Cahuzac: Opt. Commun 43, 270 (1982); J. L. Bowen, A. P. Thorne: J. Phys. B 18, 35 (1985); T. J. McIlrath, J. L. Carlsten: J. Phys. B6, 697 (1973).
25. B. Carré, F. Roussel, P. Breger, G. Spiess: J. Phys. B14, 4271 (1981).
26. W. T. Hill, III: J. Phys. B 19,359 (1986).
27. R. M. Measures, P. G. Cardinal: Phys. Rev. A23, 804 (1981); R. M. Measures, P. G. Cardinal, G. W. Shinn: J. Appl. Phys. 52, 1269 (1981); R. M. Measures, N. Drewell, P. Cardinal: J. Appl. Phys. 50, 2662 (1979).
28. W. L. Morgan: Appl. Phys. Lett. 42, 790 (1983).
29. P. G. Cardinal: Ph.D. Thesis, University of Toronto (1985).
30. W. T. Hill, III: Opt. Commun 54, 283 (1985).
31. J. Sugar, T. B. Lucatorto, T. J. McIlrath, A. W. Weiss: Opt. Lett 4, 109 (1979).
32. T. B. Lucatorto, T. J. McIlrath, G. Mehlman: Appl. Opt. 18, 2916 (1979).
33. J. M. Bridges, C. L. Cromer, T. J. McIlrath: Appl. Opt. 25, 2205 (1986).
34. C. L. Cromer, J. M. Bridges, J. R. Roberts, T. B. Lucatorto: Appl. Opt. 24, 2996 (1985).
35. K. Yoshino: private communication based on data of [36] (1986/7).
36. K. Yoshino, D. E. Freeman: JOSA B2, 1268 (1985).
37. K. T. Lu, U. Fano: Phys. Rev A 2, 81 (1970); K. T. Lu: Phys. Rev. A 4, 579 (1971).
38. W. E. Cooke, C. L. Cromer: Phys. Rev. A 32, 2725 (1985); A. Giusti-Suzor, U. Fano: J. Phys. B 17, 215 (1984).

39. C. L. Cromer: to be published (1987).
40. K. T. Cheng: private communication based on the procedure of [41] (1986/7).
41. W. R. Johnson, K. T. Cheng, K.-N. Huang, M. Le Dourneuf: Phys. Rev. A <u>22</u>, 989 (1980).

Two-Photon Resonant Parametric and Wave-Mixing Processes in Atomic Sodium

Pei-Lin Zhang and Shuo-Yan Zhao

Department of Physics, Tsinghua University, Beijing, China

The generation of coherent radiation in the UV and VUV region through four wave mixing processes has been the subject of several recent reports. In the first, the Na 4d D_J level is populated by two-photon excitation and a cascade of IR stimulated emission ensues. Photons from these cascades of frequency ω_{IR} can then mix with two laser photons with total energy related to $2\omega_L$, yielding UV photons at a frequency $\omega_{UV} = \omega_{IR} + 2\omega_L$. Using various members of the cascade, HARTWIG [1] generated radiation at 330 nm and 333nm, while Wu and Chen [2] have produced radiation at 383 nm and 388nm using this procedure.

Coherent UV radiation may also be generated using parametric oscillations [1],[3]. In this process the signal and the idler wave grow together following two-photon excitation and the coherent radiation produced undergoes a frequency shift because of the presence of atomic transitions. Thus in parametric generation more UV frequencies may be attained.

In this paper, we report results obtained in a study of UV generation using sum and difference-frequency four-wave mixing processes. We also report on the UV outputs obtained in parametric oscillation experiments using the Na 3P and 4P levels as near resonances. The wavelength and the intensity of the parametric oscillations are reported; the dependence of these parameters on pump-laser wavelength and intensity, on the oven temperature and other factors have been investigated. A theoretical analysis based on a wave equation and on atomic polarization is given for these two methods of coherent generation of UV radiation.

2. Theoretical Background

Theoretical analysis of four-wave mixing process and parametric oscillation process requires an analytical expression for the atomic nonlinear polarization. For this purpose we first use a density matrix method to describe the interaction between atom and electric field [4][5], then we use Maxwell equations to relate electric field to electric polarization. The details are described elsewhere.

For four-wave mixing processes, we obtain

$$I_{IR} = I_{IR}^0 \exp(gz), \tag{1}$$

$$I_{UV}=256\pi^4\omega_{UV}^2|\chi^{(3)}(\omega_{UV})|^2 I_L^2 I_{IR}/c^4 n_{IR}n_L^2 n_{UV}[g^2+4(\Delta k)^2], \tag{2}$$

where $g=-4\pi\omega_{IR}\mathrm{Im}\chi^{(1)}(\omega_{IR})/cn_{IR}$ is the gain coefficient; $\Delta k=2k_L-k_{IR}-k_{UV}$ is the phase mismatch of the processes; $\chi^{(1)}$, $\chi^{(3)}$ are the first- and third-order electric susceptibilities,

$$\chi^{(1)}(\omega_{IR})=-N|(j|\mu|i)|^2(\rho_{jj}-\rho_{ii})/\hbar D_{ji}^*, \tag{3}$$

$$\chi^{(3)}(\omega_{UV})=\frac{N(0|\mu|k)(k|\mu|n)(n|\mu|m)(m|\mu|0)}{\hbar^3(\omega_m-\omega_L)}\left[\frac{\rho_{oo}-\rho_{nn}}{D_{k0}^*D_{n0}^*}+\frac{\rho_{nn}-\rho_{kk}}{D_{k0}^*D_{nk}}\right], \tag{4}$$

for difference-frequency mixing process, and

$$\chi^{(3)}(\omega_{UV})=\frac{N(0|\mu|k)(k|\mu|n)(n|\mu|m)(m|\mu|0)}{\hbar^3(\omega_m-\omega_L)}\left[\frac{\rho_{oo}-\rho_{nn}}{D_{k0}^*D_{n0}^*}-\frac{\rho_{nn}}{D_{k0}^*D_{kn}^*}\right] \tag{5}$$

for sum-frequency mixing process; with $|0)$, $|n)$ denoting ground level and two-photon excited level respectively, N density of Na atom, $D_{ji}=\omega_j-\omega_i-\omega_{IR}+i\Gamma_{ji}$, $D_{nk}=\omega_n-\omega_k-\omega_{IR}+i\Gamma_{nk}$, $D_{k0}=\omega_k-\omega_0-\omega_{UV}+i\Gamma_{k0}$, and $D_{n0}=\omega_n-\omega_0-2\omega_L+i\Gamma_{n0}$.

For parametric oscillation process, the intensities of the idler wave and the signal wave are given by

$$I_I=I_I^0\exp\left\{[-\alpha+g+\mathrm{Re}\sqrt{(\alpha+g-2i\Delta k)^2+B^2}]z/2\right\}, \tag{6}$$

$$I_S=I_S^0\exp\left\{[-\alpha+g+\mathrm{Re}\sqrt{(\alpha+g-2i\Delta k)^2+B^2}]z/2\right\} \tag{7}$$

with α absorption coefficient of the signal wave,

$$B^2=(8\pi/c)^2(\omega_I\omega_S/n_I n_S)\chi^{(3)}(\omega_I)\chi^{(3)*}(\omega_S)|E_L|^4, \tag{8}$$

$$\chi^{(3)}(\omega_I)=\frac{N(0|\mu|k)(k|\mu|n)(n|\mu|m)(m|\mu|0)}{\hbar^3(\omega_m-\omega_L)}\left[\frac{\rho_{oo}-\rho_{kk}}{D_{nk}^*D_{k0}}-\frac{\rho_{oo}-\rho_{nn}}{D_{nk}^*D_{n0}^*}\right] \tag{9}$$

and $\chi^{(3)}(\omega_S)$ similar to (4).

3. Experimental Setup

The experimental apparatus is similar to that described previously [6]. A pulsed dye laser pumped by a frequency-doubled Nd:YAG laser is used to reach $4d^2D_J$ level of sodium by two-photon resonance excitation. Mixed R590 and R610 dyes in methanol are used for producing laser wavelength 578.73nm. The laser beam is focused into the center of a heat-pipe oven through a lens (f=300mm). The dimensions of the stainless steel heat-pipe oven are 22mm in diameter and

460mm in length. Coherent radiation is detected by a SPEX
0.75m spectrometer followed by a EMI9656QB photomultiplier and a
boxcar. The wavelengths of the laser and the generated coherent
radiation are carefully calibrated by Hg, Na, and Ne spectral
lamps with precision of 0.01nm near the atomic transition lines.

4. Results and Discussions

We divide the coherent radiation lines into three groups according to the generation mechanism.

4.1 Coherent lines through 4D–3P–3S parametric oscillation

We report on lines obtained through parametric oscillation with 3P
as a near resonance level. Two-photon excitation produces a population inversion between 4D and 3P levels, which causes stimulated emissions at the same wavelengths as atomic transitions
4 $D_{3/2}$-3$P_{1/2}$, 4$D_{5/2}$-3$P_{3/2}$. (Energy level shift due to optical
Stark effect is small as the pumped laser energy is less than
1mJ/pulse). At the oven temperature 280-450°C besides the stimulated emissions we observed two coherent radiation lines with
a little broader width. Their wavelengths are shorter than those
of the corresponding stimulated emissions.   The wavelength
shifts from atomic transitions increase with increasing oven temperature as shown in Table 1. The parametric oscillation lines
near atomic 3P-3S transition are also shown in this table. The
wavelengths of the latter are longer than those of atomic transitions and the wavelength shifts increase with increasing oven
temperature too. As the laser wavelength is detuned from two-
photon resonance to the longer wavelength side by about 0.05nm,
the intensities of stimulated emissions decrease rapidly while
the intensities of parametric oscillation vary relatively slowly.
The wavelength shifts of parametric oscillation lines (to
shorter side for 4D-3P transition, and to longer side for 3P-3S
transition) increase as the laser wavelength increases.

Table 1. Temperature dependence of 4D–3P–3S parametric
         oscillation wavelength (with 4 Torr of Ar)

Temperature [°C]	Wavelength [nm]			
332	568.263	568.805	589.030	589.620
356	568.240	568.790	589.047	589.635
380	568.225	568.770	589.072	589.650
Atomic transition	568.263	568.820	588.995	589.592

We have also investigated the intensities versus the oven
temperature. As an example, the thermal dependence of parametric
oscillations near 3$P_{3/2}$-3$S_{1/2}$ transition is shown in Fig. 1.
There exists an optimum temperature which yields maximum output
intensities. This can be explained if we consider the parametric
oscillations as noncollinear phase-matched lines. By utilizing
Sellmeier equation and the oscillator strength of sodium atom
[7][8],

$$n-1 = \frac{Nr_e}{2} \sum_{ij} \frac{\rho_{ii} f_{ij}}{(\gamma_{ij}^2 - \gamma^2)} \tag{10}$$

with the classical electron radius $r_e = 2.818 \times 10^{-13}$ cm, ν being the energy in cm^{-1}, and fij the oscillator strength of the transition from level i to j; we have calculated phase-matched angles θ and found they are proportional to $\sqrt{N}$. Then we calculate I_s by substituting the intensity distribution of the laser beam

$$I_L = I_L^0 \exp(-\theta^2/\theta_0^2) \qquad (11)$$

in (7), where θ_0 is the beam divergence. The calculated intensity versus oven temperature curve is also shown in Fig. 1.

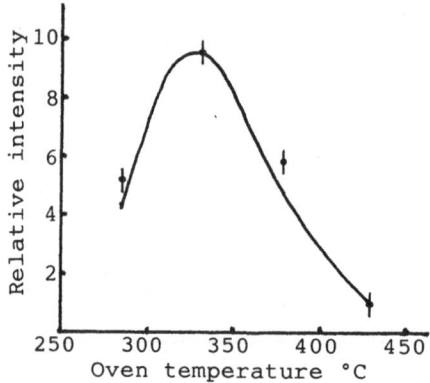

—— Theoretical
⌇ Experimental

Fig. 1. Intensity of 589nm parametric oscillation line versus oven temperature

When the oven temperature increases to 400°C or higher, the foregoing parametric oscillation decreases rapidly, at the same time another pair of parametric oscillation lines begins to appear. Their wavelengths are 568.446nm and 589.407nm. The main characteristic is that their wavelengths do not depend upon oven temperature, and their intensities increase as oven temperature increases. Therefore it is clear that these two lines are a pair of collinear phase-matched parametric oscillations. The calculated wavelengths are 568.445nm and 598.404nm which are in agreement with the experimental results (pumped laser wavenumber $\nu_L = 17274.2$ cm^{-1}, being 0.1cm^{-1} less than the resonance wavenumber). As the wavelength shift is as large as 0.18nm, they take place only if noncollinear phase-matched lines are in unfavourable condition due to their large phase-match angles at higher oven temperature.

4.2 Coherent lines through 4P-3S parametric oscillation

For parametric oscillation resonant enhanced by 4P level we have only observed UV coherent radiation near 4P-3S transition. Using a spectrometer of 0.01nm resolution and calibrating the output wavelengths by the same atomic transition 4P$_{3/2}$-3S$_{1/2}$ 330.237nm and 4P$_{1/2}$-3S$_{1/2}$ 330.298nm, we have studied the generated coherent radiation carefully. It is shown in Table 2 that the output coherent radiation has four components. Two of them correspond to collinear phase-match condition. Their wavelengths do not depend upon the oven temperature and their intensities

increase with increasing oven temperature. The other two components correspond to noncollinear phase-match condition. Their
wavelengths increase as oven temperature increases. From the
relation of the refractive index versus the wavenumber it can be
shown   theoretically that the coherent radiation line should
split into four components and that the wavelengths of noncol- •
linear phase-matched components shift to the longer side as
temperature increases[9].

Table 2. Temperature dependence of 4P-3S parametric
         oscillation wavelength (with 4 Torr of Ar)

Temperature [°C]	Wavelength [nm]			
284		330.237		330.298
428	330.229	330.249		330.309
452	330.228	330.252	330.298	330.314
476	330.227	330.260	330.299	330.320
Theoretical λ of collinear phase-matched lines	330.220		330.293	

4.3 Coherent lines through four-wave mixing processes

The observed coherent lines through four wave-mixing processes
are tabulated in Table 3. In order to obtain stronger signals
the pumped laser energy is increased to about 25mJ/pulse at
578.73nm with linewidth 0.02nm. The heat-pipe oven is heated to
temperature 405°C at sodium density $5.9 \times 10^{15} cm^{-3}$. We have observed five lines, 255.78, 257.51, 257.54, 280.47, and 280.51nm,
through processes of the type $\omega_{UV} = 2\omega_L + \omega_{IR}$. For difference-
frequency mixing processes, we have observed two lines, 298.83nm
and 298.87nm, through processes $\omega_{UV} = 2\omega_L - \omega(4P_J - 3D_{J'})$, as well as
two lines, 330.04nm and 333.11nm, through $\omega_{UV} = 2\omega_L - \omega(4P_J - 4S_{1/2})$.

Table 3. Wavelength, relative intensity, and coupling scheme
         of four-wave mixing lines.

λ [nm] Exptl.	log I	λ [nm] Theor.	coupling scheme	θ [mrad]	L_C [nm]
255.78	0.5	255.80	$2\omega_L + \omega(4P_{3/2} - 4S_{1/2})$		5.94
		255.84	$2\omega_L + \omega(4P_{1/2} - 4S_{1/2})$		5.94
257.51	0.5	257.45	$2\omega_L + \omega(4D_{3/2} - 4P_{1/2})$		5.94
257.54	0.5	257.49	$2\omega_L + \omega(4D_{5/2} - 4P_{3/2})$		5.94
280.47	1	280.44	$2\omega_L + \omega(4P_{3/2} - 3D_{5/2})$		5.93
280.51	1	280.48	$2\omega_L + \omega(4P_{1/2} - 3D_{3/2})$		5.93
298.83	1	298.82	$2\omega_L - \omega(4P_{1/2} - 3D_{3/2})$	1.27	
298.87	1.5	298.87	$2\omega_L - \omega(4P_{3/2} - 3D_{5/2})$	1.27	
333.04	2	333.00	$2\omega_L - \omega(4P_{1/2} - 4S_{1/2})$	2.73	
333.11	2.5	333.06	$2\omega_L - \omega(4P_{3/2} - 4S_{1/2})$	2.73	

Except the group of the shortest wavelength, observed lines of
all other groups are doublets, because cascade stimulated emission of either D-P or P-S transition is doublet. We can qualitatively explain the intensities of generated UV lines by the
gain coefficient of stimulated emission,nonlinear susceptibility,

and phase-match condition. For example, lines through sum-frequency mixing processes are weak in general because they can not satisfy the phase-match condition. We have calculated the coherence lengths

$$L_c = \pi / |\Delta k| \tag{12}$$

instead of phase-matched angles in Table 3.

This project has been supported by Science Fund of the Chinese Academy of Sciences.

References

1. W.Hartig: Appl. Phys. 15, 427 (1978)
2. C.Y.R.Wu and J.K.Chen: Opt. Commun. 50, 317 (1984)
3. A.V.Smith and J.F.Ward: IEEE J. Quantum Electron. 17, 525 (1981)
4. Y.R.Shen: The Principles of Nonlinear Optics (John Wiley and Sons, New York, 1984)
5. Yu.Malakyan: Sov. J. Quantum Electron. 15, 905 (1985)
6. P.-L.Zhang, Y.C.Wang, and A.L.Schawlow: J. Opt. Soc. Am. B1, 9 (1984)
7. R.B.Miles and S.E.Harris: IEEE J. Quantum Electron. 9, 470 (1973)
8. E.M.Anderson and V.A.Zilitis: Opt. Specktrosk. 16, 99 (1964)
9. P.-L.Zhang and A.L.Schawlow: Canadian J. Phys. 62, 1187 (1984)

On the Nature of Hochheim Alloy

G. W. Series

Clarendon Laboratory, Parks Road, Oxford, OX1 3PU, UK

It is not, I think, widely known that Art has access to sources of information that are denied to most of us. Since his earliest days in research - and possibly earlier than that, but I have no knowledge of anything earlier - the cerebrations of Art have transcended those of your ordinary mortal.

The story came from Fred Kelly who was a research student with Art at Toronto in the early nineteen fifties. (As this note was being revised the sad news came of Fred's death on 29 July 1986). Fred and Art were part of a team under the direction of M. F. Crawford applying to the study of hyperfine structure and isotope shifts in the spectra of magnesium the latest techniques in high resolution spectroscopy - emission from an atomic beam combined with Fabry-Perot interferometry. And because one of the lines they wished to study was in the u-v, the high reflectivity available by coating the interferometer plates with silver was useless. What should they use to maximize the finesse at 2,796 Å?

This was long before the days of multi-layer dielectric films. Remember, too, that one generally used interferometer plates of 6 or 7 cm diameter in those days, not the tiny little buttons you mount in laser beams nowadays. Silver was by far the best material to use for coating over most of the visible spectrum, though its absorption increases towards the blue. There the alternative was aluminum which, though definitely superior in the u-v is, even at its best, nothing like as good as is silver in the red.

But was that all the choice there was? Copper had been used as a mirror coating, and gold, and there was something else which turned up occasionally in the literature of the late nineteen-thirties: Hochheim alloy. You find it referred to in some of the articles and text-books on high resolution spectroscopy. It certainly deserved looking into - but there was a problem: what was Hochheim alloy, and what quantitative information was there about its alleged superiority to aluminum for use in the u-v? I quote from a well-known text-book: 'Unfortunately no data appear to be available concerning the numerical performance of the Hochheim alloy, in fact the nature of the alloy does not seem to be generally known. These reflecting alloys are prepared personally by Hochheim, apparently by an evaporation method. According to Murakawa (review of hyperfine structure, 1940, in Japanese) this alloy consists of aluminum and silver. The films made by Hochheim for work with Fabry-Perot interferometers are very thick and almost opaque in the visible region but behave excellently in the u-v. In the region between 2,000 Å and 4,000 Å they are much superior to aluminum and are capable of yielding fringes which are quite fine and indeed comparable with those given by silver in the longer wavelength regions.

A challenge indeed, and one which had not been overlooked in other laboratories. There was widespread activity in the nineteen-fifties on the best methods of depositing reflecting films for use in interferometry, and on their treatment after deposition, and - pending the re-discovery of Hochheim alloy (for Professor Hochheim was no longer on the scene after World War II) - one had to do the best one could with aluminum, for the u-v. In the Clarendon Heini Kuhn and his colleagues - Bradley, Wilson, Pery (now Thorne), and Burridge had carried out systematic studies and had come to the conclusion (for aluminum) that the key to obtaining the best films was to trap as little gas as possible in the film as is was being laid down. Thus, one aimed to have the best possible vacuum in the evaporation tank, especially while the pellet of aluminum was being evaporated, and to take the shortest possible time in securing a film of the desired density. If the pressure went up a bit, you had to speed up the evaporation to get a film of comparable quality.

At about this time Fred Kelly joined us in the Clarendon as a post-doc. He set about studying hyperfine structure in the resonance lines of gold. They, too, are in the u-v, and Fred brought to our research group in Oxford the wisdom of North America. Hochheim alloy: yes, of course we in Toronto were puzzled. Yes, indeed, we managed to sort it out; things like that happen when Art's on the scene. The literature - hopeless: Art has private ways and means. He came into the lab one morning and he said, 'What d'you think? - I know about Hochheim alloy.'

'You're crazy.'

'Maybe I'm crazy, but I met Professor Hochheim last night.'

'You're crazy.'

'Maybe I'm crazy, but it was him all right. It must have been. I spoke to him. I said, 'Professor Hochheim, I should like to ask you a question.'

And Professor Hochheim replied, 'Go ahead, my boy, what is it?'

'Professor Hochheim, what is Hochheim alloy?'

And Professor Hochheim replied, 'It's aluminum chum, it's aluminum - but put it on faaast !'

Part III

Solid State Spectroscopy

*Anything worth doing
is worth doing twice –
the first time quick and dirty,
and the second time the
best way you can*

Optical Spectral Linewidths in Solids

R.M. Macfarlane

IBM Almaden Research Center, 650 Harry Road,
San Jose, CA 95120, USA

1. INTRODUCTION

Physicists and chemists have long realized that the information about material systems which can be obtained by spectroscopic studies is strongly dependent on the resolution which can be achieved. This may be limited by the physical system itself or by the measurement process. A number of early papers studied sharp-line spectra in solids. One of the pioneers was Jean Becquerel [1], who worked extensively with rare earth materials in Paris in the early 1900's. He recognized that at low temperatures, spectral lines could become quite sharp, and in 1908 took his samples and spectrograph to Leiden and collaborated with Kammerlingh Onnes [2] on a measurement of the absorption spectra of rare earth ions in naturally occurring crystals such as tysonite (LaF_3) and xenotime (YPO_4) at the temperatures of liquid and solid hydrogen. The resulting spectra, recorded with a theoretical resolving power of $\sim 10^4$, were extremely rich but certainly contained many lines which were instrumentally broadened. In the 1930's, Otto Deutschbein working in Marburg [3] carried out extensive measurements of the spectra of chromium ions in natural crystals of many materials including ruby, spinel, alexandrite and tourmaline, as well as synthetic materials. He found that at liquid nitrogen temperatures, a number of spectral lines were ~ 3 cm^{-1} wide, but again, since he was working with a prism spectrograph with a linear dispersion of 50Å/mm, many of these lines were broadened by the measurement process. In much of the work on the line spectra of ions in solids, even a half century or more after the work of Becquerel and Deutschbein, the sharpest lines are still broadened by the resolution of the spectrometer, which even in very favorable cases is 2-3 GHz.

Although sharp spectral lines are more or less universal in gas phase systems, their occurrence in solids is much less widespread. At the second Quantum Electronics Conference in Berkeley in 1961 [4], Schawlow noted that "It seems possible that under some conditions, very sharp lines may be obtained (in solids), perhaps even rivalling atomic spectra in sharpness." This has certainly been borne out in recent years in solid-state rare-earth spectroscopy and to some extent also for transition metal ions. However as we will see below, the measurement of these narrow linewidths had to await the development of single-mode tunable lasers.

The study of the narrow-line spectra of solids containing Cr^{3+}, specifically the R-lines arising from the $^4A_2 \leftrightarrow {}^2E$ transition and the B-lines from the $^4A_2 \leftrightarrow {}^2T_2$

transition, received great stimulus from the invention of the laser by Schawlow and Townes [5] and the operation of the first laser using the R-lines of ruby by Maiman [6]. Schawlow and his collaborators in the 1960's contributed substantially to this field with elegant studies of the effects of external perturbations such as magnetic fields [7] and stress [8] and isotope shifts [4] in materials containing trivalent chromium, using the narrow Cr^{3+} lines as a probe of the local environment. The sensitivity of such probes depends strongly on their spectral linewidths. This focussed attention on ways of measuring these widths and the factors which controlled them. The advent of single frequency tunable cw dye lasers has had an enormous impact on our knowledge and understanding of the sources of optical line broadening in solids. Here, I will discuss some examples which illustrate this progress, without attempting a comprehensive review of the subject which is given elsewhere [9]. These examples use rare earth ion impurities in single crystals, but the techniques used, and the mechanisms for line broadening which have been demonstrated, are much more generally applicable. Art Schawlow was always fascinated by the occurrence of sharp spectral lines and the origin of their width, so it is fitting that the brain-child of Schawlow and Townes should have enabled this field to progress so rapidly.

Before proceeding to individual examples, a few general remarks on inhomogeneous and homogeneous broadening are in order. Inhomogeneous broadening, in solid-state spectroscopy, is the spread of resonance frequencies produced by the range of static micro-environments in which atoms, ions or molecules are found. It is inhomogeneous in the sense that different optical centers have different resonant frequencies. The sensitivity of an optical transition frequency of an ion in a crystal to strain fields tells us something about the likely magnitude of the inhomogeneous broadening [4]. The ability to specify the resonance frequency for a given center depends on its homogeneous width. This is the width exhibited by all ions (ideally independent of where their resonance falls in the inhomogeneous profile), due to dynamical perturbations from phonons or spin fluctuations or lifetime effects. The homogeneous linewidth (Γ_h) can be expressed in terms of the optical dephasing time T_2:

$$\Gamma_h = (1/\pi T_2) = (1/2\pi T_1) + \left(1/\pi T_2'\right)$$

where T_1 is the ("longitudinal") population decay time and T_2' is the ("transverse") pure dephasing term due essentially to frequency modulation of the optical transition. Figure 1 illustrates schematically the case where inhomogeneous broadening dominates homogeneous broadening. This is often the case for transitions to metastable levels at low temperatures. In this context, metastable levels are those with lifetimes of approximately 1 μsec-1 msec. These have relatively large energy gaps below them which reduce spontaneous phonon emission processes and the associated lifetime broadening (Fig. 2). For a group of levels such as illustrated in Fig. 2, optical transitions to the upper levels are usually homogeneously broadened by spontaneous phonon emission, and for transitions to the lowest one, inhomogeneous broadening dominates the small homogeneous part. This will be illustrated with an example below. It has been implied that this

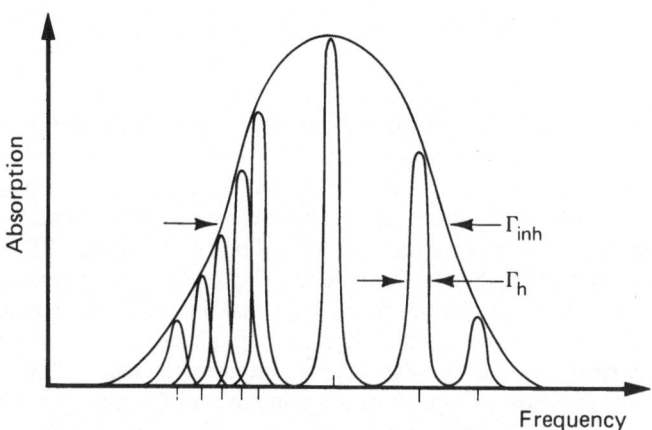

Figure 1. Schematic illustration of a spectral line profile in which inhomogeneous broadening (Γ_{inh}) dominates homogeneous broadening (Γ_h).

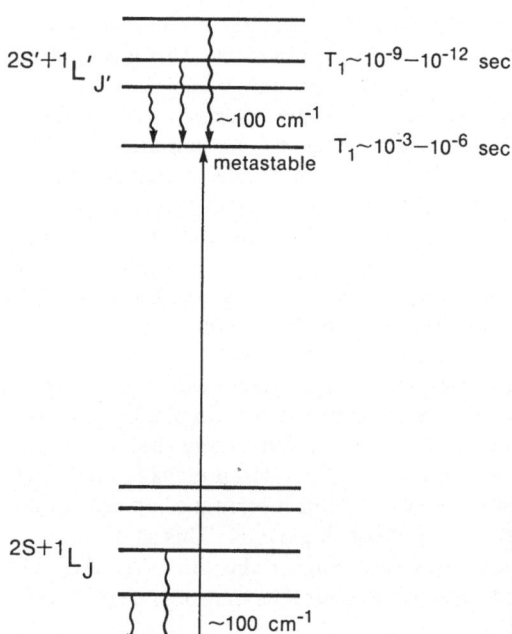

Figure 2. Schematic energy level diagram for trivalent rare earth ions in crystals showing ground and excited state J-manifolds split by the crystal field.

distinction between inhomogeneous and homogeneous broadening is clean and simple.   Unfortunately, this is not always the case.   For example, the "homogeneous" width may depend on the position in the inhomogeneous line, or what is considered homogeneous and what inhomogeneous may depend on the timescale of the measurement. Generally, however, the separation is a clear and useful one.

207

2. INHOMOGENEOUS BROADENING

There are several reasons to be interested in inhomogeneous broadening apart from finding ways to eliminate its effect on spectral resolution. In the first place, it can provide a measure of single crystal perfection and strains. It can also be used to monitor the damage produced by ion implantation, the strains at interfaces and in superlattices, or the reduction of strain by thermal annealing for example. There may even be geological applications, using naturally occurring probe ions in crystals of known geological origin. It has been proposed [10] that information can be stored in the frequency domain using a laser addressed optical memory in which permanent but reversible spectral holes are bleached in an inhomogeneous line of a material held at low temperatures. This frequency domain multiplexing has the potential to increase storage densities by the ratio Γ_{inh}/Γ_h which can be a factor of 10^4 or greater [11]. Finally, of course, an understanding of the origins of inhomogeneous broadening may make it possible to control it and significantly influence the optical properties of materials especially at low temperatures.

Very little is known about the microscopic origins of inhomogeneous broadening in solids which, in general, are exceedingly complex. This is in contrast to the case of gases where the distribution of atomic or ionic velocities determines inhomogeneous broadening, and a single parameter – the velocity – describes the position in the line. In solids, it is usually assumed that a Gaussian distribution of resonance frequencies results from the random strain fields due to dislocations and point defects. Many variables control the inhomogeneous width and it is generally assumed that Γ_{inh} is all that we need to specify, indeed is all that we can specify. In some cases, such a simple description fails. Figure 3, for example, shows three traces of the inhomogeneous profile of the $^7F_0 \leftrightarrow {}^5D_0$ transition of EuP_5O_{14} at 2K taken at different parts of the crystal separated by less than 1 mm. Not only are the profiles not Gaussian, but they vary with position in the crystal and also the spot size of the probe laser. These are macroscopic inhomogeneities. Non-Gaussian lineshapes also arise from ions in special environments, for example, distant pairs of ions that are spatially correlated and have an interaction energy that shifts their resonance by a small amount. Vial and Buisson [12] used an elegant method to separate this contribution: they measured an excitation spectrum of up-conversion fluorescence which occurs only for pairs and not for single ions. This showed, in the case of $LaF_3:Pr^{3+}$, considerable structure in the wing of the line. As the pairs become more distant, their resonance frequency approaches that of the single ion.

Non-Gaussian distributions are also expected when the number of ions being probed is very low, such as in small volumes at low concentrations. In this situation, fluctuation in the number of centers in different homogeneous packets would occur, giving structure in the inhomogeneous profile.

The lack of a simple parametric dependence of the optical transition frequency on strain makes the specification of a unique position in the inhomogeneous line difficult. There is no equivalent of the "zero-velocity packet" which can be determined by an analog of Lamb-dip saturation spectroscopy [13]. There are,

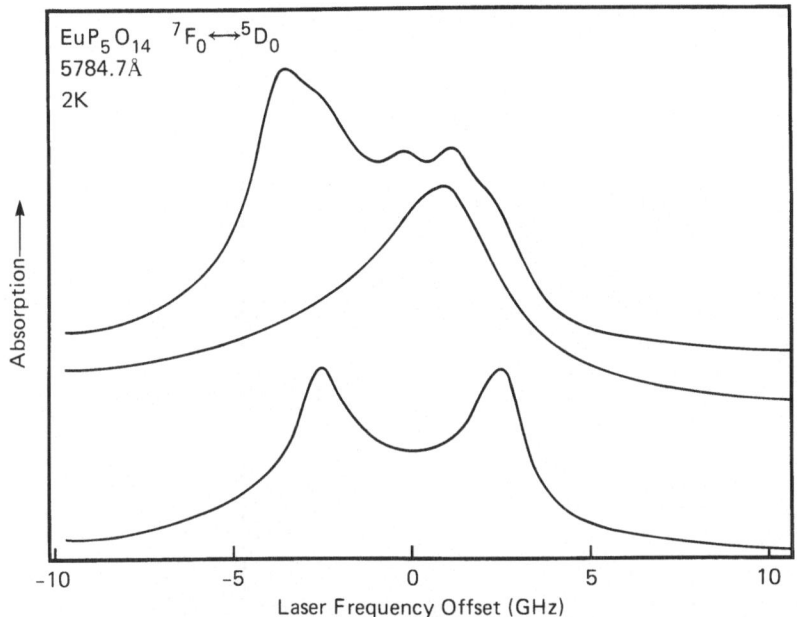

Figure 3. Inhomogeneous line profile for the $^7F_0 \leftrightarrow {}^5D_0$ transition of EuP_5O_4 measured at three different positions in the crystal showing the effect of macroscopic inhomogeneities.

however, some cases where another quantity, such as an rf-optical double resonance frequency, varies smoothly as a function of the optical resonance frequency [14] and could provide an additional piece of information to fix the optical transition frequency.

What are some of the factors which control inhomogeneous broadening? In addition to crystal defects, the introduction of the dopant ions themselves produces lattice strains, so that inhomogeneous linewidths can be quite sensitive to defect concentration. This is illustrated in Fig. 4 for Pr^{3+} in F^- compensated sites of C_{4v} symmetry in CaF_2. Three crystals were used, all with quite low concentrations of Pr^{3+} ($\approx 0.05\%$). The relative concentrations measured from absorption vary by more than a factor of 8. For the lowest doping, the hyperfine structure associated with the ^3H_4E level is clearly resolved and the inhomogeneous linewidth of the transition to the metastable $^1D_2$ level is only 700 MHz [15]. This, in itself, is quite remarkable since the ground state is a non-Kramers' doublet whose degeneracy could be lifted by strain, and this might be expected to produce large inhomogeneous broadening. The observation of this hyperfine structure also provides an example of the information which can be extracted simply by probing an inhomogeneous line profile with a narrow-band laser. This spectrum had been studied many times by conventional spectroscopy and only a single broad line observed. There are numerous cases of this. The high resolution laser can be thought of as a frequency domain "microscope" which enables new interactions and

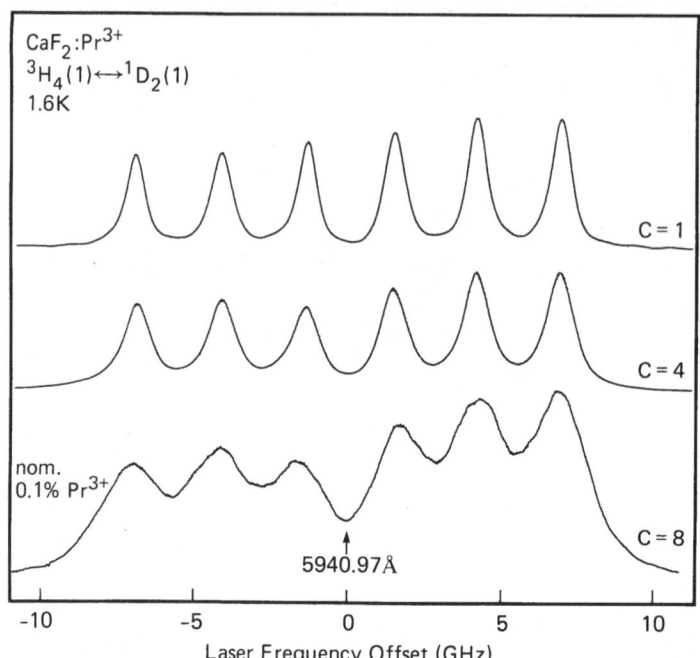

Figure 4. Resolved hyperfine structure on the absorption from the ground state to the lowest component of $^1D_2$ of $CaF_2:Pr^{3+}$ at 5941Å showing the concentration dependence of the inhomogeneous broadening. Relative concentrations are shown on the right and one nominal concentration, i.e., dopant added to the melt, is shown on the left. This is the F^- compensated C_{4v} site.

new dynamics to be discovered when its resolution is increased. Referring to Fig. 4, we see that for a higher doping of Pr^{3+}, significantly increased broadening is observed and at a nominal doping level of 0.1%, the hyperfine structure is barely resolved. This is rather qualitative but that is the nature of the subject at this time. The growth and annealing of the individual crystals (all supplied by Optovac Inc.) is believed to be rather uniform but may contribute somewhat to the variations observed. There are many examples known where there is a strong dependence of Γ_{inh} on the doping level. Very little has been written about the absolute magnitude of Γ_{inh}, however, which is appropriate because little is yet understood. At high doping levels, pair structure will become evident in the inhomogeneous profile leading both to a broadening and the appearance of complex structure [12,16]. Materials in which the optical centers are present in stoichiometric quantities can again exhibit narrow lines (see below) so as a function of concentration the inhomogeneous linewidth at first increases, and then decreases.

Two other factors controlling inhomogeneous broadening should be noted. The first is that different optical transitions may show very different strain sensitivity and, hence, inhomogeneous broadening. A study of the effect of external stress on

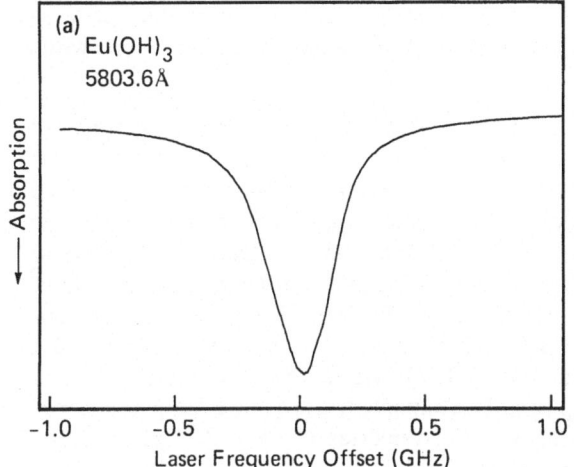

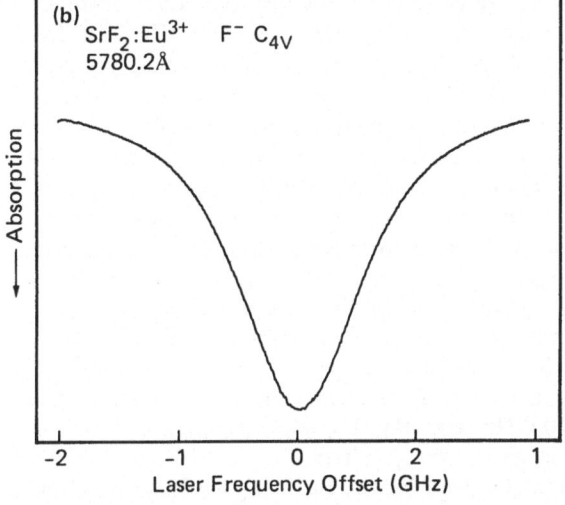

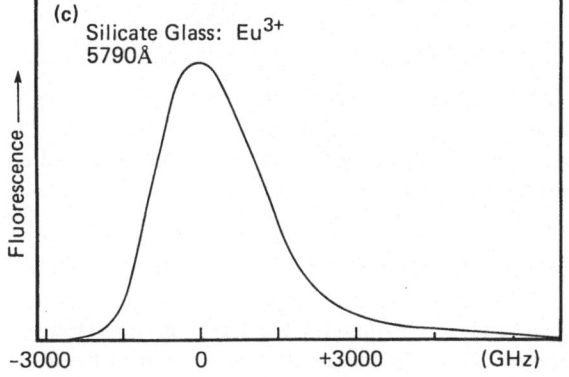

Figure 5. Inhomogeneous line profiles of the $^7F_0 \leftrightarrow {}^5D_0$ transition of Eu^{3+} in three materials.
(a) $Eu(OH)_3$ where the narrowest width exhibited by a single crystal is 170 MHz. For the sample shown here, Γ_{inh} is 260 MHz; (b) $SrF_2:Eu^{3+}$ showing a representative linewidth for a good quality doped single crystal of $\Gamma_{inh} = 1.2$ GHz; (c) a Eu^{3+} doped silicate glass showing the large inhomogeneous broadening typical of disordered materials. Here, $\Gamma_{inh} = 100$ cm^{-1} or 3000 GHz.

the optical transition frequencies of rare earth ions, similar to the studies made by Schawlow and co-workers on ruby [4] and $MgO:Cr^{3+}$ [8] would be very worthwhile in this context.

Another factor which strongly influences inhomogeneous broadening of defect absorption lines is the specific host material. For substitutional doping, the ease with which a dopant can be incorporated into the lattice without introducing local deformations of the structure, clearly influences Γ_{inh}. It also determines the distribution coefficient K, which is the ratio of dopant concentration in the crystal to that in the melt. For rare-earth and transition-metal doping, it is sometimes found that a wide range of solid solutions can be formed, e.g., $Pr_xLa_{1-x}F_3$ for $0 < x < 1$. In other cases, such as $YAG:Nd^{3+}$, K has values between 0.1 and 0:3 depending on the method of crystal growth [17]. An inverse correlation between the distribution coefficient and Γ_{inh} might be expected, and some examples support this. For example, for the lowest $^3H_4 \leftrightarrow {}^1D_2$ transition of $LaF_3:Pr^{3+}$ at 5925.2Å, the linewidth is ~5 GHz for $x = 5 \times 10^{-4}$, and in YAG, where K is small (~0.1), it is 50-60 GHz for the same concentration. This host dependence of Γ_{inh} is illustrated in Fig. 5 for the $^7F_0 \leftrightarrow {}^5D_0$ transition of some different Eu^{3+} systems. For this transition, the free-ion levels are nondegenerate so the effect of the crystalline environment is to shift the $^5D_0$-$^7F_0$ separation, which typically varies between $17200\,cm^{-1}$ and $17350\,cm^{-1}$ in different ionic solids. Inhomogeneous broadening results from the shift of this energy from site to site within a given material. Perhaps the most remarkable of the cases illustrated in Fig. 5 is that of $Eu(OH)_3$. For this material, the inhomogeneous linewidth is extremely small, varying between 170 MHz and 280 MHz in different samples [18]. This appears to be the narrowest inhomogeneous linewidth yet reported in a solid. It is not a doped system so questions of lattice matching of the Eu^{3+} ion do not arise. As we have seen in Fig. 4 for the case of EuP_5O_{14}, such narrow lines are not always found in stoichiometric europium compounds so it reflects a high degree of crystal perfection and the low strain of the hydrothermally grown $Eu(OH)_3$ crystals which were produced by Dr. Stanley Mroczkowski of Yale University. In doped single crystal systems, Γ_{inh} is typically 1-10 GHz (e.g., Fig. 5b) and in a silicate glass, for example, where disorder produces large inhomogeneities, Γ_{inh} is $\sim 100\,cm^{-1}$ [19]. Thus, the extreme range of linewidths observed for the Eu^{3+} $^7F_0 \leftrightarrow {}^5D_0$ transition in these different hosts is 2×10^4.

2.1 Inhomogeneous broadening and ground state hyperfine structure in $LaF_3:Ho^{3+}$

Before leaving the subject of inhomogeneous broadening, the example of $LaF_3:Ho^{3+}$ is introduced, and it will be further developed below. The optical spectrum associated with transitions from the $^5I_8$ ground state to the $^5F_5$ manifold very effectively illustrates the situation shown in Fig. 2.

Figure 6a shows fluorescence excitation traces made at 2K with a cw laser having a frequency width of ~1 MHz. The six absorption lines originate from transitions to the six lowest crystal field components of $^5F_5$ separated by Δ_n from the lowest one (Fig. 6b). Because of the C_2 site symmetry for Ho^{3+}, all electronic states are nondegenerate. The lowest level is metastable with a lifetime of

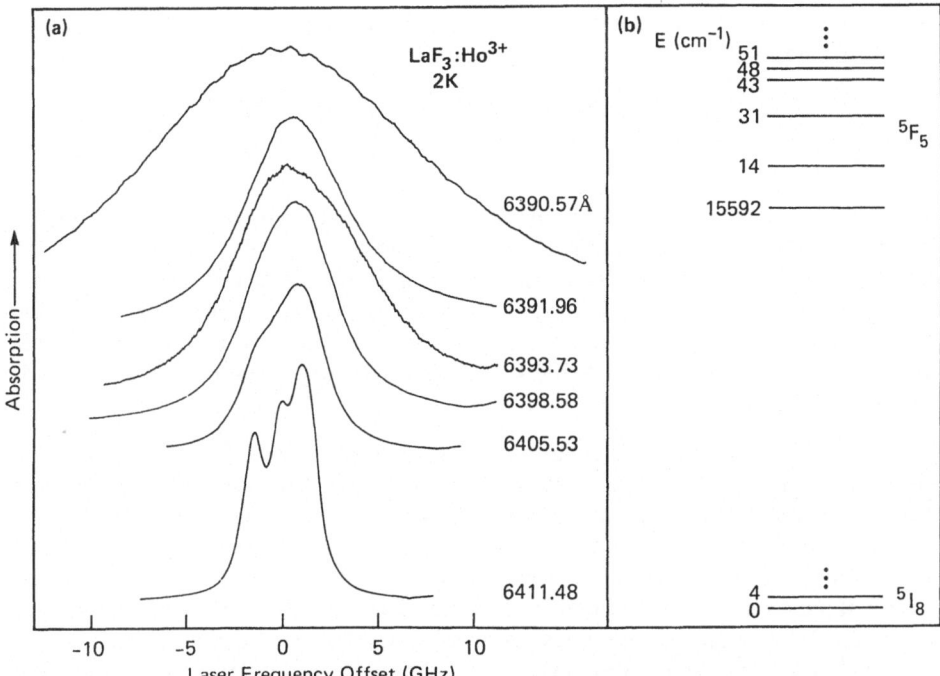

Figure 6. (a) Fluorescence excitation spectra of $LaF_3:Ho^{3+}$ at 1.6K from the ground state ($^5I_8$) to the six lowest crystal field components of $^5F_5$ at the wavelengths given. The lowest transition is inhomogeneously broadened, and for higher transitions, homogeneous broadening due to spontaneous phonon emission becomes increasingly important. Spectral resolution of the dye laser used in those scans is ~1 MHz. (b) The energy level scheme for the transitions shown on the left. At 1.6K only the ground state has appreciable population.

0.55 msec which contributes a negligible 290 Hz to the homogeneous linewidth, and at 2K, thermally induced broadening is absent. The linewidth of ~1 GHz is dominated by inhomogeneous broadening and shows structure which is assigned to the four components of the singlet ground state, coupled to the holmium nuclear spin $I = 7/2$. These hyperfine splittings are a combination of a pure quadrupole contribution and a pseudoquadrupole or second-order hyperfine contribution [20]. The latter dominates because of the close proximity of the neighboring electronic states (Fig. 6b). The pseudoquadrupole Hamiltonian has the form

$$\mathcal{H}_{pq} = D_{pq}\left[I_z^2 - I(I + 1)/3\right] + E_{pq}\left[I_x^2 - I_y^2\right]$$

with

$$D_{pq} = A_J^2[(\Lambda_{xx} + \Lambda_{yy})/2 - \Lambda_{zz}], \quad E_{pq} = A_J^2(\Lambda_{yy} - \Lambda_{xx})/2 \qquad (1)$$

where the Λ coefficients express the magnetic coupling between the singlet electronic states $|0>$, $|n>$ of the J-manifold separated by Δ_n, with

$$\Lambda_{\alpha\beta} = \sum_{n=1}^{2J+1} \frac{<0|J_\alpha|n><n|J_\beta|0>}{\Delta_n} . \tag{2}$$

For the ground state, Δ_1 is only 4 cm^{-1}. The matrix elements of J can, in principle, be obtained from the nonlinear Zeeman effect, but this has not yet been done. They will, however, be large because of the high angular momentum of the ground state. The pseudoquadrupole splittings are therefore extremely large ($\sim$1 GHz, see Table 1) and this appears to be the only case where they can be resolved outside the inhomogeneous width. In Table 1, the three splittings are labelled δ_1, δ_2 and δ_3 in order of increasing energy.

For the transition to the next level of the $^5F_5$ manifold at 6405.5A, the resolution of the hyperfine structure is reduced because of homogeneous broadening due to spontaneous emission of 14.5 cm^{-1} phonons. This emission rate is proportional to the phonon density of states at the frequency differences Δ_n, and also to the coupling strength between pairs of electronic levels and the phonons. This coupling may be significantly different for acoustic and optic modes for example. As Δ_n increases, the lines get broader and there is a dramatic increase for the level at 51 cm^{-1} where the lowest optic phonon mode contributes a peak in the density of phonon states. The linewidth of 15 GHz corresponds to a lifetime due to spontaneous phonon emission of 10 psec. Spectral holeburning, or time resolved fluorescence measurements [21] are required to obtain more precise values of the relaxation rates for each level. For values of Δ_n greater than the highest frequency lattice phonons, two-phonon processes are required and, typically, lines become narrow again. This behavior is seen quite generally in solid-state systems and, in addition to providing important information on relaxation processes, it can also be a useful probe of the effective phonon density of states. Figure 6 illustrates the transition from inhomogeneous broadening to homogeneous broadening with increasing excitation energy above the metastable level.

3. HOMOGENEOUS BROADENING

When homogeneous broadening dominates the total linewidth, its measurement presents little problem for conventional spectroscopy. This is usually the case at high temperatures ($\lesssim$50K) where phonon absorption and scattering processes [22,23] are responsible for Γ_h. Figure 7 illustrates schematically the temperature dependence of the linewidth observed for metastable levels. On cooling from room temperature, the linewidth narrows by one or two orders of magnitude and becomes inhomogeneous at the lowest temperatures. This is the regime of interest to us here, i.e., where $\Gamma_{inh} >> \Gamma_h$. A number of techniques of laser spectroscopy have been developed to measure Γ_h in the presence of inhomogeneous broadening which is often many orders of magnitude larger. As indicated in Fig. 7, this makes it possible to study new mechanisms for dephasing such as coupling to nuclear-spin fluctuations. A catalog of these techniques includes fluorescence line narrowing

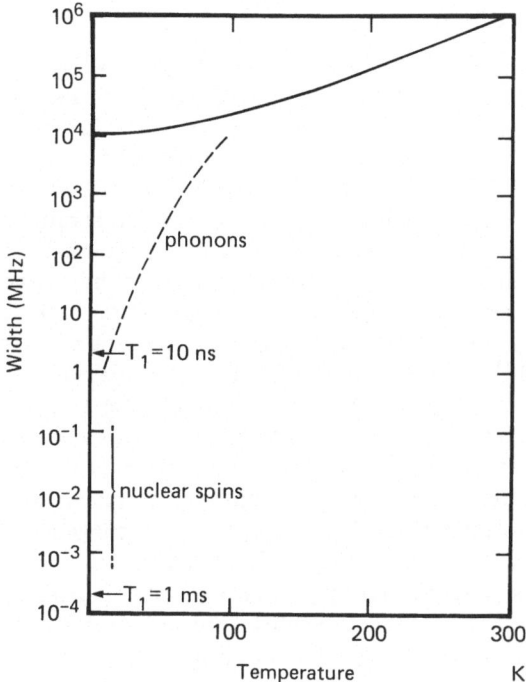

Figure 7. Schematic diagram of the temperature dependence of the homogeneous linewidth of a rare earth impurity ion. At the lowest temperatures, the measured width (solid curve) is dominated by inhomogeneous broadening. By using a variety of techniques of nonlinear laser spectroscopy, the inhomogeneous contribution can be eliminated and this enables the homogeneous width to be followed over many decades to the limit set by the population decay time T_1. For sufficiently long T_1, nuclear-spin fluctuations determine the homogeneous linewidth typically in the range of 1 kHz to 1 MHz.

[24] and spectral holeburning [25,26] in the frequency domain and photon echoes [27], optical free induction decay [28] and optical phase switched transients [29] in the time domain. For a detailed exposition of these techniques, see the original references, or the review by Macfarlane and Shelby [9].

On the subject of eliminating the effects of inhomogeneous broadening, it is interesting to quote another section from Schawlow's 1961 paper at the second Quantum Electronics Conference [4], i.e., "For example, if the lines are inhomogeneously broadened due to strain, then a sufficiently strong source can saturate the absorption of a part of the line. This technique of 'eating a hole in the line' is widely used with microwaves but could not be applied optically before the existence of maser sources." This form of holeburning was observed 14 years later by Szabo in ruby [25]. Since that time, with the availability of tunable lasers, spectral holeburning activity has mushroomed and now includes many mechanisms other than the two-level saturation envisaged by Schawlow, for example, optical pumping of hyperfine, superhyperfine or Zeeman split electronic levels and selective photochemistry of many kinds which can lead to essentially permanent holes. Recent reviews have been given by Macfarlane and Shelby [9,26]

Before the application of these techniques of laser spectroscopy, very little was known about the magnitude of Γ_h in the low temperature regime (<4K) or of the mechanisms responsible for it.

3.1 Homogeneous broadening in LaF$_3$:Pr^{3+}

A large amount of work was devoted to the study of the $^3H_4(Z_1) \rightarrow {}^1D_2(D_1)$ transition of LaF$_3$:Pr^{3+}. Out of this, a much greater understanding of the low temperature dephasing mechanisms evolved. This, and related work, has been reviewed by Macfarlane and Shelby [9], but some important results will be mentioned here. It was shown that fast, time domain, photon echo [30] and optical free induction decay [31] techniques could reliably measure Γ_h at low temperatures and the values were lower than at first expected, i.e., 56 kHz at zero magnetic field and 12 kHz in a field of 80G. Spectral holeburning by optical pumping of hyperfine levels is a much longer timescale experiment ($\sim$sec) which puts more serious demands on laser frequency stability for measurements requiring resolution $\sim$10's of kHz. In addition, it was shown [32] that there are slowly varying local fields due to spin flips on ^{19}F nuclei which are close to the Pr^{3+} ion and hence strongly perturbed. These can broaden the hole. The ^{19}F spin flips which contribute on a faster timescale to the photon echo and FID decays are further from the praseodymium ion and outside the "frozen core" of perturbed fluorine nuclei. The role of nuclear-spin flips as the dominant source of low temperature homogeneous broadening was clearly demonstrated by nuclear-spin decoupling experiments [31,33]. Here, the ^{19}F nuclear spins were coherently driven by strong rf fields to average out their local field fluctuations, or their flip rates were slowed down by "magic-angle" decoupling. It is now believed that nuclear-spin coupling is a universally important source of optical coherence loss in solids at temperatures where the phonon contributions have been frozen out. Figure 8 summarizes the progressive steps used in determining the homogeneous linewidth of the 5925Å transition of LaF$_3$:Pr^{3+}. It can be thought of as an example of the frequency domain "microscope" concept. By cooling the sample to 1.6K, homogeneous broadening due to phonons was eliminated and the line profile became dominated by inhomogeneous broadening. Spectral holeburning yielded hyperfine structure [34] but the hole width was limited by laser frequency jitter. Time domain techniques were therefore used, and photon echo measurements gave the homogeneous linewidth of 54 kHz which was narrowed to 4 kHz by decoupling the ^{19}F nuclear spins and essentially removing their contribution from the linewidth. This example demonstrates many of the factors which control coherence loss in solids at temperatures where the phonon contributions have been frozen out.

3.2 Homogeneous broadening and excited state hyperfine structure in LaF$_3$:Ho^{3+}

We return now to the illustrative example of LaF$_3$:Ho^{3+}. A spectral holeburning experiment was carried out on the 6411.5Å line using two single frequency cw dye lasers – one to "saturate" the absorption and the other to probe the resulting hole. The result is shown in Fig. 9. In addition to a hole at the pump laser frequency, a pattern of side holes was observed which measures the excited state hyperfine splittings. Again, these are due to second-order interactions, but in this case, the splittings are smaller than in the ground state. This is mainly because the separation between the two closest lying excited states is now 14.5 cm^{-1}, almost four times as large as in the ground state, but matrix elements of J are also smaller.

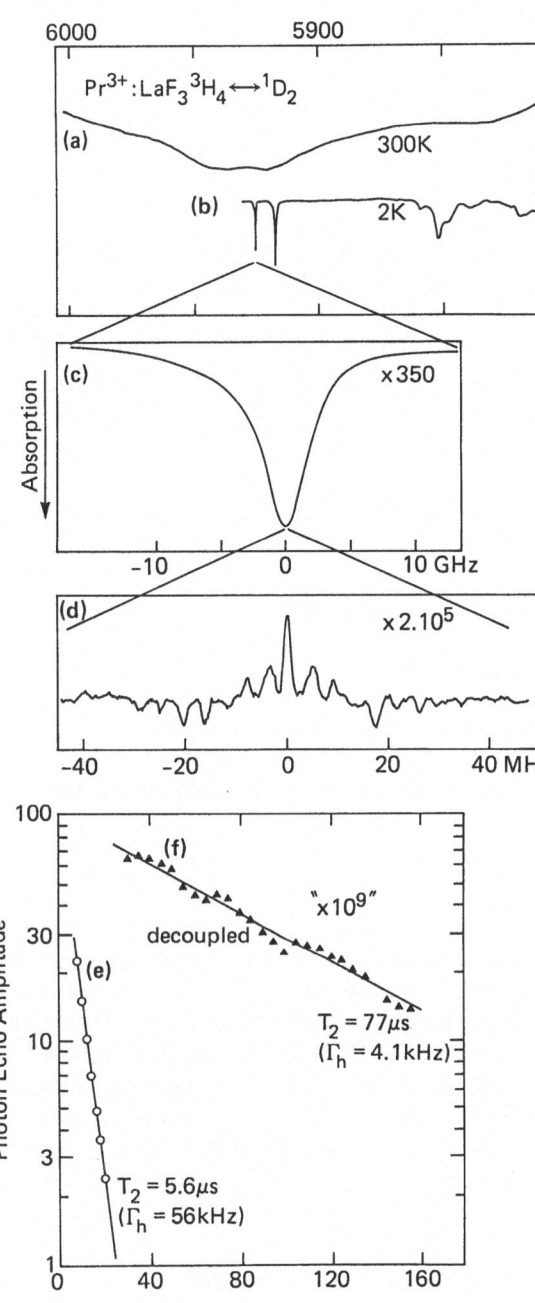

Figure 8. Progressive steps in determining the homogeneous linewidth of the $^3H_4(Z_1)$ to $^1D_2(D_1)$ transition of $LaF_3:Pr^{3+}$ at 5925.0Å. (a) The 300K spectrum broadened by phonon interactions. (b) At 2K, the linewidth of 6 GHz is limited by static inhomogeneous strain as shown in (c). In (d), this broadening is largely eliminated by holeburning, showing hyperfine structure. The hole width is limited by laser frequency jitter. Using photon echoes, the true homogeneous width $\Gamma_h = 56$ kHz is revealed, and this can be further reduced to 4.1 kHz by decoupling the ^{19}F nuclear spins, thus showing that they dominate Γ_h at very low temperatures. The magnification factors refer to the effective increase in spectral resolution.

217

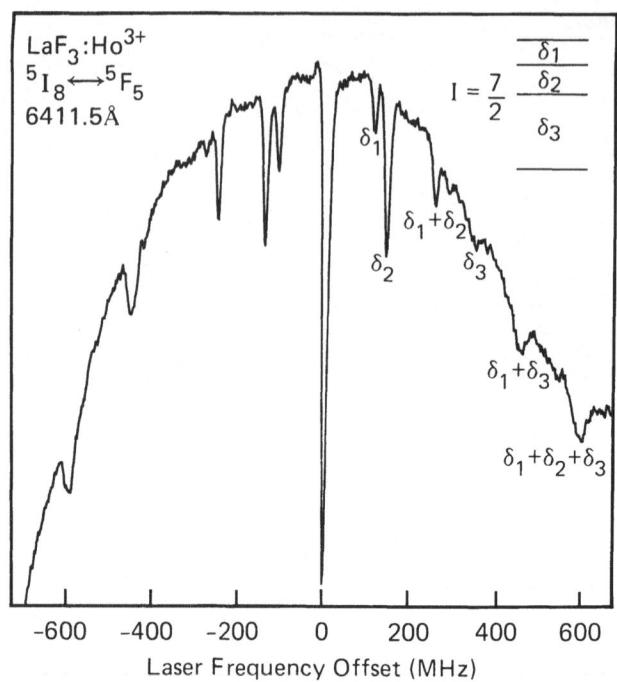

Figure 9. Holeburning spectrum measured in the 6411.5Å transition of LaF$_3$:Ho^{3+} using two cw dye lasers of resolution ~1 MHz. The side holes are due to excited state hyperfine structure. It is thought that holeburning occurs due to population storage in the ^{5}I$_7$ level.

The mechanism for holeburning in this case appears to be population storage in the metastable ^{5}I$_7$ level which has a lifetime of 10's of msec. This allows depletion of the ground state population for those ions whose environment puts them into resonance with the laser. From the fluorescence excitation spectrum and the holeburning, we have obtained approximate but reasonably good values for ground and excited state hyperfine splittings (see Table 1). The pseudoquadrupole splittings are very approximately in the axial ratio of 1:2:3, which is consistent with a dominant contribution coming from hyperfine interactions with a single low-lying level.

Table 1. Pseudoquadrupole Splittings in LaF$_3$:Ho^{3+}

	δ_1(MHz)	δ_2(MHz)	δ_3(MHz)
^{5}I$_8$(Z$_1$)	400	800	1500
^{5}F$_5$(D$_1$)	110	150	350

If we expand the frequency scale and look at the width of the central hole, we find that it is $W = 18$ MHz (Fig. 10). This is broader than the 2-3 MHz contribution from laser frequency jitter, but does it give the homogeneous linewidth (i.e., $\Gamma_h = 9$ MHz)? To further investigate this question, we measured the optical dephasing time using optical-free induction decay (Fig. 11) and found that $T_2 = 380$ nsec [9], i.e., Γ_h from this measurement is 0.83 MHz. The resolution

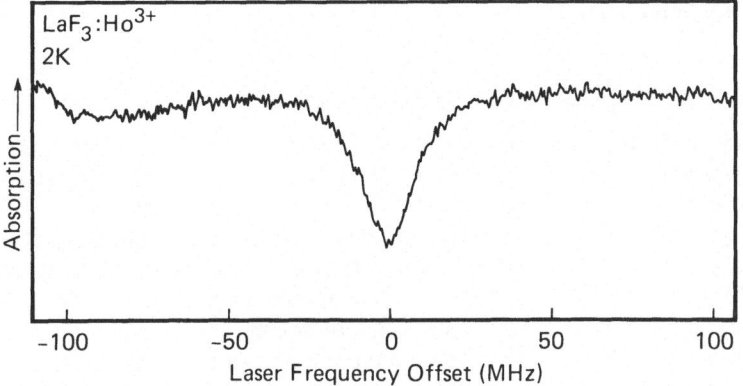

Figure 10. Expanded trace of the central hole of Fig. 7 showing a hole width, $W = 18$ MHz.

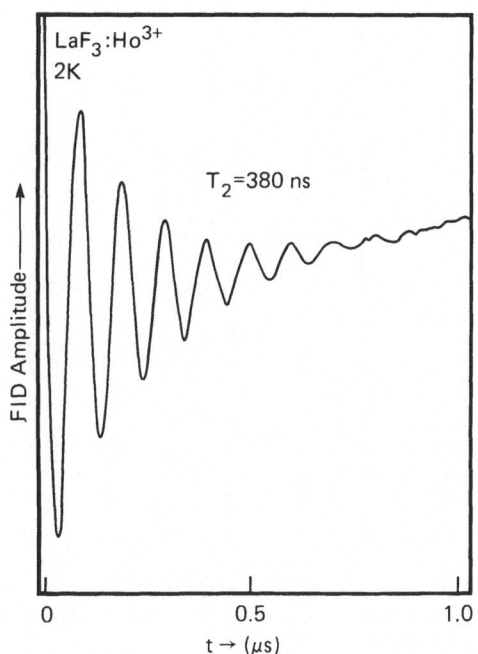

Figure 11. Optical free induction decay measurement of the homogeneous linewidth of the $^5I_8 \leftrightarrow {}^5F_5$ transition at 6411.5Å. From the dephasing time of 380 nsec, $\Gamma_h = 0.83$ MHz. The measurement shown here was made in collaboration with Dr. Richard Meltzer.

of this dilemma, as discussed already, for $LaF_3:Pr^{3+}$, almost certainly arises from the timescale on which the two measurements were carried out. For the FID, it is ~1 μsec, and for holeburning, it is at least as long as the storage time of $^5I_7$ which is a factor of 10^4 longer. The homogeneous width and the hole width are broader here than was the case for $LaF_3:Pr^{3+}$ because of the much larger nuclear magnetic moment of the Ho^{3+} hyperfine levels. The second-order hyperfine interaction which gave rise to such large pseudoquadrupole splittings also induces an enhanced nuclear moment [35], which for the ground state is ~1 MHz/G or almost as large as a typical electronic moment.

3.3 $Y_2O_3:Eu^{3+}$ – The narrowest optical linewidth in a solid

We have already seen that the $^7F_0 \leftrightarrow {}^5D_0$ transition of Eu^{3+} provides the narrowest inhomogeneous linewidth in a solid (170 MHz) due in part to the relative insensitivity of the optical transition frequency to crystal strains. This transition is also noteworthy in that the ground and excited states have no electronic magnetic moment, and the ground state nuclear moment is typically quenched by hyperfine coupling to $^7F_1$ [36]. For this reason, homogeneous broadening due to nuclear-spin fluctuations is generally very small. This depends, of course, on the nature of the host material. In Y_2O_3, only the yttrium ions have nuclear spin and the magnetic moment of yttrium is small (–0.137 nuclear magnetons), contributing approximately 200 Hz to the optical linewidth. The homogeneous linewidth of the

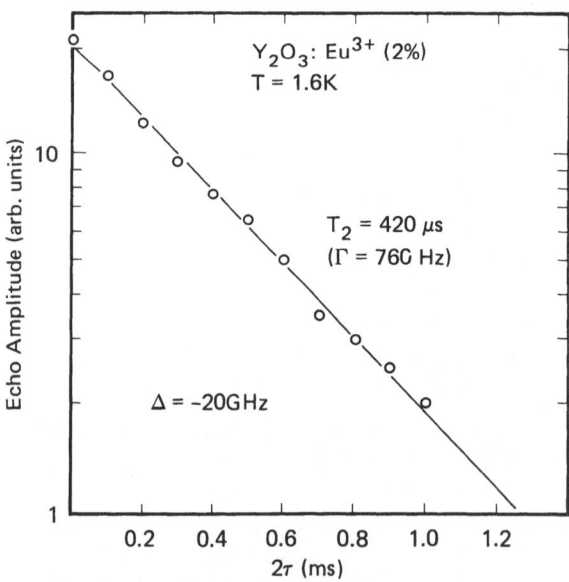

Figure 12. Decay of the photon echo amplitude measured for the $^7F_0 \leftrightarrow {}^5D_0$ transition of $Y_2O_3:Eu^{3+}$ at 1.6K. The homogeneous linewidth measured in this experiment has the very small value of $\Gamma_h = 760$ Hz.

$^7F_0 \leftrightarrow {}^5D_0$ transition of $Y_2O_3:Eu^{3+}$ was measured by Macfarlane and Shelby [37] from the photon echo decay. They found a concentration dependent linewidth, and also a dependence of Γ_h on position in the inhomogeneous line. Because the inhomogeneous line profile is essentially a plot of concentration versus frequency, a lower effective concentration of Eu^{3+} ions can be found in the wing of the line. A width of 760 Hz was found at 2K, in a 2% $Y_2O_3:Eu^{2+}$ sample at a frequency 20 GHz off the peak of the inhomogeneous line (Fig. 12). This shows that very narrow homogeneously broadened resonances can be observed in solids at low temperatures, and many more examples await study.

4. CONCLUSION

As techniques for measuring very narrow linewidths become refined and more understanding of the sources of optical homogeneous broadening is obtained, it is natural to ask how narrow these linewidths can become and how to measure them. The simple answer to the ultimate limit on the linewidth is just that it is $(1/2\pi T_1)$, where T_1 is the population decay time. For metastable levels, this may only contribute several Hz to hundreds of Hz. Eliminating the effect of nuclear-spin coupling is clearly necessary, and as linewidths become narrower, interaction between the optical centers themselves may lead to spectral diffusion and line broadening. These ideas have been illustrated principally by the examples of $LaF_3:Pr^{3+}$, $LaF_3:Ho^{3+}$ and $Y_2O_3:Eu^{3+}$. The case of the trivalent europium ion has the distinction of providing both the narrowest inhomogeneous linewidth (170 MHz in $Eu(OH)_3$) and the narrowest homogeneous linewidth (760 Hz in $Y_2O_3:Eu^{3+}$) in solid-state spectroscopy. There is every reason to believe that narrower lines will be observed, providing more and more detailed probes of the nature of crystalline materials and of the complex magnetic interactions between nuclear and electron spins.

When Art Schawlow presented his paper at the Quantum Electronics Conference 26 years ago, and projected that the "optical maser" would enable us to measure narrow spectral lines in solids, did he envisage just how far this field would develop? Probably he did.

ACKNOWLEDGEMENTS

I wish to acknowledge the invaluable contributions of my collaborators in previously published work, particularly Robert Shelby, and also Art Schawlow for his encouragement and guidance during the "impressionable years."

REFERENCES

1. J. Becquerel: Compt. Rend. Acad. Sci. (Paris) 142, 775 (1906); ibid. 145, 413 (1907); ibid. 145, 1150 (1907); Le Radium 4, 49 (1907); ibid. 4, 328 (1907); ibid. 5, 5 (1908).
2. J. Becquerel and H. Kammerlingh Onnes: Le Radium 5, 227 (1908).
3. O. Deutschbein: Ann. Phys. 14, 712 (1932).

4. A.L. Schawlow: In Advances in Quantum Electronics, ed. by J.R. Singer (Columbia University Press 1961) p.50.
5. A.L. Schawlow and C.H. Townes: Phys. Rev. 112, 1940 (1958).
6. T.H. Maiman: Nature 187, 493 (1960).
7. S. Sugano, A.L. Schawlow and F. Varsanyi: Phys. Rev. 120, 2045 (1960).
8. A.L. Schawlow, A.H. Piksis and S. Sugano: Phys. Rev. 122, 1469 (1961).
9. R.M. Macfarlane and R.M. Shelby: In "Spectroscopy of Crystals Containing Rare Earth Ions," Modern Problems in Condensed Matter Sciences, ed. by A.A. Kaplyanskii and R.M. Macfarlane (North-Holland, Amsterdam 1987) in press.
10. G. Castro, D. Haarer, R.M. Macfarlane and H.P. Trommsdorff, "Frequency Selective Optical Data Storage System," U.S. Patent No. 4,101,976 (1978).
11. W.E. Moerner, W. Lenth and G.C. Bjorklund: In "Persistent Spectral Holeburning: Science and Applications," Topics in Current Physics, ed. by W.E. Moerner (Springer-Verlag, Heidelberg 1987) in press.
12. J.C. Vial and R. Buisson: J. Physique Lett. 43, L339 (1982).
13. L.S. Vasilenko, V.P. Chebotayev, A.V. Shishaev: JETP Lett. 12, 113 (1970); B. Cagnac, G. Grynberg and F. Biraben: Phys. Rev. Lett. 32, 643 (1974); M.D. Levenson and N. Bloembergen: Phys. Rev. Lett. 32, 645 (1974).
14. L.E. Erickson and K.K. Sharma: Phys. Rev. B24, 3697 (1981).
15. R.M. Macfarlane, R.M. Shelby and D.P. Burum: Opt. Lett. 6, 593 (1981).
16. P. Kisliuk, N.C. Chang, P.L. Scott and M.H.L. Pryce: Phys. Rev. 184, 367 (1969).
17. R.F. Belt, R.C. Putthach and D.A. Lepore; L. Pryce; J. Cryst. Growth 13/14, 268 (1972); A.G. Petrosyan, Kh.S. Bagdasarov, T.I. Butaeva, A.M. Kevorkov and A.A. Shakhnazaryan: Kristallogr. 20, 1089 (1975) [Sov. Phys. Crystallog. 20, 665 (1975).
18. M.S. Otteson, R.L. Cone, R.M. Macfarlane and R.M. Shelby: J. Opt. Soc. Am. 73, P1391 (1983).
19. C. Brecher and L.A. Riseberg: Phys. Rev. B 13, 81 (1976).
20. J.M. Baker and B. Bleaney: Proc. Roy. Soc. (Lond.) A 245, 156 (1958).
21. R. Bayerer, W. Schneider, J. Heber and D. Mateika: Z. Phys. B 64, 195 (1986).
22. D.E. McCumber and M.D. Sturge: J. Appl. Phys. 34, 1682 (1963).
23. W.M. Yen, W.C. Scott and A.L. Schawlow: Phys. Rev. 136, A271 (1964).
24. Yu. V. Denisov and V.A. Kizel: Opt. i Spektr. 23, 472 (1967) [Opt. Spectr. 23, 251 (1967)]; A. Szabo: Phys. Rev. Lett. 25, 924 (1970).
25. A. Szabo: Phys. Rev. B11, 4512 (1975).
26. R.M. Macfarlane and R.M. Shelby: In "Persistent Spectral Holeburning: Science and Applications" Topics in Current Physics, W.E. Moerner (Springer-Verlag, Heidelberg), in press.
27. N.A. Kurnit, I.D. Abella and S.R. Hartmann: Phys. Rev. Lett. 13, 567 (1964).
28. R.G. Brewer and R.L. Shoemaker: Phys. Rev. A 6, 2001 (1972).
29. A.Z. Genack, D.A. Weitz, R.M. Macfarlane, R.M. Shelby and A. Schenzle: Phys. Rev. Lett. 45, 438 (1980).
30. R.M. Macfarlane, R.M. Shelby and R.L. Shoemaker: Phys. Rev. Lett. 43, 1726 (1979).

31. S.C. Rand, A. Wokaun, R.G. DeVoe and R.G. Brewer: Phys. Rev. Lett. 43, 1868 (1979).
32. R.M. Shelby, C.S. Yannoni and R.M. Macfarlane: Phys. Rev. Lett. 41, 1739 (1978).
33. R.M. Macfarlane, C.S. Yannoni and R.M. Shelby: Opt. Commun. 32, 101 (1980).
34. L.E. Erickson: Phys. Rev. B16, 4731 (1977).
35. B. Bleaney: Physica 69, 317 (1973).
36. R.J. Elliott: Proc. Phys. Soc. (Lond.) Sec. B 70, 119 (1957); R.M. Shelby and R.M. Macfarlane: Phys. Rev. Lett. 47, 1172 (1981); K.K. Sharma and L.E. Erickson: Phys. Rev. Lett. B23, 69 (1981); R.M. Macfarlane and R.M. Shelby: Opt. Commun. 39, 169 (1981).
37. R.M. Macfarlane and R.M. Shelby: Opt. Commun. 39, 169 (1981).

Spectroscopy of Solid-State Laser Materials

S. Sugano

The Institute for Solid State Physics, The University of Tokyo,
Roppongi, Tokyo 106, Japan

1. Before the Dawn of Laser History

After finishing the work "On the Absorption Spectra of Complex Ions"/1/,
Y. Tanabe and myself were looking for clear-cut experimental evidence to
justify our energy level diagrams for d^N (N=2, 3, $\cdots$, 8) electron configurations in a cubic field. As is well known, the diagram predicts the
co-existence of gaslike narrow lines and broad bands in the absorption spectra of transition-metal ions in cubic environments. We thought that detailed studies of the gaslike lines could justify the diagram as they could
provide us with much information such as fine structure, Zeeman effects,
and so on. We noticed that the gaslike lines were first reported in 1893
by Lapraik/2/ and that many detailed spectroscopic studies had been done on
the narrow lines arising from the $t_{2g}^3\,{}^4A_{2g} \rightarrow t_{2g}^3\,{}^2E_g$, ${}^2T_{1g}$, and ${}^2T_{2g}$
transitions of Cr^{3+} ions in complex salts/3/ and oxide crystals such as
spinel, alexandrite, ruby etc../4/

Shortly after examining these experimental data, we arrived at the conclusion that detailed studies of the ruby spectrum would be the most appropriate for our purpose. The reasons are as follows; (1) The crystal
structure is simple with uniaxial trigonal symmetry, and the crystal is
stable undergoing no phase transition at low temperatures as in the case of
complex salts. (2) The spectral intensity is relatively high because of
the absence of inversion symmetry at the Cr^{3+} site, and the spectral widths
of the gaslike lines are so narrow at low temperatures that Zeeman experiments may be performed with a readily available magnetic field.

Soon we finished the calculation of the fine structure of the R, R', and
B absorption lines arising, respectively, from the $t_{2g}^3\,{}^4A_{2g} \rightarrow t_{2g}^3\,{}^2E_g$,
${}^2T_{1g}$, and ${}^2T_{2g}$ transitions by using the even-parity trigonal field and the
spin orbit interaction. We also calculated the optical anisotropy, or the
optical polarization, of the absorption lines and the broad U and Y absorp-
tion bands arising, respectively, from the $t_{2g}^3\,{}^4A_{2g} \rightarrow t_{2g}^2 e\,{}^4T_{2g}$ and ${}^4T_{1g}$
by using the odd-parity trigonal field and the spin-orbit interaction. We
created symbols U and Y for the broad bands to form RUBY in the order of
increasing wavenumbers, where symbols R and B had already been used for the
narrow lines. The results seemed to explain all the observed features of
the spectrum in the absence of a magnetic field./5/

When we proceeded to the study of Zeeman patterns of the absorption
lines, however, we encountered a serious difficulty. We could not find any
Zeeman splitting of g $\sim$2 of the ground state, which had been observed by the
paramagnetic resonance/6/, in the Zeeman patterns of the R_1 and R_2 absorption lines observed by H. Lehmann/7/ with a magnetic field (27,000 Gauss)

parallel to the trigonal axis and the polarization perpendicular to it:
the Zeeman patterns consisted of three split components and the apparent
g-values were reported to be ∿1.47. As far as I remember, we had no difficulty in explaining the other features of the Zeeman patterns. The difficulty just described, however, was so serious that we were eager to find
some experimentalist who could repeat the Zeeman experiment to examine the
extraordinarily small g-value reported by Lehmann.

Meanwhile we met I. Tsujikawa, who had just finished Zeeman experiments
on chrome alums in collaboration with Madame L. Couture at Bellevue and was
constructing a new grating spectrograph in Tohoku University. Examining,
with Tsujikawa, very carefully the propriety of performing the Zeeman
experiment of ruby by using his new experimental facility, we finally
decided that Tsujikawa and myself should start the Zeeman experiment of
ruby as soon as his spectrograph was completed. At that time I was a research associate in an experimental laboratory of optics.

Tsujikawa's spectrograph in Eagle mounting with dispersion 2.5 A/mm in
the first order was located on the premier étage and a large magnet on rez-
de-chaussée. In spite of this separation, adjustment of the optical path
was not so difficult. The magnet was used by several groups, so that our
experiment was mostly done at night. Liquid Helium and liquid hydrogen
were supplied once a week on different days, and I had enough time to
enjoy exploring an attractive old city, Sendai, in the north-east (Tohoku)
part of Japan. Furthermore, since it took several hours to take a photograph of the emission spectrum, my ability in playing flute showed good
progress. In some preliminary experiment, we placed an unpolished pink
ruby in a Dewar vessel. We were very excited to see the R_1 emission and
the R_2 absorption lines on a photographic plate, as shown in Fig.1.

The experiment with Tsujikawa was successful. We obtained the Zeeman
patterns of the R lines at 20 K with a magnetic field parallel to the trigonal axis and the polarization perpendicular to it as shown at the top of
Fig.2. The qualitative features of the patterns are quite similar to those
observed at T=-190°C by Lehmann: they consist of three components, being
asymmetric for R_1 and symmetric for R_2 in shape. Only a difference may be
found in the magnitude of the apparent g-value. In our patterns, the averaged apparent g-values for R_1 and R_2 are 1.78 (±0.04) and 1.74 (±0.04),
respectively, while they are 1.47 in Lehmann's experiment. Comparison
showed that the effective magnetic field was 20% smaller than the value
reported in his experiment. The existence of two unresolved lines at the
central component was confirmed by observing the relative shift of the cen-

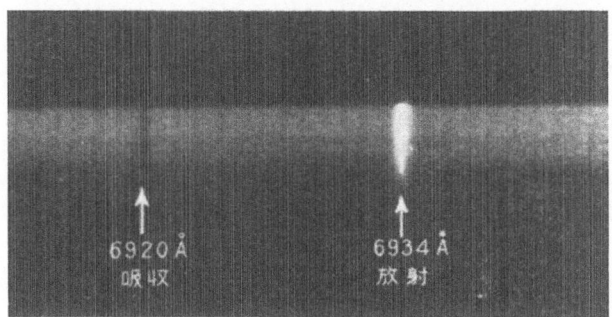

6920 Å
吸収

6934 Å
放射

Fig.1. Photograph show-
ing simultaneous obser-
vation of the R_1 emis-
sion line of 6934 Å
and the R_2 absorption
line of 6920 Å of an un-
polished ruby at 20 K.
The observation is
made in the same direc-
tion as the incident
light.

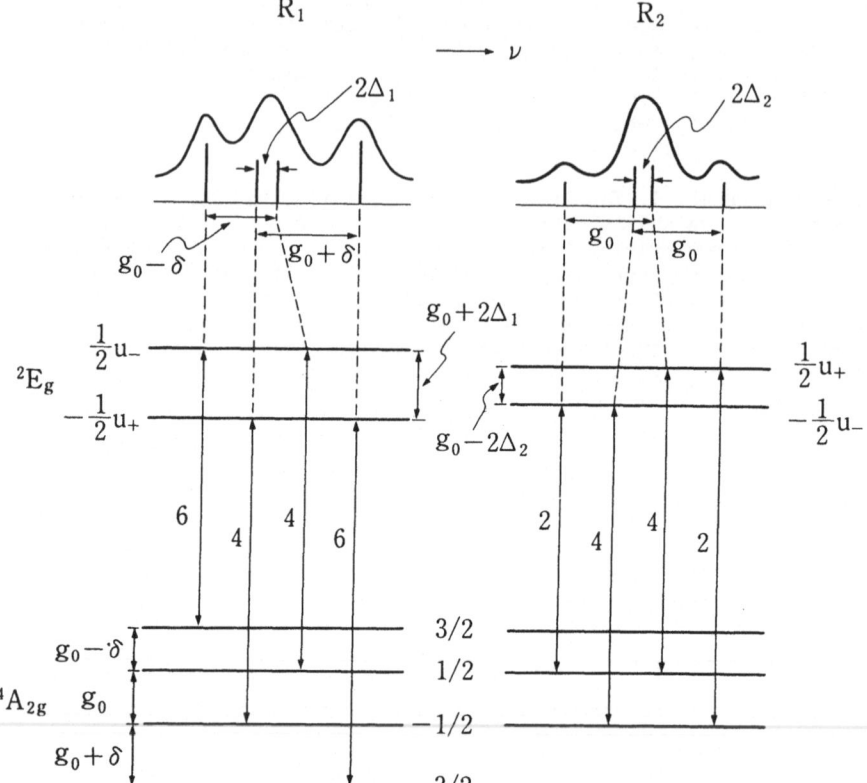

R_1 R_2

Fig.2. The observed Zeeman patterns of the R_1 and R_2 absorption lines at 20K and the corresponding transition diagrams for a magnetic field parallel to the trigonal axis and the polarization perpendicular to it. (g_0= 1.98, δ=0.34, Δ_1=0.22, Δ_2=0.26)./5/

tral component at 4K. As shown in Fig.2, we can find in our patterns the Zeeman splitting of the ground state observed by paramagnetic resonance. Analyzing our Zeeman patterns, we obtained the g-shifts in the R_1 and R_2 excited Kramers doublets as Δ_1=0.22 and Δ_2=0.26.

Shortly after publishing the papers on ruby in August, 1958/5/, we received a preprint of short communication on Maser Action in Ruby by G. Makhov, C. Kikuchi, J. Lambe and R. W. Terhune from C. Kikuchi of the University of Michigan in October, 1958./8/ This preprint brought the following idea to my notice. As easily seen in Fig.2, the transitions from the R_2 excited state populate the excited Zeeman levels of the ground state. Therefore, if one could find a trigonal Cr^{3+} system in which the R_2 excited state is lower than the R_1 excited state contrary to the case of ruby, one may expect to induce maser action by optical pumping as long as non-radiative transitions could be ignored. I wrote about this idea to Andy Liehr of Bell Telephone Laboratories who was trying to invite me to Bell Labs. At the end of that year, I read the epoch-making paper of Schawlow and Townes on "Infrared and Optical Masers"/9/ with strong excitement.

On February 19 of 1959, I received a letter from Andy, in which the following statement was made: [I have taken the liberty of showing your letter to Al Clogston, Art Schawlow, Martin Peter, Stan Geschwind, and Derek Scovil, all of whom have had ideas concerning optical pumping in Ruby. They have asked me to include the following paragraph concerning the connection between your ideas and theirs: "Since reading your very interesting paper on the ruby spectrum, we have been aware that there are several different kinds of optical pumping experiments which might be done on ruby. Several of our people are thinking about this and some of these experiments will probably be tried fairly soon. However, it would be interesting to know more about the one you propose."] Later Art Schawlow told me that a trigonal crystal is available in which the sign of the trigonal field is opposite to that in ruby. However, I am unaware if the optical pumping experiment has ever been performed on that crystal.

When I arrived at Bell Telephone Laboratories at the end of summer in 1959, Art Schawlow showed me many big ruby crystals of different shapes. It took some time for me to understand the purpose these crystals were grown for. On some day soon after my arrival, Art pushed me into his car. When we finally got off the car, I realized that we were attending the First Quantum Electronics Conference held at Shawanga Lodge. Listening to several talks, I could easily understand the purpose of this conference. Speakers seemed to be insisting that all the variety of the optical transitions they were studying could be used for possible optical masers to be realized in the near future.

Many interesting experiments on ruby were conducted at Bell Labs. in 1959. Among them, the experiment on "Self-Absorption and Trapping of Sharp-Line Resonance Radiation in Ruby" by Frank Varsanyi, Dar Wood and Art Schawlow/10/ reminded me of the spectrum in Fig.1, where the R_1 emission and the R_2 absorption lines are seen on a photographic plate. I suspected that this might be due to the relaxation, within the ^2E state, of the trapped photons at the rough surface. However, I did not pursue this problem further.

I would like to comment upon the experiment performed in this period by the group of Stan Geschwind on "Optical Detection of Paramagnetic Resonance in an Excited State of Cr^{3+} in Al_2O_3" /11/. They detected paramagnetic resonance in the R_1 excited levels $(\bar{E}:\pm1/2u_\mp)$ by observing change of the emission intensities from these levels. The presence of the resonance signal with a static magnetic field parallel to the trigonal field indicated the existence of non-zero linear Zeeman splitting with a magnetic field perpendicular to the trigonal axis. Later, we studied the Zeeman

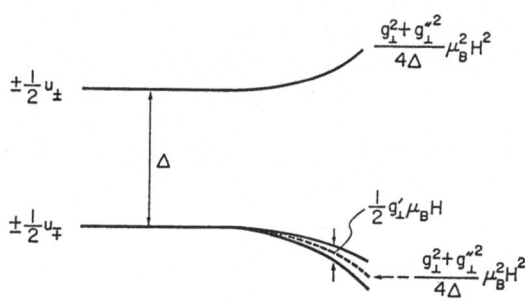

Fig.3. Zeeman splitting of the t_{2g}^3 ^2E excited states of ruby in a magnetic field perpendicular to the trigonal axis /12/.

splittings of the R_1 and R_2 excited states by using the method of effective Hamiltonian giving the exact results /12/. The result for a magnetic field perpendicular to the trigonal field is shown in Fig.3, which indicates that the linear Zeeman splitting exists in the R_1 excited state but not in the R_2. This means that, if the sign of the trigonal field was reversed making the R_1 excited state higher than the R_2 in energy, we could not observe the paramagnetic resonance at low temperatures where only the lower excited state is populated. This conclusion is very interesting when compared with that given before for the optical pumping of the excited Zeeman levels of the ground state.

2. Spectroscopic Work at Bell Labs

In 1960, I collaborated in a series of the spectroscopic work on the cubic-field line of Cr^{3+} in a MgO crystal developed in Art's group. The line has to arise from either an electric dipole transition slightly released by the asymmetric distortion of the system due to lattice vibration, or by a magnetic dipole one. Zeeman studies can decide which transition is responsible for the line. As an example, the experimental and the calculated Zeeman patterns for the electric dipole and magnetic dipole transitions with a magnetic field parallel to the [001] axis are shown in Fig.4/13/, which makes one conclude that the transition is of a magnetic dipole nature. As far as I know, no Zeeman effect has been reported on crystalline-field lines allowed by the coupling with odd-parity vibrations. This would be due to the broadening of these lines by the coupling of the local modes of vibration.

At the beginning of this work, we used a Bausch and Lomb Dual Grating Spectrograph. The photographic plates used were calibrated and photometered for intensities. During the course of the work, however, we got a Jarrell-Ash high-resolution photoelectric spectrometer. It was very nice to find that the results obtained with this new instrument agreed well with those obtained photographically in the cases where comparisons were made. In this period Frank Varsanyi carried me between Bell Labs and my apartment in his car every day, so that we had enough time to exchange new ideas during the ride.

The study of the longitudinal Zeeman effect of the same emission line was more interesting. As an example, we show the experimental and theoretical longitudinal Zeeman patterns with a magnetic field parallel to the [111] axis for circular polarization. In Fig. 5 the central component is linearly polarized as it is given by superposition of the opposite circular polarizations with a fixed relative phase. However, this phase difference would vary from train to train of the emitted light, so that the linear polarization would take any direction in the plane perpendicular to the [111] axis. Unfortunately we did not confirm this experimentally.

The degenerate cubic-field line may be split by applying uniaxial pressure. The experiment on such a splitting was performed by Art Schawlow and A. H. Piksis./14/ I joined the analysis of the experiment. Piksis was also enthusiastic in making some calculations by himself. The stress, simple pressure P, was applied normal to the (001), (110), and (111) planes. The strain-induced change of the crystal field V linear to the strain can be written in the form,

$$V = \sum_{\Gamma\gamma} C_{\Gamma\gamma} V_{\Gamma\gamma}(\vec{r}) e_{\Gamma\gamma},$$

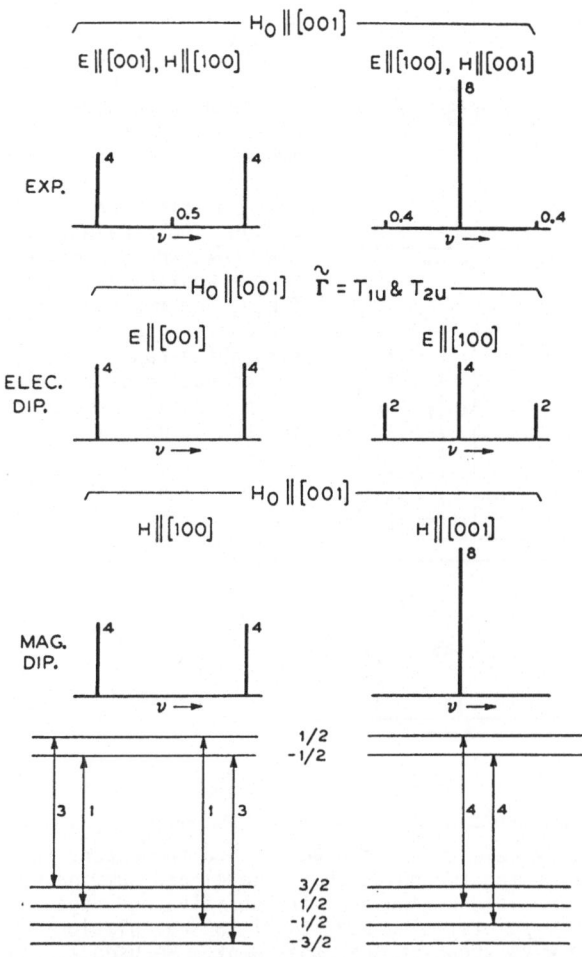

Fig.4. The experimental and the calculated Zeeman patterns of the electric dipole and magnetic dipole transitions for a cubic-field emission line $(t_{2g}^3 \, {}^2E_g \rightarrow t_{2g}^3 \, {}^4A_{2g})$ of Cr^{3+} in MgO with a magnetic field H_0 parallel to the [001] axis. E and H stand for the oscillating electric and magnetic fields of the emitted light, and $\tilde{\Gamma}$ for the vibrational mode coupled with the electric dipole transitions. The intensities are relative and not normalized /13/.

where $\Gamma\gamma$ runs over the irreducible representation Γ ($\Gamma=A_1$, E, T_2) and their components γ of the cubic symmetry group. $V_{\Gamma\gamma}(\vec{r})$ is a function of electron coordinates $\vec{r}$ transforming as the γ basis of the Γ representation. $e_{\Gamma\gamma}$ is the strain component transforming in the same way as $V_{\Gamma\gamma}$;

$$e_{A_1} = e_{xx} + e_{yy} + e_{zz}, \qquad e_{T_2\xi} = e_{yz},$$

$$e_{Eu} = 2e_{zz} - e_{xx} - e_{yy}, \qquad e_{T_2\eta} = e_{zx},$$

$$e_{Ev} = e_{xx} - e_{yy}, \qquad e_{T_2\zeta} = e_{xy},$$

229

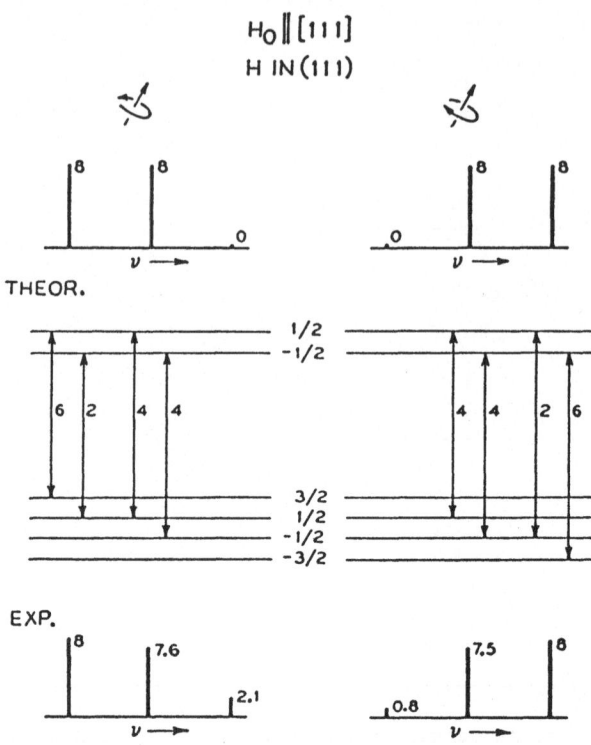

$H_0 \parallel [111]$

H IN (111)

THEOR.

1/2
-1/2

6 2   4 4   4 4   2 6

3/2
1/2
-1/2
-3/2

EXP.

Fig.5. The experimental and the calculated longitudinal Zeeman patterns of the R cubic-field emission line of Cr^{3+} in MgO with a magnetic field H_0 parallel to the [111] axis for circular polarization. The straight arrow shows the direction of the applied magnetic field, and the circulating one around the straight arrow indicates the direction of the rotating magnetic vector of the radiation /13/.

wnere e_{ij}'s are the strain components ordinarily used. Further, $V_{\Gamma\gamma}(\vec{r})$ is expanded in a power series of the electron coordinates, and the expansion coefficients are calculated by using the point-charge model and summing up appropriately the electrostatic field coming from the positive and negative charges at all the lattice points. The results of the calculation are given in Table 1 together with those of the experiment. We used Watson's analytical Hartree-Fock wavefunction of a free Cr^{3+} ion/25/ to calculate the average values;

$$<r^2/R^2> = 0.0918, \qquad <r^4/R^4> = 0.0179,$$

$$R = 3.97 \text{ (a.u.)},$$

where R is the lattice constant.

In Table 1, the sign of the observed splitting was determined to give agreement between the calculated and observed polarization. The agreement between the calculated and observed splitting is surprisingly good. One might suspect that the agreement is accidental, but the fact that it is uniformly good in all the cases seems to suggest something significant. It should be noticed that, if only the point charges at the nearest neighbor are taken into account in the calculation of the strain-induced change of the crystal field, the sign of the calculated splitting turns out to be wrong in some case. It is well known that the calculation of the cubic-field strength by the use of the point-charge model and the Hartree-Fock wavefunction gives a too small value; here we have to take into account the

Table 1. The strain-induced splitting, shifts, and the polarization of the split components of the R cubic-field emission line of Cr^{3+} in MgO. The polarization is defined as the ratio of the magnetic dipole strength of the shorter-wavelength split component to that of the longer-wavelength one when the splitting is positive. Δ and λ are given in cm^{-1} and P in dynes/cm^2 (P < 0) /14/.

Case	Calc. split. $\Delta/(-P)$	Obs. split. $\Delta/(-P)$	Calc. pol.	Obs. pol.	Obs. shift $-\lambda/(-P)$
	$\times 10^{-10}$	$\times 10^{-10}$			$\times 10^{-10}$
$P_\perp(001)$	6.8	6.3±0.2	H⊥P 3	2.4	2.9
			H∥P 0	0.4	
$P_\perp(110)$	-3.8	-(3.6±0.1)	H∥[1$\overline{1}$0] 1.7	1.2	2.9
			H∥[00$\overline{1}$] 0.05	0.01	
			H∥[110] 4.6	1.5	
$P_\perp(111)$	-2.0	-(2.8±0.1)	H⊥P 0.6	0.61	2.6
			H∥P 3	1.6	

non-orthogonality and the covalency effects in addition to the point-charge crystal field./12/ However, the contribution of these effects to the low-symmetry crystal field may be expected to be small, as the contribution of the crystal field from distant point-charges is appreciable. This would be a reason why the agreement is so good. In Table 1, we also see that the observed shifts λ of the center of the split components are almost the same in the three cases. This is reasonable, as the shifts are caused by the strain-induced change of the spherically symmetric and cubic crystal fields which should be the same in all the three cases. Later, such a shift will be discussed again more quantitatively.

In 1960, T. H. Maiman reported the first observation of laser action of the R lines of ruby /16/, and Art was busy in observing the laser action of the satellite lines of concentrated ruby./17/ On January 31, 1961, a continuously-operating He-Ne gas laser built by A. Javan, W. R. Bennett Jr., and D. R. Herriott/18/ was demonstrated at a press conference of Bell Labs. In the 1961 spring meeting of the Optical Society of America, the red spot of a ruby laser was projected on a screen in Ballroom of Penn-Sheraton Hotel when Charles Townes counted down to "zero". In this meeting Art gave an invited talk on Solid-State Optical Masers, and I gave an invited talk on Spectroscopy for Solid Optical Masers./19/ The title of the present article comes from the title of my talk: at that time, as far as I remember, Bell people used the term "Optical Masers" instead of "Lasers". When my talk ended, A. Kastler, chairman, asked me the question "Do you think that a new type of laser could be designed by use of chemical reaction?" My answer "May be" induced an explosion of laughter. Later, Kastler asked me to spend one year at Ecole Normale Supérieure.

In 1961, which was the last year of my stay at Bell Labs., I was very busy in studying the electronic structure of $KNiF_3$ with Bob Shulman. One day in the spring, W. Kaiser and Dar Wood came to my office and showed me the preliminary results of their experiment on the electric field

effect on the R emission line of ruby. The result seemed to show clearly
the splitting of the R line, but it was not clear whether the splitting was
quadratic or linear to the field strength. Since I had been working on
ruby so long, I knew the existence of two sites of Cr^{3+} in ruby at which
the directions of the odd-parity crystal fields are opposite. It did not
take much time for me to arrive at the conclusion that the apparent splitting is due to the superposition of the linear spectral shifts of the opposite signs. Since the odd-parity crystal field E_{cryst} at the Cr^{3+} site
with C_3 symmetry may transform like an electric field along the trigonal
axis, the application of an external electric field E_0 along the trigonal
axis induces the spectral shift proportional to E_0 as shown by

$$(E_{cryst} + E_0)^2 \sim E_{cryst}^2 + 2E_{xryst}E_0. \qquad (E_{cryst} \gg E_0)$$

In Fig.6, we show the final result of the experiment/20/, which ver-
ifies our reasoning. It is easy to show that such a linear shift does not
exist if the site symmetry is D_3./12/ The apparent splitting thus observed
is called "Pseudo-Stark Splitting". Later, an apparent linear splitting
with the same origin was also observed in the paramagnetic resonance of ruby
in Bloembergen's group.

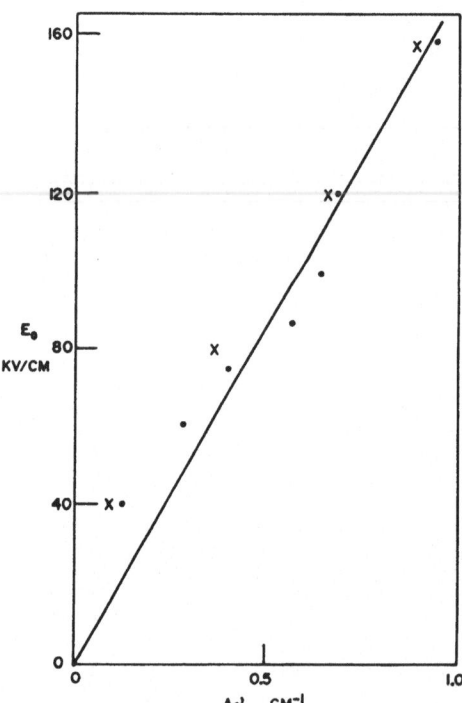

Fig.6. Pseudo-Stark splitting $\Delta\nu$ of the R emission
lines of ruby vs applied
electric field E_0. Data
are obtained from the $R_1(\cdot)$
and $R_2(\times)$ lines /20/.

3. Spectroscopy with a Strong Pulsed Magnetic Field

When I came to the Institute for Solid State Physics (ISSP), I found that
some experimental facilities for producing a strong pulsed magnetic field
were available in ISSP. I asked K. Aoyagi and A. Misu, who were working

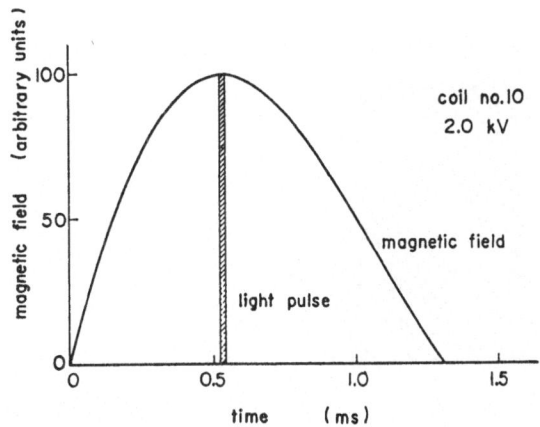

Fig.7. The principle of spectroscopic experiment by using a strong pulsed magnetic field./21/

coil no.10
2.0 kV

magnetic field

light pulse

in the experimental laboratory of optics and spectroscopy where I worked on ruby before my stay at Bell Labs., if they were interested in the research project of solid state spectroscopy by using a strong pulsed magnetic field. Our project started in the summer of 1962.

The principle of our experiment is simply illustrated in Fig.7. Let us assume that the field is a simple sine curve such as $H_0 \sin \omega t$, and the pulsed light is of a rectangular shape having the width 2Δ. Then, the fluctuation of the magnetic field, H_0, during an exposure is given by $\delta H_0/H_0 \sim (\omega\Delta)^2/2$, where $\omega\Delta \ll 1$ is assumed. For example, by assuming $T \equiv \pi/\omega$ =1 ms and Δ=1 μs, we obtain $\delta H_0/H_0 \sim 0.05\%$ which is negligible compared with the inhomogeneity of the field inside a sample. The maximum field of 230 kOe was produced by a copper wire coil surrounded by a cylindrical ring of berryllium-copper alloy immersed in liquid nitrogen.

Our experiment was very useful in determining the g-values of the B_1,B_2 excited states arising from the $t_{2g}^3 \, ^2T_{2g}$ state. The values obtained are given in Table 2. In our strong field, we could observe the onset of Paschen-Back effect in the Zeeman patterns of the R lines with a magnetic field perpendicular to the trigonal axis. Instead of showing this Zeeman pattern, however, I would like to show in Fig.8 the change of the Zeeman pattern when the field is increased up to 140 T (1.4 MOe). This data was recently obtained by N. Miura's group in ISSP./23/ This change shows the perfect Paschen-Back effect as plotted in Fig. 9.

Table 2. g-values of the excited states of the B lines of ruby

	Exp.	Theor./22/
$g_\perp (B_1)$	< 0.5	0.01
$g_\perp (B_2)$	< 0.6	0.13
$g_{/\!/} (B_1)$	0.69 ± 0.09	0.96
$g_{/\!/} (B_2)$	-2.97 ± 0.15	-2.97

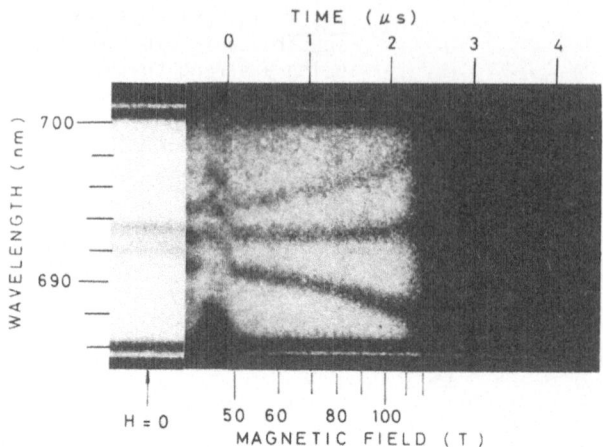

Fig.8. The observed change of the Zeeman splitting of the R lines of Ruby at T=150 K when a pulse magnetic field is applied perpendicular to the trigonal axis./23/

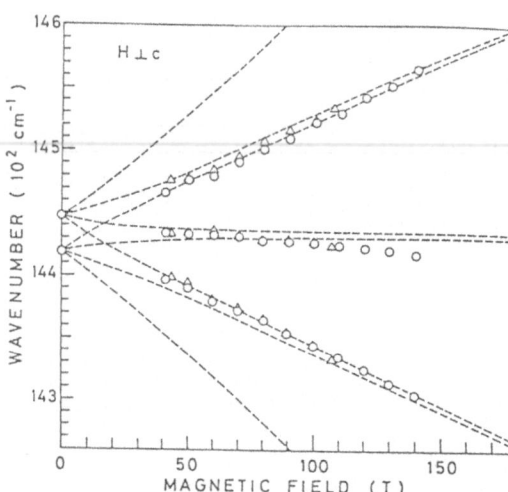

Fig.9. Plot of the Zeeman splitting of the R lines of Ruby with a magnetic field perpendicular to the trigonal axis. The broken curves are theoretical curves taking into account the Paschen-Back effect. The circles and triangles represent data for two different samples. T=150-160 K./23/

Besides observing Zeeman patterns, we were successful in observing optically the process of electron spin relaxation in the ground state of ruby in the magnetic field of ∿100 kOe at low temperatures, by using a pulsed magnetic field and a light flash./24/ The principle of the experiment is illustrated in Fig.10. The Zeeman pattern is observed at time t measured from the time when the field was applied. We vary t in the repeated experiments making the field strength at t to be constant. Then, we found that the relative intensities of the Zeeman components vary as t varies. This is shown in Fig.11, where the Zeeman pattern of the B_1 and B_2 lines is observed at t=0.27, 0.67, and 1.13 ms in a magnetic field, $H_0(t)$=110 kOe, as shown in Fig.10. The transition diagram giving the Zeeman pattern of Fig.11 is shown in Fig.12.

234

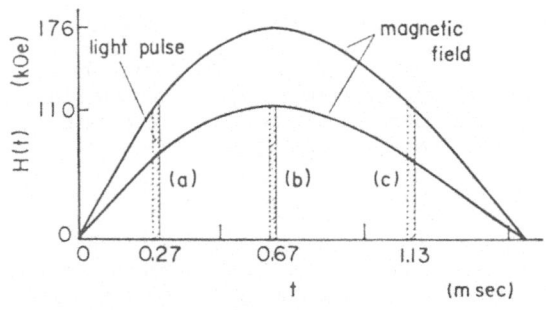

Fig.10. The combinations of a pulsed magnetic field and a light flash. /24/

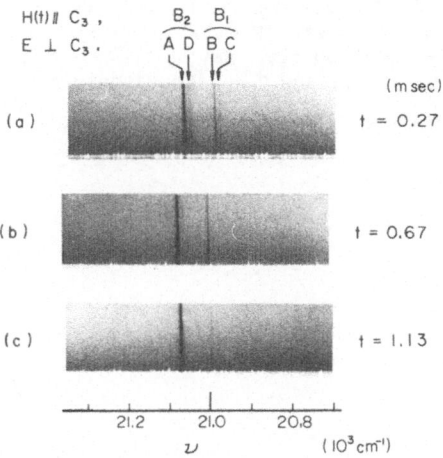

Ruby No 5 (concentration 0.28 %),
H(t) = 110 kOe, 4.2°K.

Fig.11. The Zeeman patterns taken by light flash (a), (b), and (c) shown in Fig. 10. /24/

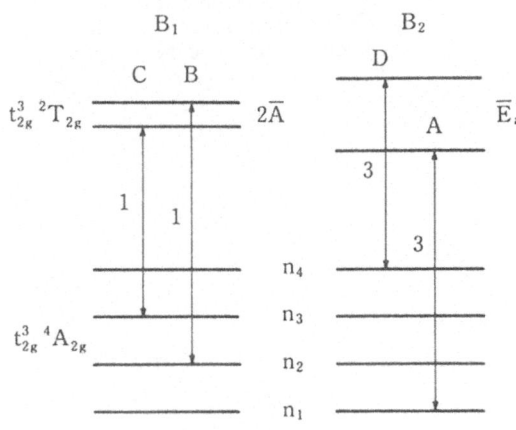

$H(t) // C_3$, $E \perp C_3$

Fig.12. The transition diagram giving the Zeeman pattern of Fig.11. n_i (i=1, 2, 3, 4) are populations of the Zeeman levels of the ground state.

Since we know the relative transition probabilities of A, B, C, and D components as indicated in Fig.12, we can determine from the observed relative intensities, the population n_i (i=1, 2, 3, 4) of the Zeeman levels of the ground state. The populations thus determined are plotted in Fig.13 against t. Since the energy separations between the neighboring Zeeman levels are almost equal, we may expect that n_i's are equally spaced at a fixed value of t if an effective temperature of the spin system can be defined. Fig.13 shows that it is not the case.

Another spectroscopic experiment by using the pulse property of the magnetic field was done on rare earth compounds. In this experiment, we applied a light flash at delay time t and took the photograph of the spectrum in a magnetic field H(t) in the same way as was done in the spin relaxation experiment of ruby. At first we observed the Zeeman spectrum of $Dy_3Al_5O_{12}$(DAG) at 4.2K as shown in Fig.14. It is surprising to see that the d_{21} and d_{22} lines, which arise from the transitions from the Z_2 excited state 70.3cm^{-1} above the ground state as shown in Fig.15, begin to appear when the magnetic fields are above 30 kOe even at 4.2K. If a pulse field is replaced by a static field, no d_{21} and d_{22} lines appear at 4.2K as shown in Fig.16, since the population of the Z_2 excited state is negligibly small.

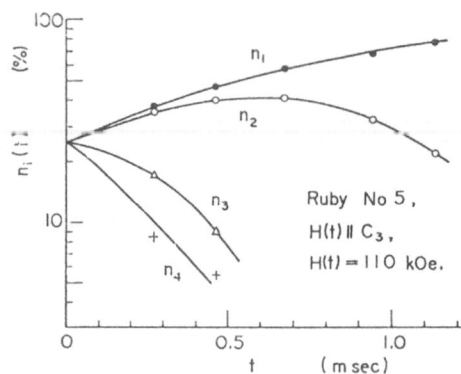

Fig.13. The population of the Zeeman levels of the ground state versus t./24/

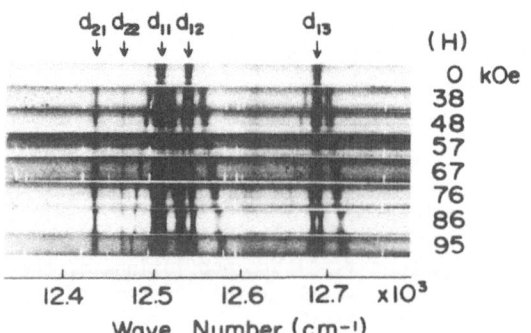

Fig.14. The Zeeman spectra of DAG in the axial polarization at 4.2 K. The pulsed magnetic fields are applied along the (001) cubic axis. d_{ij} indicate the transitions shown in Fig.15. /25/

236

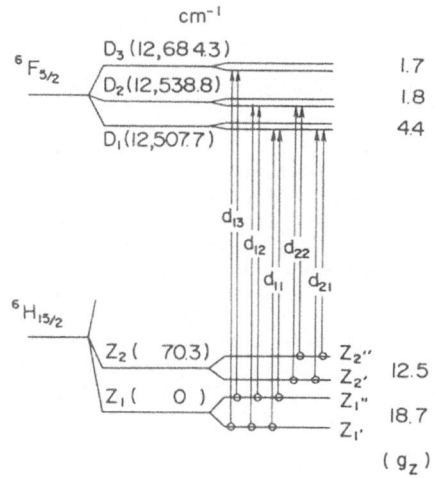

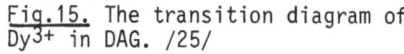

Fig.15. The transition diagram of Dy^{3+} in DAG. /25/

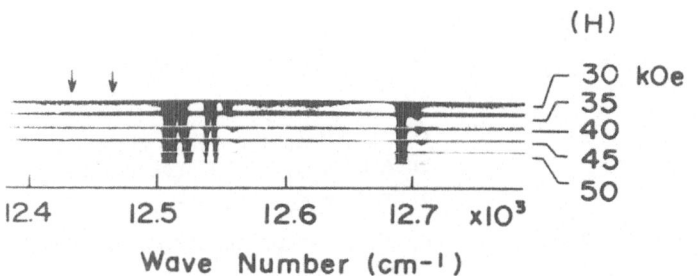

(H)

30 kOe
35
40
45
50

12.4   12.5   12.6   12.7 $\times 10^3$

Wave Number (cm^{-1})

Fig.16. The Zeeman spectra of DAG in the axial polarization with the static magnetic field along the <001> cubic axis. /25/

Using the relative transition probabilities determined from the absorption measurement at 77K, we can determine population N_i (i=1", 2' and 2") of the Z_i level. Then, assuming a Boltzmann distribution, we can determine the effective temperature T_{eff} to satisfy $N_i/N_{1'} = \exp(-E_i/kT)$, where E_i is the energy difference between Z_i and $Z_{1'}$. The effective temperatures thus determined from $N_{2'}$ and $N_{2''}$ are almost the same, but T_{eff} from $N_{1''}$ is a little higher as shown in Fig.17. The important result is the observation that T_{eff} does not change when t is changed as long as H(t) is fixed. This suggests that there is no energy flow from the electron system to the heat bath surrounding the crystal within the duration of the pulsed magnetic field.

Summarizing the experimental results, we may conclude that (i) the electron system is thermally isolated from the heat reservoir in the duration of the pulse field and (ii) electrons in the $Z_{2'}$ and $Z_{2''}$ levels are in thermal equilibrium allowing us to define an effective electron temperature. These facts seem to show that the anomalous effect we observed may be discussed from a purely thermodynamic point of view, i.e. an effective temperature higher than that of the heat reservoir may be interpreted as due to heating by adiabatic magnetization.

To calculate the entropy of the electron system, we assume that the electron system has only two Kramers doublets Z_1 and Z_2 with g-values, $g_1 = 10.8$ and $g_2 = 7.2$, and energy separation $E_2 = 70.3$ cm^{-1}. The calculated temperature dependence of the entropy is given in Fig.18 for magnetic fields up to 100 kOe. A broken line in the figure shows the path taken by the system when the field is increased from zero to H. The crossing points of this path with the entropy curves determine the T_{eff} versus H curve which is shown in Fig.17. Fig.17 shows that the calculated T_{eff} are in fair agreement with the experimental ones.

Thermal isolation of the electron system from the heat bath may be due to either a long lattice-bath relaxation time τ_ℓ or a long spin-lattice

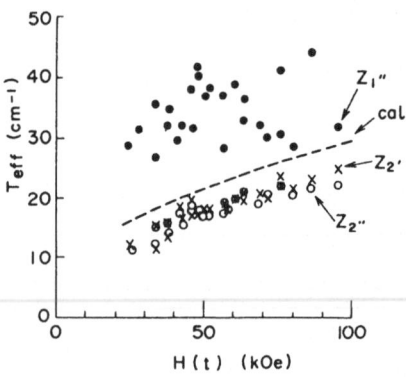

Fig.17. Magnetic field dependence of T_{eff}. The broken line is T_{eff} numerically calculated by assuming adiabatic magnetization. /25/

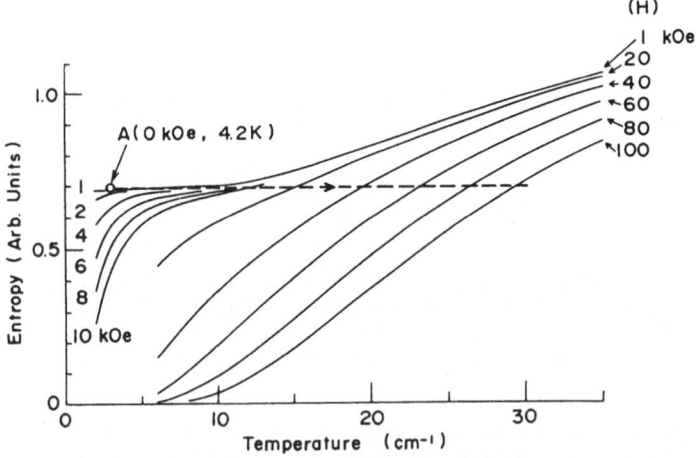

Fig.18. The calculated temperature dependence of the entropy of the electron system. The broken line shows the path taken by the system when the magnetic field is increased from zero to H. Point A indicates the starting point of the path. /25/

relaxation time $\tau_{s\ell}$ as compared with the delay time of the light flash. The former possibility, however, may be excluded by considering the fact that T_{eff} suddenly decreases when the temperature of the heat bath decreases below the Néel temperature of DAG, $T_N=2.5$ K, as shown in Fig.19: there is no reason why τ_ℓ should change abruptly at T_N. Thus, our conclusion may be summarized as

$$\tau_s, \tau_\ell < t < \tau_{s\ell} \qquad \text{at} \qquad T > T_N,$$

$$\tau_s, \tau_\ell, \tau_{s\ell} < t \qquad \text{at} \qquad T < T_N,$$

where τ_s is the cross relaxation time of the effective spins and t the delay time of the light flash. The abrupt change of $\tau_{s\ell}$ at T_N might be due to the crossing of the dispersion curves of effective-spins in the magnetic field with those of phonons below T_N. The adiabatic magnetization effect mentioned here was also observed in the optical spectra of $DyAlO_3$, where $E_2=54$ cm^{-1} and the maximum g-value of the ground state is 13.2, and in $Er_3Ga_5O_{12}$, where $E_2=46$ cm^{-1} and $g_{max}=11.3$ in the ground Kramers doublet.

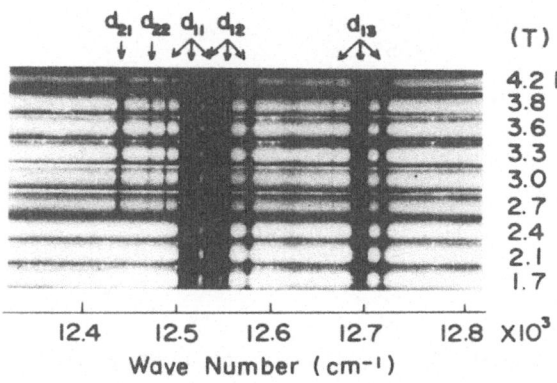

Fig.19. Temperature variation of the Zeeman spectra of DAG in the axial polarization with the magnetic field parallel to <001>. H_{max} and t are 76 kOe and 0.7 ms, respectively. The d_{21} and d_{22} lines suddenly disappear at $T < T_N=2.5$ K. /25/

4. Absorption Spectrum of Optically Pumped Ruby/26/

In 1965, T. Kushida of Central Research Laboratory of Tokyo Shibaura Co. showed me beautiful absorption spectra of ruby optically pumped by a xenon flash lamp. M. Shinada, my research associate, and myself immediately joined Kushida to help in the analysis of his data.

The energy level diagram of ruby is shown in Fig.20, in which the locations of the spin doublets of the $t_{2g}^2 e_g$ electron configuration had not been experimentally confirmed. Kushida's experiment measuring the absorption transitions from the t_{2g}^3 ^2E_g and $^2T_{1g}$ states could determine these locations, as the transitions are spin-allowed: for the conventional absorption in the ground state, they are spin-forbidden and difficult to be observed being masked by the spin-allowed transitions. In the experiment, it is reasonable to assume that thermal equilibrium is established among the five Kramers doublets of t_{2g}^3 ^2E_g and $^2T_{1g}$, as the relaxation time between

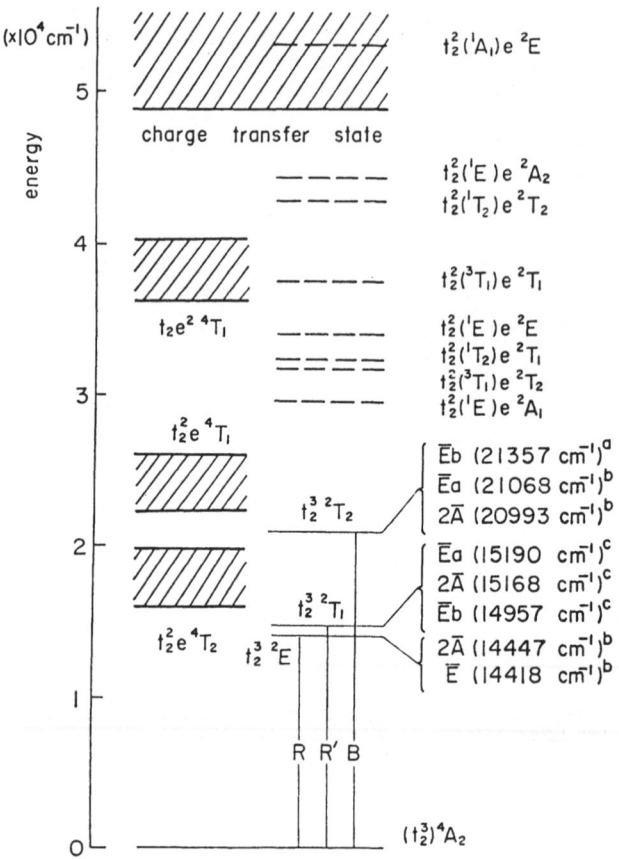

Fig.20. The energy level diagram of ruby. /26/
a. O. Deutschbein/4/, b. S. Sugano and I. Tsujikawa/5/, c. J. Margerie/27/.

these doublets is shorter than a fraction of 1 μs while the lifetime of these levels is as long as ∿3.4 ms at room temperature. Thus, the transition probabilities from these levels to the $t_{2g}^3 \, {}^2T_{2g}$ Kramers doublets can also be determined in Kushida's experiment.

Let us first discuss the transitions to the $t_{2g}^3 \, {}^2T_{2g}$ Kramers doublets in the optically pumped ruby. The transitions lie in the infrared region. The experimental spectrum due to these transitions is compared with the theoretical one in Fig.21, where we use the locations of the relevant Kramers doublets indicated in Fig.20. The agreement of the relative intensities is surprisingly good between theory and experiment. The important points are as follows; (1) The theory predicts no absorption due to the $\bar{E}_b({}^2T_{1g}) \to \bar{E}_b({}^2T_{2g})$ transition expected at ∿6,400 cm^{-1} and the experimental absorption curve shows a minimum: (2) The theory predicts very weak π intensities of the $2\bar{A}({}^2T_{1g}) \to 2\bar{A}({}^2T_{2g})$ and $\bar{E}_a({}^2T_{1g}) \to \bar{E}_a({}^2T_{2g})$ transitions expected at ∿5,900 cm^{-1} and no π-absorption is observed. The un-

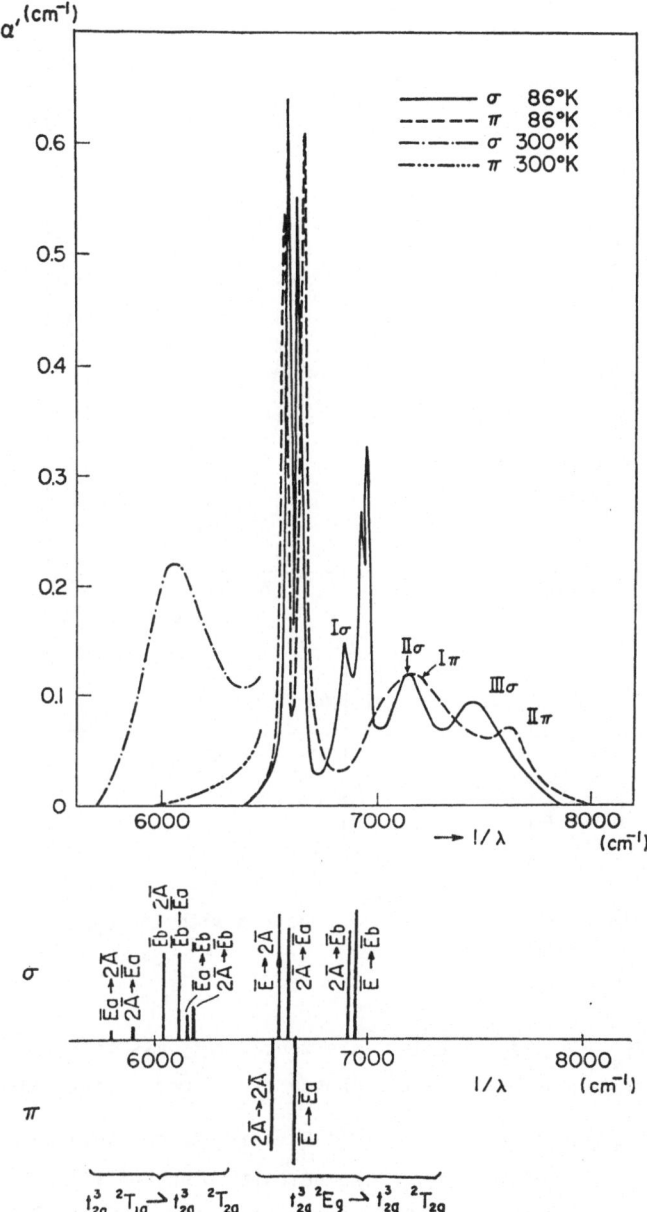

Fig.21. Comparison of the calculated absorption spectrum at 300 K with the experimental one of the optically pumped ruby in the infrared region./26/

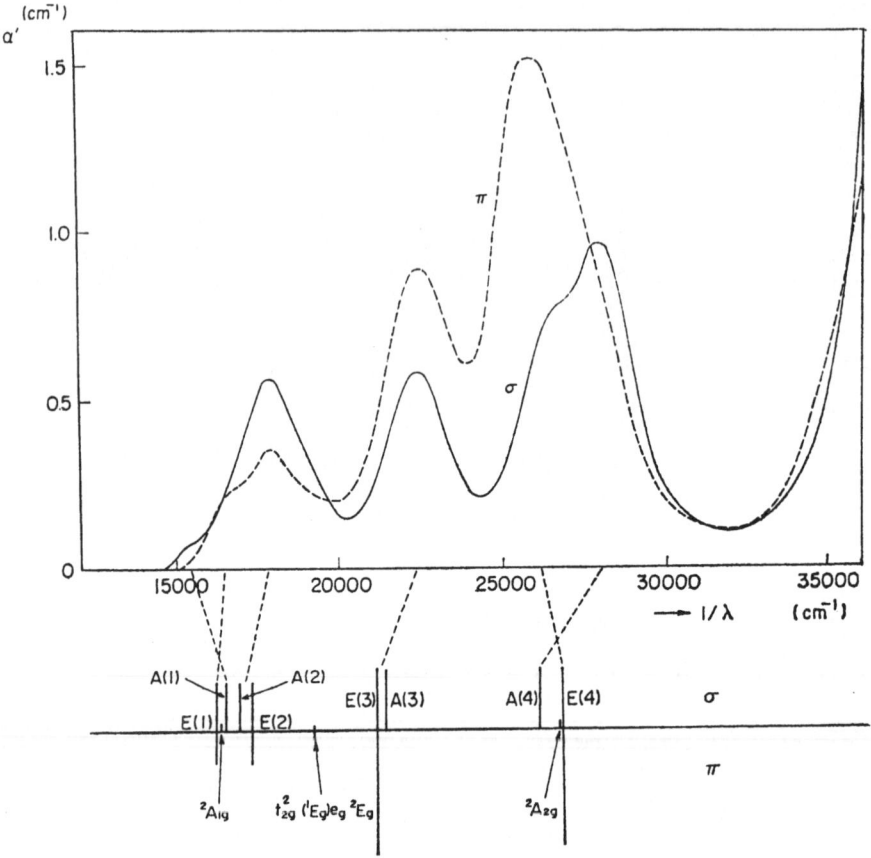

Fig.22. Comparison of the predicted absorption spectrum with the experimental one in the visible region./26/

identified peaks, I_σ, II_σ, III_σ, I_π, and II_π, are tentatively assigned to the electric dipole transitions exciting one phonon of the odd-parity mode.

Now, we discuss the transitions from the t_{2g}^3 ^2E_g state to the spin doublets of the $t_{2g}^2 e_g$ electron configuration. The calculated energies/22/ of these spin doublets measured from the t_{2g}^3 ^2E_g state are given in Table 3. Comparison is made of the calculated absorption spectrum with the observed one in the visible region in Fig.22. For A(i) and E(i) (i=1, 2, 3, 4), see Table 3, the agreement between theory and experiment is fair. In this way, we confirmed the applicability of the ligand field theory to such highly excited states as those lying 40,000 cm^{-1} above the ground state.

Table 3. The calculated energies of the $t_{2g}^2 e_g {}^2\Gamma_g$ states measured from the $t_{2g}^3 {}^2E_g$ excited state./22/ Here, the spin-orbit splitting is averaged and the centers of gravity of the spin-orbit split components are indicated. The indicated splittings are due to the trigonal field: A's are mainly M=0, and E's are mainly M=±1 components.

Excited states	Calculated energies
$t_{2g}^3 {}^2E_g$	0 cm^{-1}
$t_{2g}^2({}^1E_g)e_g {}^2A_{1g}$	16,300
$t_{2g}^2({}^3T_{1g})e_g {}^2T_{2g} \begin{cases} E(1) \\ A(1) \end{cases}$	16,200 16,500
$t_{2g}^2({}^1T_{2g})e_g {}^2T_{1g} \begin{cases} A(2) \\ E(2) \end{cases}$	16,900 17,300
$t_{2g}^2({}^1E_g)e_g {}^2E_g$	19,200
$t_{2g}^2({}^3T_{1g})e_g {}^2T_{1g} \begin{cases} E(3) \\ A(3) \end{cases}$	21,100 21,400
$t_{2g}^2({}^1T_{2g})e_g {}^2T_{2g} \begin{cases} A(4) \\ E(4) \end{cases}$	26,100 26,800
$t_{2g}^2({}^1E_g)e_g {}^2A_{2g}$	26,700
$t_{2g}^2({}^1A_{1g})e_g {}^2E_g$	38,600

5. High-Pressure Effects on the Optical Spectrum of Ruby/28/

In 1978, S. Ohnishi of the Geophysics Department joined our group. He was naturally interested in high-pressure effects on solids. We started studies of high-pressure effects on the optical spectrum of ruby.

In the ligand field theory, the most important physical parameters are cubic field splitting parameter 10Dq and Racah parameter B: parameter C is approximately related to B. The dependence of these physical parameters upon the metal-ligand distance, R, has been one of the main subjects of the optical studies of transition-metal compounds under high-pressure. It was embarrassing that the pressure experiments show/29/ the dependence of 10Dq upon R to be fairly close to that given by the point charge model. However, it is now believed that this coincidence is rather accidental and more elaborate quantum mechanical treatments including the covalency in the metal-ligand bond are necessary to explain this observed pressure dependence. Moreover, the pressure experiments show a decrease of parameter B with pressure increase as observed in the spectral red shift of the R emission line of ruby on the pressure application. The same red shift was also observed in the R cubic field line of Cr^{3+} in MgO as seen in Table 1. The red shift may be interpreted to be due to an increase of the covalency with a decrease of the metal-ligand distance.

To calculate the high-pressure effects, we used a cluster model, $[CrO_6]^{9-}$ in a flat potential well. The use of this cluster model seems to be a fair approximation/30/ as long as we are concerned with the localized d-electrons. Further, we assumed that the cluster has O_h symmetry. It was carefully determined that the pressure-induced change of the trigonal field gave negligible spectral shifts of the U band and the R lines as compared with those due to the change of 10Dq and B./28/

For the calculation, we applied the discrete-variational Xα method/30/ taking into account the spin-polarization in the transition-state scheme. Spectral positions of the U band peak, E_U, and the R line, E_R, were calculated for transition-states $(t_{2g\uparrow})^{2.5}(e_{g\uparrow})^{0.5}$ and $(t_{2g\uparrow})^{2.5}(t_{2g\downarrow})^{0.5}$, respectively. Nine cases of the Cr-O distance from 2.000 Å through 1.754 Å were examined: they are enough to cover the pressure range of more than 500 kbar for ruby. The distance in the absence of pressure is assumed to be 1.9 Å. The calculation was made self-consistent and the degree of convergence was within 0.1 % of the Mulliken charge.

The peak energy of the U band is given by

$$E_U = \varepsilon(e_{g\uparrow}) - \varepsilon(t_{2g\uparrow}),$$

where $\varepsilon(e_{g\uparrow})$ and $\varepsilon(t_{2g\uparrow})$ are the spin-polarized $e_{g\uparrow}$ and $t_{2g\uparrow}$ eigenvalues of the transiton-state $(t_{2g\uparrow})^{2.5}(e_{g\uparrow})^{0.5}$. The energy of the R line is given by

$$E_R = \frac{4}{5}\{\varepsilon(t_{2g\downarrow}) - \varepsilon(t_{2g\uparrow})\},$$

where $\varepsilon(t_{2g\downarrow})$ and $\varepsilon(t_{2g\uparrow})$ are the spin-polarized $t_{2g\downarrow}$ and $t_{2g\uparrow}$ eigenvalues of the transition-state $(t_{2g\uparrow})^{2.5}(t_{2g\downarrow})^{0.5}$. The numerical factor 4/5 comes from the fact that the $\{\varepsilon(t_{2g\downarrow}) - \varepsilon(t_{2g\uparrow})\}$ is supposed to give the excitation energy, 15(3B+4)/4, of the center of gravity of the multiplets, $t_{2g}^3\,^2E_g$, $^2T_{1g}$, and $^2T_{2g}$, while the excitation energy of the R line in the strong crystal field limit is 3(3B+C)./1/

The R dependence of E_U and E_R in our calculation turns out to be

$$\delta(\ln E_U)/\delta(\ln R) = -4.5,$$

$$\delta(\ln E_R)/\delta(\ln R) = 0.375.$$

The value, -4.5, in the first equation should be compared with -5.0 of the point-charge model. Sato and Akimoto/31/ have given two sets of values of isothermal bulk modulus K_T and its pressure derivative K_T' for α-Al$_2$O$_3$: (A) $K_T = 2.26$, $K_T' = 4.0$, (B) $K_T = 2.39$, $K_T' = 0.9$. If one assumes the incompressibilities of the $[CrO_6]^{9-}$ cluster in ruby to be that of α-Al$_2$O$_3$, one obtains from the above-given second equation the wavelength (Å) derivative of the pressure (kbar) at p=0 as follows;

$$dp/d\lambda\big|_{p=0} = 2.60 \quad \text{for set (A),}$$

$$= 2.75 \quad \text{for set (B).}$$

244

Those values should be compared with the linear relationship /32/ between pressure p(kbar) and the wavelength shift $\Delta\lambda(\text{Å})$ from λ_0=6942 Å for the pressure calibration;

$$p = 2.74 \, \Delta\lambda.$$

The relation between λ/λ_0 and p was calculated as shown in Fig.23.

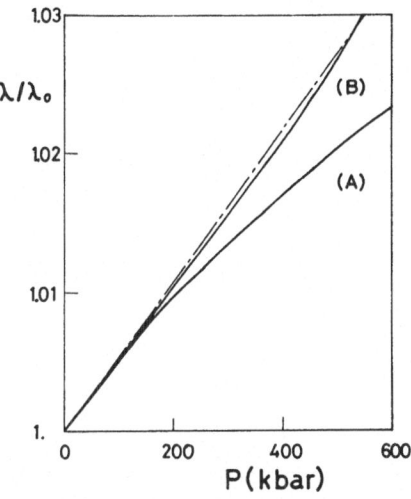

Fig.23. The calculated pressure dependence of wavelength λ of the R line of ruby for K_T=2.26, K_T'=4.0 (A), and K_T=2.39, K_T'=0.9 (B). The chain line shows the linear relationship used for the pressure calibration./28/

6. Further Developments

In the present article, I have not mentioned the spectroscopy of magnetically ordered materials. The research group of Art Schawlow at Stanford University is one of the pioneers of this field. When R. L. Greene, D. D. Sell, W. M. Yen, A. L. Schawlow, and R. M. White published a famous paper on the magnon sideband of MnF_2/32/. We, Y. Tanabe, T. Moriya and myself, immediately responded by publishing a theory on the magnon-induced electric dipole transition moment./33/ We have many interesting stories of research developments before and after the dawn of the history of this field. It is my great regret, however, that the space left for me in this volume is too small to accommodate the stories. I would like just to mention that I enjoyed the collaboration/34/ with K. Tsushima, K. Aoyagi and many others to perform a research project on the spectroscopy of anti-ferromagnetic rare-earth orthochromites at Broadcasting Science Research Laboratories of NHK where I worked on the absorption spectra of complex ions/1/ in 1952 - 56. In rare-earth orthochromites, the R lines of Cr^{3+} played as usual an important role in guiding us through the maze of data.

In concluding this article, I would like to point out an additional theoretical development of the spectroscopy of transition-metal compounds including ruby, which has not been confirmed by experiments. This is on a light-induced change in multiplet satellites of 3p-photoelectron spectra of transition-metal compounds. In this work, I collaborated with Y. Miwa and T. Yamaguchi who preformed careful algebraic manipulation as well as tedious numerical calculations./35/ The idea is as follows: The energy

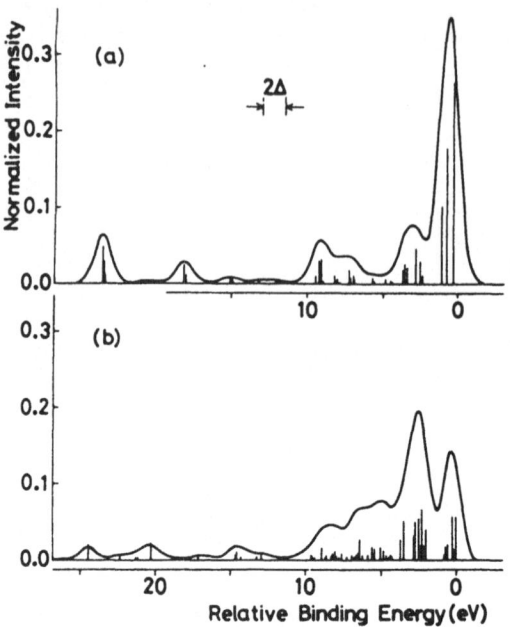

Fig.24. The calculated 3p-photoelectron spectra of a Cr^{3+} compound for the two cases of the $t_{2g}^3 \, ^4A_{2g}$ initial state (a), and the 100% excited initial state of the $t_{2g}^3 \, ^2E_g$ (b)./35/

spectrum of photoelectrons emitted by exciting 3p electrons of a transition-metal compound consists of a number of multiplet satellites due to the interaction between the incomplete d-shell and the 3p-hole. This spectrum depends upon the initial state of the compound. Therefore, if the initial state is changed by optical pumping, the photoelectron spectrum would be changed. As an example, the calculated 3p-photoelectron spectra of a Cr^{3+} compound are shown in Fig.24(a) and (b) for the two cases of the $t_{2g}^3 \, ^2A_{2g}$ initial state and the 100% excited initial state of the $t_{2g}^3 \, ^2E_g$, respectively. The change of the spectum can be seen clearly.

Finally the author would like to express his hearty thanks to the following friends and collaborators for allowing him to use the figures and quote the essence of the joint papers; Professor A. L. Schawlow, Professor Y. Tanabe, Professor I. Tsujikawa, Professor H. Kamimura, Dr. F. Varsanyi, Dr. A. H. Piksis, Dr. D. L. Wood, Professor W. Kaiser, Professor G. Kuwabara, Professor K. Aoyagi, Professor A. Misu, Professor T. Kushida, Professor M. Shinada, Dr. S. Ohnishi, Dr. K. Tsushima, and Professor T. Yamaguchi. The author would also like to express his sincere thanks to Professor N. Miura for providing him with the beautiful photograph (Fig.8) and the plot of the Zeeman splitting (Fig.9) of the R lines of ruby in ultra-high magnetic fields showing the perfect Paschen-Back effect. The author is also indebted to Miss T. Oto for her careful preparation of this camera-ready manuscript.

References

1. Y. Tanabe and S. Sugano: J. Phys. Soc. Japan 9, 753 (1954); ibid. 9, 766 (1954)
2. Lapraik: J. prakt. Chem. 47, 305 (1893)
3. F. H. Spedding and G. C. Nutting: J. Chem. Phys. 3, 369 (1935)

4. O. Deutschbein: Ann. d. Phys. [5] 14, 712 (1932); ibid. 14, 729 (1932); ibid. 20, 828 (1934)
   B. V. Thosar: Phys. Rev. 60, 616 (1941); J. Chem. Phys. 10, 246 (1942)
5. S. Sugano and Y. Tanabe: J. Phys. Soc. Japan 13, 880 (1958)
   S. Sugano and I. Tsujikawa: J. Phys. Soc. Japan 13, 899 (1958)
6. J. E. Geusic: Phys. Rev. 102, 1252 (1956)
7. H. Lehmann: Ann. d. Phys. [5] 19, 99 (1934)
8. G. Makhov, C. Kikuchi, J. Lambe and R. W. Terhune: Maser Action in Ruby, The University of Michigan 2616-1-T, June 1958
9. A. L. Schawlow and C. H. Townes: Phys. Rev. 112, 1940 (1958)
10. F. Varsanyi, D. L. Wood, and A. L. Schawlow: Phys. Rev. Lett. 3, 544 (1959)
11. S. Geschwind, R. J. Collins, and A. L. Schawlow: Phys. Rev. Lett. 3, 545 (1959)
12. S. Sugano, Y. Tanabe, and H. Kamimura: Multiplets of Transition-Metal Ions in Crystals (Academic Press, 1970) p.187
13. S. Sugano, A. L. Schawlow, and F. Varsanyi: Phys. Rev. 120, 2045 (1960)
14. A. L. Schawlow, A. H. Piksis, and S. Sugano: Phys. Rev. 122, 1469 (1961)
15. R. E. Watson, Mass. Inst. Technol., Solid-State and Molecular Theory Group, Tech. Rept. No.12 (June 15, 1959)
16. T. H. Maiman: Nature 187, 493 (1960)
    R. J. Collins, D. F. Nelson, A. L. Schawlow, W. Bond, C. G. B. Garrett, and W. Kaiser: Phys. Rev. Lett. 5, 303 (1960)
17. A. L. Schawlow: Quantum Electronics edit. by C. H. Townes (Columbia University Press, New York, 1960)
18. A. Javan, W. R. Bennett Jr., and D. R. Herriott: Phys. Rev. Lett. 6, 106 (1961)
19. S. Sugano: Applied Optics 1, 295 (1962)
20. W. Kaiser, S. Sugano, and D. L. Wood: Phys. Rev. Lett. 6, 605 (1961)
21. K. Aoyagi, A. Misu, and S. Sugano: J. Phys. Soc. Japan 18, 1448 (1963)
22. S. Sugano and M. Peter: Phys. Rev. 122, 381 (1961)
23. N. Miura and F. Herlach: Strong and Ultrastrong Magnetic Fields and their Applications edit. by F. Herlach (Springer Verlag, Berlin, Heidelberg, 1985) p.330
24. K. Aoyagi, A. Misu, G. Kuwabara and S. Sugano: J. Phys. Soc. Japan 19, 412 (1964)
25. K. Aoyagi and S. Sugano: J. Phys. Soc. Japan 45, 837 (1978)
26. T. Kushida: J. Phys. Soc. Japan 21, 1331 (1966)
    M. Shinada, S. Sugano, and T. Kushida: J. Phys. Soc. Japan 21, 1342 (1966)
27. J. Margerie: CR Acad. Sci. (Paris) 255, 1598 (1962)
28. S. Ohnishi and S. Sugano: Japanese J. Appl. Phys. 21, L309 (1982)
29. H. G. Drickamer: Solid State Physics edit. by F. Seitz and D. Turnbull (Academic Press, New York, 1965) vol.17, p.1
30. H. Adachi, S. Shiokawa, M. Tsukada, C. Satoko and S. Sugano: J. Phys. Soc. Japan 47, 1528 (1979)
31. Y. Sato and S. Akimoto: J. Appl. Phys. 50 (8), 5285 (1979)
32. R. L. Greene, D. D. Sell, W. M. Yen, A. L. Schawlow, and R. M. White: Phys. Rev. Lett. 15, 656 (1965)
33. Y. Tanabe, T. Moriya, and S. Sugano: Phys. Rev. Lett. 15, 1023 (1965)
34. S. Sugano, K. Aoyagi, and K. Tsushima: J. Phys. Soc. Japan 31, 706 (1971)
35. S. Sugano, Y. Miwa, and T. Yamaguchi: Phys. Rev. Lett. 44, 1527 (1980)

Ruby – Solid State Spectroscopy's Serendipitous Servant

G.F. Imbusch[1] *and W.M. Yen*[2]

[1]Department of Physics, University College, Galway, Ireland
[2]Department of Physics and Astronomy, The University of Georgia,
  Athens, GA 30602, USA

1. Introduction

Ruby is a handsome gemstone whose color can vary from pale pink to a deep
purple-red, depending on the percentage of chromium present in the crystal.
Large naturally-occurring good quality specimens are rare and highly-prized.
The hardness and high refractive index of the material gives the cut gemstone
a particular brilliance.   In addition, the "cold fire" of ruby - its deep
red luminescence - adds to the mystique and aura of this beautiful crystal.

Ruby is also a most interesting scientific material.   It is crystalline
Al_2O_3 (sapphire) with a dilute doping of Cr^{3+} ions (usually much less than
1%) replacing some Al^{3+} ions.   Its spectroscopic properties have been
studied for over one hundred years;   the early work has been described briefly
by MOLLENAUER /1/.   In a treatise on light (*La Lumière, Ses Causes et Ses
Effets*) published in 1867 Edmond BECQUEREL /2/ devoted nearly a full chapter
to *Alumine et ses Combinaisons*, whose luminescence he studied with the aid of
his "phosphoroscope" (Fig. 1).   In this device the sample was excited by a
beam of sunlight and its luminescence viewed through a prism spectrometer a
controllable time after the exciting sunlight beam had been shut off.   With
this apparatus Becquerel was able to observe the two intense sharp
luminescence lines of ruby and to correctly estimate the lifetime at around
4 ms.   Because the luminescence could be observed from apparently "pure"
samples of alumina Becquerel claimed that the emission was the result of some
intrinsic property of the Al_2O_3 crystal and that chromium merely played the
role of activator.   This view was challenged by BOISBAUDRAN /3/, and a
protracted debate ensued which is well chronicled in the literature /1/.

The first high-resolution study of the ruby spectrum was that of DU BOIS
and ELIAS /4/ in 1908 who measured the absorption and emission spectra.   It
was they who first used the label R ("rot") to designate the sharp luminescence
lines.   These workers, as well as J. BECQUEREL /5/ and MENDENHALL and WOOD
/6/, observed and attempted to analyze the Zeeman splitting of the R lines.
In addition, du Bois and Elias reported the presence of neighboring sharp
lines (N lines) in the vicinity of the R lines.   In a series of papers
beginning in 1932 DEUTSCHBEIN /7/ reported extensive studies of chromium
luminescence in a number of crystalline host materials.

The proper theoretical analysis of the ruby spectrum awaited a number of
major steps.   First the method of quantum mechanics had to be developed to
permit a proper description of the interaction between radiation and atoms.
The method of determining the splitting of atomic energy levels by the
electrostatic fields of neighboring ions in a solid was explained by BETHE /8/

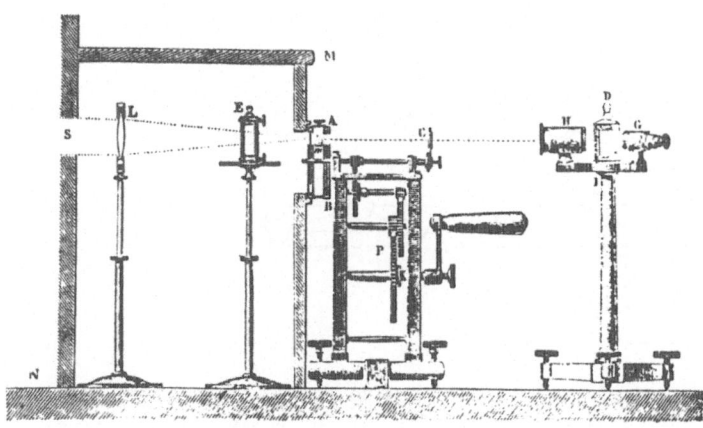

Fig. 1. Becquerel's phosphoroscope. Sunlight enters at S and is focussed
onto the sample through a filter at E. By turning the handle a pair of
segmented disks, one on either side of the sample, are rotated, which
alternatively block the sunlight and the path of the luminescence to the
spectrometer.

in 1929. Finally, the systematic analysis of the spectroscopic properties of
the Cr^{3+} ion in ruby was included in a series of papers, beginning in 1954, by
SUGANO, TANABE, and KAMIMURA /9/ which have provided a solid basis for modern
studies of the spectroscopy of transition metal ions in solids. As a check
on the theory in the case of ruby SUGANO and TSUJIKAWA /10/ made precise
measurements on the anisotropy and Zeeman effect in ruby. These experiments,
incidentally, were carried out under difficult conditions, as Sugano relates
in his review paper ("Spectroscopy of Solid-State Laser Materials") in this
volume.

Electron paramagnetic resonance was observed in the ground state of Cr^{3+} in
ruby in 1955 by MANENKOV and PROKHOROV /11/ and maser operation was
successfully achieved in ruby by MAKHOV et al./12/. In thinking about
materials which might exhibit optical maser action Art Schawlow was attracted
to ruby by virtue of its sharp intense luminescence lines, to which his
attention had been drawn by the experiments of Sugano and Tsujikawa. The
credit for producing the first operational laser goes to MAIMAN /13/; the
active medium was ruby.

Mainly because of its importance as a laser material methods have been
perfected for growing large single-crystal specimens of ruby of excellent
quality, and the availability of these crystals has greatly assisted research
into its spectroscopic properties. We both began our training in solid state
optical spectroscopy in Art Schawlow's laboratory at Stanford University where
we were introduced by him to this unique material which was then, and which
continues to be, the testing ground for many of the ideas of solid state
spectroscopy. We will describe some of these studies in this paper.

2. The Spectroscopy of the Cr^{3+} Ion in Ruby

Part of the α-Al_2O_3 crystal structure is shown in Fig. 2(a), the smaller dark
circles represent Al^{3+} ions while the larger open circles represent O^{2-} ions.

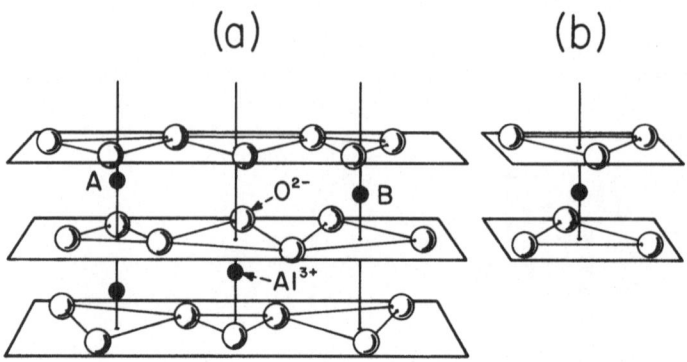

(a)   (b)

Fig. 2(a). Part of the α-Al$_2$O$_3$ crystal showing the A and B sites of the Al^{3+} ions.   The Al site has trigonal symmetry about the optic axis (vertical line through the Al ion).
Fig. 2(b). Al^{3+} ion in a site of octahedral symmetry.

There is a similar arrangement of oxygen ions about each Al^{3+} ion - a large triangle and a small triangle of oxygen ions, one above and one below the Al^{3+} ion - but these arrangements are rotated relative to each other on the different Al^{3+} sites, as Fig. 2(a) shows.   The vertical lines through the Al^{3+} ions in Fig. 2(a) give the direction of the optic axis of the crystal.

In ruby a small fraction of the Al^{3+} ions are substituted by Cr^{3+} ions. When present in very dilute amounts these ions can be considered too far apart to interact with each other.   We then have the equivalent of a dilute "gas" of Cr^{3+} ions in an inert sapphire host matrix.   The interesting spectroscopic properties of the ruby crystal derive from these Cr^{3+} ions.   In contrast to the case of a normal gas, whose particles are constantly in translational motion, the constituents of the Cr^{3+} gas are confined to specific Al^{3+} sites in the sapphire lattice.   All of these ions are, in principle at least, in identical sites and are subject to the same kind of electrostatic field (the *crystal field*) of neighboring O^{2-} and Al^{3+} ions.   Because they are identical and isolated from each other in dilute ruby we need only confine our attention to a single representative Cr^{3+} ion and calculate its spectroscopic properties. Then, using suitable statistical considerations, we can predict the behaviour of a macroscopic sample of dilute ruby.   Although each chromium ion is confined to a specific site in the sapphire host it is not rigidly fixed in space since it participates in the vibrational motion of the host lattice. However, it is convenient, initially at least, to ignore this vibrational motion and to imagine that the representative Cr^{3+} ion and its neighboring Al^{3+} and O^{2-} ions are at rest.

In this paper we are concerned with optical and, to a lesser extent, paramagnetic resonance studies of ruby.   The properties of interest derive from the interaction of the electromagnetic radiation with the outer unpaired 3d electrons of the Cr^{3+} ion.   The inner core electrons do not actively participate in the interaction and so can be ignored.   This greatly simplifies the analysis of the behaviour of the Cr^{3+} ions.   The filled inner core of electrons can be considered to form a time-average spherically-symmetric charge distribution.   This and the nucleus can be regarded as creating an electrostatic *central field* potential $V_{central}(\vec{r})$, in which the outer electrons move.   In such a potential orbital and spin angular momenta of the individual outer electrons are good quantum numbers, and these one-electron states are classified by their n, ℓ values. The outer three electrons of Cr^{3+} are in n = 3, ℓ = 2 states, so Cr^{3+} has the configuration $3d^3$.

Next we must take into account the interaction among these outer 3d electrons. This strong Coulomb interaction is described by the energy term $V_{Coulomb} = \sum e^2/(4\pi\varepsilon_0 r_{ij})$, where $r_{ij} = |\vec{r}_i - \vec{r}_j|$, i and j being indices labelling the outer electrons. This Coulomb interaction and the requirement of the Pauli principle split the energy of the $3d^3$ configuration into a number of distinct states (called *terms*) classified by the total orbital angular momentum L and total spin angular momentum S. For the three outer electrons S can have values 1/2 or 3/2, so the terms are divided into spin doublets (S = 1/2) and spin quarters (S = 3/2). The lowest energy term is ^4F, in conformity with Hund's rule. The wavefunctions of these free-ion LS terms are of even parity and, to very good approximation, are derived solely from single-electron 3d states.

When the Cr^{3+} ion is substituted for an Al^{3+} ion in the Al_2O_3 crystal the outer 3d electrons of the chromium ion are affected by the electrostatic crystal field of the neighboring O^{2-} and Al^{3+} ions, which for the present we regard as rigidly fixed in position in the Al_2O_3 lattice. As Fig. 2(a) shows the arrangement of O^{2-} ions about each Al^{3+} site is seen to have three-fold symmetry around the optic axis through the Al^{3+} ion. This would be much more symmetrical if the two parallel triangles of O^{2-} ions (above and below the Cr^{3+} ion) were identical and symmetrically arranged, as in Fig. 2(b). In this symmetrical arrangement the perpendicular separation between the two triangles is $\sqrt{2}/\sqrt{3}$ times the length of the side of the oxygen triangle, and the Al^{3+} ion is midway between the triangles. This arrangement has octahedral (O_h) symmetry. The crystal field potential in this case is written $V_{O_h}^{(g)}(\vec{r})$, where the (g) superscript indicates that this potential function has even parity. We find that the actual arrangement of oxygen ions in ruby has a distorted octahedral arrangement; the crystal field potential is predominantly octahedral, but two weaker potential terms with lower symmetry (trigonal) also occur. We write these as $V_{trig}^{(g)}(\vec{r})$ and $V_{trig}^{(u)}(\vec{r})$. The first is of even parity, the second is of odd parity, as denoted by the superscript (u) and is a reflection of the fact that the Al^{3+} site in ruby lacks inversion symmetry. (Dopant Cr^{3+} ions in all insulating materials occupy sites with six negatively-charged anions nearby. The crystal field potential sometimes has full octahedral symmetry, but more often additional smaller terms of lower symmetry also occur.)

To solve for the electronic energy levels of the Cr^{3+} ion in ruby we can start by neglecting the small terms in the Hamiltonian, such as $V_{trig}^{(g)}$, $V_{trig}^{(u)}$, and spin-orbit coupling, and calculate the effect of $V_{O_h}^{(g)}$ on the LS terms. This octahedral crystal field term splits the free ion terms into crystal field *multiplets*. The energy levels of the crystal field multiplets for all $3d^n$ configurations have been calculated by Sugano et al./9/, and the separations between these multiplet levels are given in terms of three parameters (i) two Racah parameters, B and C, which are a measure of the strength of the Coulomb interaction among the outer electrons, and (ii) a single parameter, Dq, which describes the strength of the octahedral crystal field. Since $C \simeq 4B$ for all $3d^n$ systems the separations between the energy levels can be approximately described by two parameters, B and Dq. The splittings of LS terms for Cr^{3+} in an octahedral crystal field are shown in Fig. 3, where E/B is plotted against Dq/B. All of these wavefunctions have even parity, as they are formed from even-parity free-ion LS terms by the even-parity octahedral crystal field term. Odd-parity crystal field states,

251

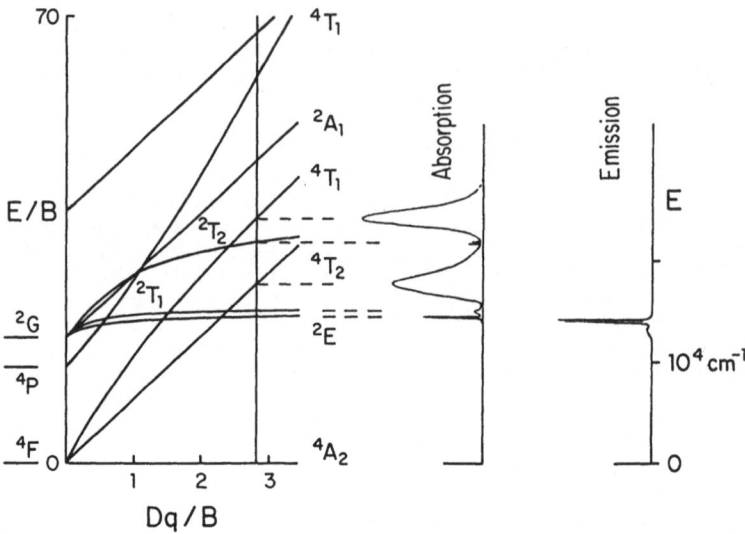

Fig. 3. Splitting of the Cr^{3+} free-ion levels in an octahedral crystal field. Low-temperature absorption and emission spectra are shown on the right.

formed from $3d^2 4p$ orbitals, occur some 50,000 cm^{-1} higher in energy. The solid vertical line drawn at $Dq/B \approx 2.8$ shows the value of Dq/B for the average octahedral crystal field in ruby.

Since we are neglecting spin-orbit coupling the total spin S is a good quantum number and the states are still labelled by their spin multiplicity (2S+1). Now, however, the orbital states are labelled according to the irreducible representations of the octahedral symmetry group. These representations are A_1, A_2, E, T_1, T_2, and this labelling scheme is used to describe the orbital levels in Fig. 3. On the right-hand side of the figure the low-temperature absorption and luminescence spectra of ruby are drawn. There is good agreement between the predicted energy levels and the absorption peaks. One observes that the spin-allowed transitions, $^4A_2 \rightarrow {}^4T_2, {}^4T_1,$ are much more intense than the spin-forbidden transitions, $^4A_2 \rightarrow {}^2E, {}^2T_1, {}^2T_2,$ as expected.

Next, the smaller terms, $V_{\text{trig}}^{(g)}$ and spin-orbit coupling, must be taken into account. These cause a splitting of all the octahedral levels, and we are particularly interested in the 29 cm^{-1} splitting of the ^2E level and the 0.38 cm^{-1} splitting of $^4A_2$. Since much of the subsequent discussion will involve the optical transitions between these two states, we show in Fig. 4 how these octahedral levels are split by the lower-symmetry crystal field and spin-orbit coupling, and how these levels are further split by a magnetic field applied parallel to the optic axis.

Our analysis so far predicts *sharp* electronic energy levels and, hence, sharp optical transitions between the levels. However, as the absorption spectrum in Fig. 3 shows, some of the observed transitions are quite broad. This broadening is a consequence of the vibrational motion of the Al_2O_3 lattice including the dopant Cr^{3+} ions. These vibrations are quite large; even at the absolute zero of temperature the root-mean-square lattice displacement of an ion is around 5% of the inter-ion separation. As a result the Cr^{3+} ions

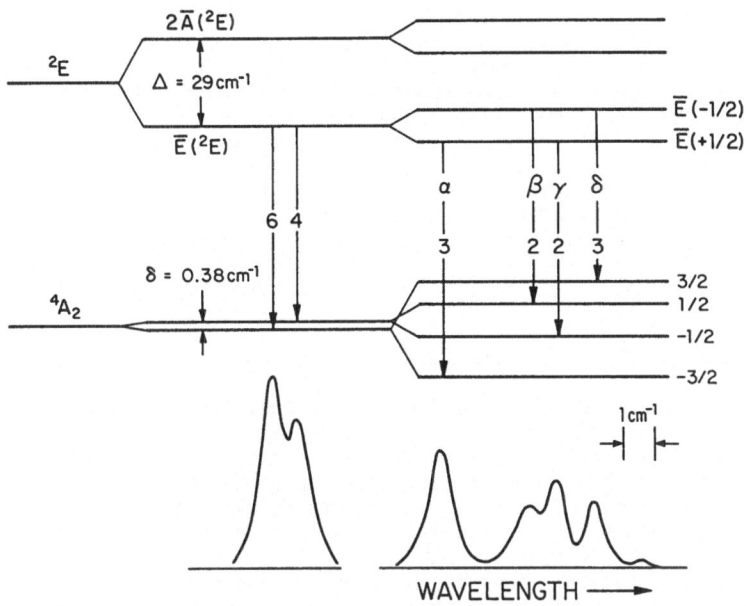

WAVELENGTH ⟶

Fig. 4. Under the trigonal crystal field and spin-orbit interaction the ^2E and $^4A_2$ levels exhibit splittings of 29 cm^{-1} and 0.38 cm^{-1}, respectively. These levels are further split by a magnetic field, B, along the optic axis. The R_1 line ($\bar{E}(^2E) \to {}^4A_2$) in emission at 2 K is shown for B = 0 and B ≃ 1.2 T.

experience a *dynamic* crystal field as well as the average *static* crystal field we have discussed earlier. Since the crystal field is modulated by the lattice vibrations the optical transitions are *frequency modulated* by the lattice vibrations. Consequently, the optical transitions appear as sharp zero-phonon lines accompanied by FM sidebands. As is clear from Fig. 3, a variation of around ±5% in Dq/B significantly changes the energies of the $^4T_1$ and $^4T_2$ levels relative to the $^4A_2$ ground state. Hence the $^4A_2 \to {}^4T_1$, $^4T_2$ transitions are strongly modulated and are dominated by broad FM sidebands. On the other hand, the $^4A_2 \leftrightarrow {}^2E$ transition, which appears in absorption and in emission, is seen to be affected to a much smaller extent by the variation in crystal field, and the zero-phonon lines are the dominant feature of this transition.

Finally we take account of the odd-parity crystal field term, $V_{trig}^{(u)}$. This has a negligible effect on the energy and splitting of levels; its most important consequence is a mixing of some odd-parity wavefunctions with the even-parity $3d^n$ wavefunctions. As a result the $^4A_2 \leftrightarrow {}^2E$ zero-phonon lines in ruby occur by a weak electric dipole process, which is an order of magnitude stronger than the analogous magnetic dipole process.

Ions raised to the $^4T_2$ and $^4T_1$ levels by optical absorption lose some energy to the lattice (as lattice vibrations) and quickly drop to the ^2E level, which is the lowest excited state and is about 14,000 cm^{-1} above the ground state. The radiative transition between the ^2E and $^4A_2$ levels is spin-forbidden and has a relatively long lifetime (≃ 3.6 ms at low temperatures). Despite its small transition probability the radiative emission process is the most effective decay process out of the ^2E level, the luminescence process at room temperature and lower temperatures being almost 100% efficient.   Two sharp zero-phonon lines dominate this transition: the R_2 line between the upper ^2E level ($2\bar{A}(^2E)$) and the $^4A_2$ ground state, and the R_1 line between the lower ^2E level ($\bar{E}(^2E)$) and the ground state.   The strong broad absorption bands of ruby in the green and blue permit efficient optical pumping of the ^2E state.   The long lifetime of this state, then, allows a sizeable population to be maintained there by continuous optical pumping. These properties were put to good use when the first laser was constructed using ruby as the active medium, with laser action taking place in the R_1 line.

3. The Zero-phonon R_1 and R_2 Luminescence Lines

A matter of some importance in the early days of the ruby laser was to gain an understanding of the origin of the broadening of the R_1 line.   That this broadening could be due in part to strains in the crystal was suspected by Schawlow who noticed that the more perfectly-grown specimens exhibited the narrowest lines.   This was confirmed by him and others when the R lines were found to shift in wavelength if the material was subjected to static stresses in controlled laboratory experiments.   Random strain fields occur in all ruby crystals which lead to variations in the time-average crystal field among the Cr^{3+} sites and, consequently, to a distribution of R_1 frequencies. The situation is illustrated schematically in Fig. 5. Cr^{3+} ions in sites of identical strain have a small homogeneous broadening at low temperatures and emit a very sharp R_1 line.   The observed R_1 line from a ruby crystal at low temperatures is a composite of such emissions from Cr^{3+} ions in a range of different crystal field sites; the observed line exhibits *inhomogeneous* broadening.   The fluorescence line-narrowing (FLN) experiment of SZABO /14/ illustrates this inhomogeneous broadening very clearly.   In this experiment a narrow band laser, tuned to a frequency within the R_1 line, excites a subset of the Cr^{3+} ions - those in resonance with the laser - and these excited Cr^{3+} ions emit their luminescence in a sharp line whose width is determined by the width of the laser beam and by the *homogeneous* width of the transition, not by the width of the inhomogeneous line.   In his initial FLN experiments Szabo achieved a linewidth of around 30 MHz, a factor of 100 smaller than the normal inhomogeneous linewidth.   Szabo's experiments in ruby were the first demonstration of fluorescence line narrowing in a solid. Subsequent experiments employing FLN and hole-burning techniques by MURAMOTO et al./15/, and JESSOP, MURAMOTO and SZABO /16/ have succeeded in reducing the observed low-temperature homogeneous linewidth of the R_1 line to a few MHz.

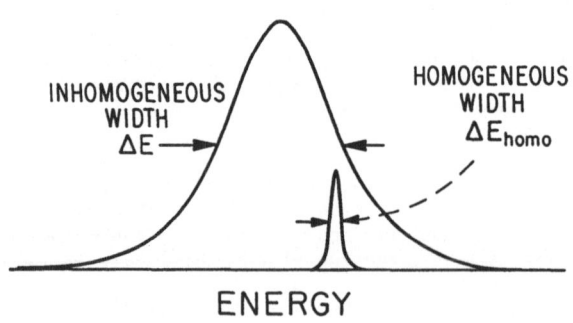

Fig. 5.   The inhomogeneously broadened emission line is a composite of narrower emission lines from ions in different strain environments.

As the temperature is raised the homogeneous broadening of the R lines increases significantly, as SCHAWLOW's early measurements showed /17/. The origin of the broadening is a second-order effect due to interaction with lattice vibrations, in which two lattice vibrations of the same frequency interact with the Cr^{3+} ion to cause a broadening of the zero-phonon line. In a quantum description this broadening is ascribed to a Raman scattering of phonons, and a theoretical formula for the broadening was first developed by McCUMBER and STURGE /18/. They also made accurate measurements of the temperature-dependent widths of the R_1 and R_2 lines of ruby and found excellent agreement with the theoretical formula. There is also a variation with temperature of the central frequency of the zero-phonon line which comes about because the mean crystal field energy varies slightly as the lattice vibrational energy increases. The calculations of McCumber and Sturge show that the frequency of the zero-phonon line varies as the mean-square lattice displacement. Fig. 6 shows the good agreement between their measurements (open circle) of the shift of the R_1 line of ruby and the temperature-dependence predicted theoretically (solid curve). McCumber and Sturge included some measurements of the variation of the R_1 line with temperature (the triangles in Fig. 6) made by GIBSON /19/ in 1916. The temperature-dependent shift and broadening of sharp optical transitions, first elucidated in ruby, are general effects in the spectroscopy of doped transition metal and rare earth ions.

The occurrence of an energy shift proportional to the mean-square lattice displacement has an interesting consequence. Chromium occurs in four

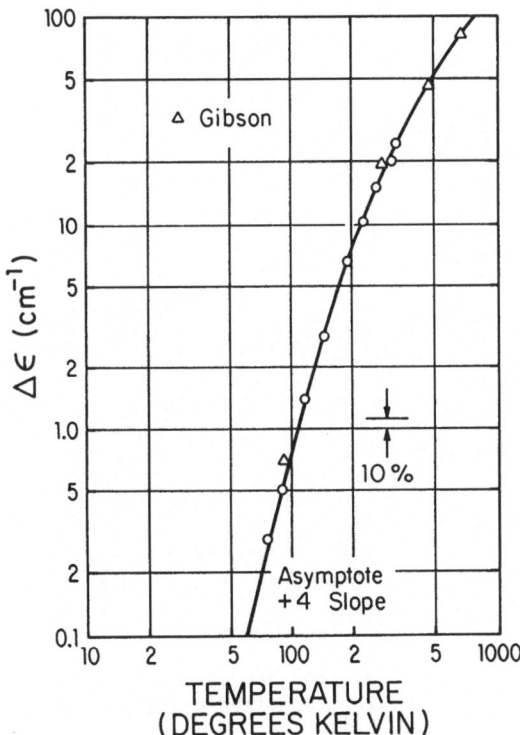

Fig. 6. Shift of the R_1 line as a function of temperature (from McCumber and Sturge /18/).

reasonably abundant isotopes: $Cr^{50}(4.3\%)$, $Cr^{52}(83.6\%)$, $Cr^{53}(9.6\%)$, and $Cr^{54}(2.4\%)$. Now the lighter isotopes are expected to vibrate with larger amplitude than the heavier isotopes, so the mean-square lattice displacement, and consequently the energy shift, is greater for the lighter isotopes. This should lead to a slight variation of the position of the zero-phonon line with isotopic mass. Such an isotope shift was originally found in the R_1 line of ruby by SCHAWLOW /20/ and his spectrum is seen in Fig. 7(a). This spectrum was taken at helium temperatures, where the homogeneous broadening is a minimum but where there is still a sizeable mean-square lattice displacement due to zero-point vibrations. Figure 7(a) shows the two components of the R_1 line due to the 0.38 cm^{-1} ground state splitting, and the isotope shift is seen as barely resolvable structure on these components. JESSOP and SZABO /21/ utilized FLN techniques to resolve the isotope shift with much greater clarity. Their spectrum is shown in Fig. 7(b) and illustrates the great experimental advantages of laser excitation spectroscopy.

Even since the development of the first laser spectroscopists have employed this unique tool in diverse ways in their laboratory. In the beginning, when only the ruby laser was available, it was natural to study the resonant interaction of the ruby laser beam with a target ruby sample. The phenomena of *photon echoes* analogous to the concept of spin echoes originally demonstrated by HAHN /22/, was observed in ruby by ABELLA, HARTMANN, and KURNIT /23/. This is but one example of a class of phenomena which can be classified as *coherent radiation spectroscopy*.

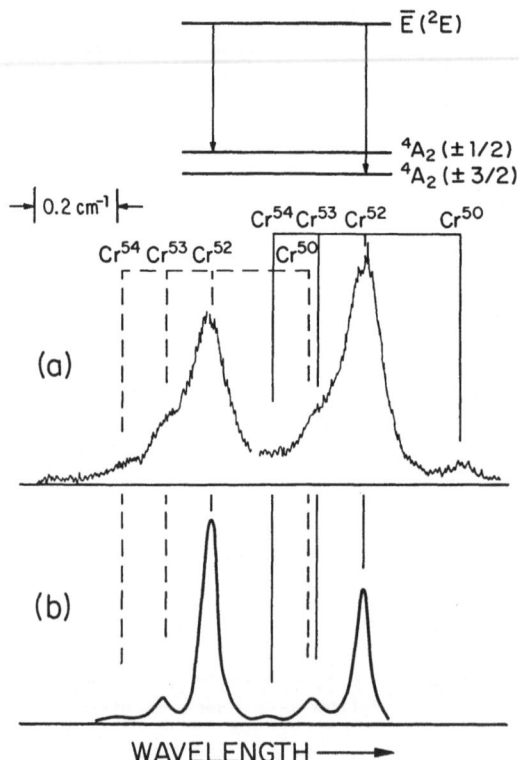

Fig. 7. Structure on the R_1 line due to ground state splitting and isotope shift: (a) Schawlow's original spectrum, (b) the spectrum of Jessop and Szabo obtained by laser excitation.

Because of the narrow width of the R_1 and R_2 lines at low temperatures and the high density of Cr^{3+} ions present in a typical ruby sample, the peak absorption in the R lines can be quite strong. Thus when an excited Cr^{3+} ion decays with emission of R_1 or R_2 radiation the emitted photon may be reabsorbed ("trapped") by an unexcited Cr^{3+} ion before it leaves the crystal. That this phenomenon was occurring in ruby was first recognized by VARSANYI, WOOD, and SCHAWLOW /24/ and was the first example of this effect, well known to occur in gaseous spectroscopy, to be reported in a solid. Trapping acts as a mechanism for the radiative transfer of resonant optical excitation throughout the full ruby crystal. Its most noticeable consequence is a lengthening of the observed radiative lifetime, from its intrinsic value of around 3.6 ms found in dilute powdered samples, to over 10 ms in large heavily-doped samples (NELSON and STURGE /25/).

Under intense optical pumping into the broad absorption bands at low temperatures the final step of the nonradiative relaxation process brings the Cr^{3+} ion across the 29 cm^{-1} gap from $2\bar{A}(^2E)$ to $\bar{E}(^2E)$ and can result in the generation of a high density of 29 cm^{-1} phonons /26/. Interesting experiments can be carried out to study the dynamics of these nonequilibrium phonons /27,28,29,30/.

4. Magnetic Resonance Studies in Ruby

Electron paramagnetic resonance (EPR or ESR) is a high-resolution technique for probing the fine structure of electronic levels. The material is irradiated with microwave power of a specific frequency, the electronic splitting can be tuned by applying a variable magnetic field, and the occurrence of microwave absorption is detected by use of a balanced bridge network. The microwave photon is a finer probe than the optical photon, but the detection of microwave photons is a much more difficult task than the detection of optical phonons, and as a result EPR is only possible when a large number of ions interacts with the microwave radiation. EPR studies, then, are restricted to ground electronic states. (Under constant optical pumping a large population can be maintained in some *excited* triplet states in some organic solids, and conventional EPR studies can be made in these states. Such situations do not occur in doped inorganic materials.) EPR in the ground state of ruby was originally achieved in 1955 /11/; a very comprehensive study of ruby ground state EPR was published by SCHULZ-DU BOIS /31/.

In order to carry out an EPR study in an excited electronic state where, even with intense optical pumping, the density of ions will be small, a much more sensitive detection technique than that employed in conventional EPR must be used. Such a technique was proposed by GESCHWIND, COLLINS, and SCHAWLOW /32/ in the case where the excited electronic state emits luminescence. The scheme is based on the fact that when one microwave photon is absorbed in the excited state the luminescence pattern is changed by one optical photon; the occurrence of microwave absorption can then be detected with much greater sensitivity by monitoring the luminescence.

We have seen how the ^2E state of Cr^{3+} in ruby is efficiently pumped by broadband optical sources and has a relatively long lifetime, so a sizeable population can be maintained in this state. Geschwind proposed to carry out his experiment for the optical detection of magnetic resonance (ODMR) in this state. A number of detection schemes were outlined by him, one of which we can illustrate with the aid of Fig.4, which shows the Zeeman splitting of the $\bar{E}(^2E)$ level, from which the R_1 line originates. At helium temperatures the lower Zeeman level ($\bar{E}(\frac{1}{2})$) has the greater population and, as a consequence, the α line, which originates on this level, is the most intense optical Zeeman component.   If resonant microwave power is strongly absorbed in the $\bar{E}(^2E)$ state the populations in the two Zeeman levels become equal and the α line is reduced in intensity.   Thus the onset of EPR in the $\bar{E}(^2E)$ state is detected by monitoring the intensity of the α component.   The pioneering ODMR experiments of Geschwind et al./32/ were carried out in ruby in 1959.   Since then the technique of ODMR has been developed into a very useful and sensitive probe, particularly in studies of doped semiconductor materials.

When the resonant microwave power is switched off the intensity returns to its equilibrium value with a time constant which is determined in part by the spin-lattice relaxation between the two Zeeman levels, $\bar{E}(1/2)$ and $\bar{E}(-1/2)$. ODMR measurements of this spin-lattice relaxation time (T_1) were made by GESCHWIND et al./33/ and these are shown in Fig. 8.   The temperature dependence shows that this is clearly a two-phonon Orbach process, proceeding via the $2\bar{A}(^2E)$ level 29 cm^{-1} higher in energy, the Orbach process predicting

$$T_1 = C \exp(\Delta/kT), \tag{1}$$

where Δ is the energy gap (29 cm^{-1}) between $2\bar{A}(^2E)$ and $\bar{E}(^2E)$.

At 3 K the spin lattice relaxation time in the $\bar{E}(^2E)$ state has the same value as the radiative decay time of the $\bar{E}(^2E)$ state;   below this temperature the spin-lattice relaxation rate is slower than the radiative decay rate. This fact could have a bearing on the optical detection of magnetic resonance in ruby.   In order for magnetic resonance to occur there must be a greater population in the lower $\bar{E}(1/2)$ level than in $\bar{E}(-1/2)$ so that microwave power will be absorbed.   How is this population difference achieved?   The optical pumping process involves first raising the ions to the $^4I_2$ or $^4T_1$ levels, after which the ions relax by phonon emission until the $\bar{E}(^2E)$ state is

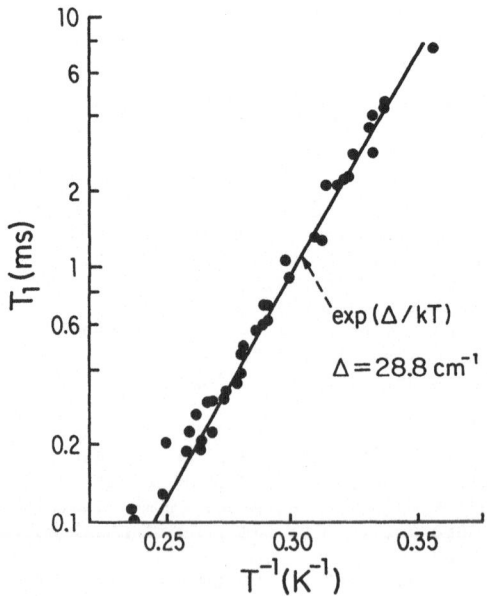

Fig. 8. Spin-lattice relaxation time in the $\bar{E}(^2E)$ state of Cr^{3+} in ruby (after Geschwind et al./33/)

reached. At the conclusion of this indirect pumping process it might be assumed that both the $\bar{E}(1/2)$ and $\bar{E}(-1/2)$ levels are equally populated. Next spin-lattice relaxation in the $\bar{E}$ state causes an adjustment of population ("thermalization"), giving a larger population in the lower level. However, thermalization can only occur if the spin-lattice relaxation rate is faster than the radiative decay rate (which, as Fig. 8 shows, occurs above 3 K in ruby); if there is not enough time for thermalization to occur the populations stay equal, no microwave absorption occurs, and no ODMR signal should be found. Such, however, was not found experimentally; the ODMR signal became stronger as the temperature was reduced below 3 K. Clearly the optical pumping process itself leads *directly* to a population difference between the $\bar{E}(-1/2)$ and $\bar{E}(1/2)$ levels. This would occur if there were spin selection rules in the broadband ($^4A_2 \rightarrow {}^4T_2$, $^4T_1$) optical pumping of the ^2E state. At low temperatures the Cr^{3+} ions in the ground $^4A_2$ state are preferentially in the lower spin states, under broadband optical pumping they retain ground state *spin memory*, and at the end of the pumping cycle they preferentially populate the lower spin level of the $\bar{E}(^2E)$ state. That such a spin memory occurs in the optical pumping of ruby was experimentally verified by changing the population in the *ground* spin states during optical pumping and detecting the resultant change in population in the ^2E spin states (IMBUSCH and GESCHWIND /34/). Although it was unambiguously demonstrated for the first time in ruby, the phenomenon of spin memory is of very general occurrence and has been demonstrated in many other luminescence systems.

5. Interaction between Cr^{3+} Ions in Ruby - Excitation Transfer

The picture of ruby so far presented is of isolated Cr^{3+} ions randomly distributed in Al^{3+} sites in the sapphire crystal. If the ions are close enough for a weak interaction to occur between them new effects will manifest themselves. For example, an excited Cr^{3+} ion may transfer its excitation *nonradiatively* to a nearby Cr^{3+} ion with which it interacts, and there may be a number of such transfers before the excitation is released as a photon. We can try to distinguish *resonant* nonradiative transfer - in which excitation is transferred between ions whose excited levels coincide to within the homogeneous linewidth - from *nonresonant* nonradiative transfer between nearby ions whose excited levels are sufficiently different that a phonon must be included in the process to compensate for the energy mismatch.

The occurrence of nonresonant transfer among the Cr^{3+} ions in ruby was clearly demonstrated in the laser excitation studies of SELZER et al. /35/. Using a pulsed narrowband laser tuned to a frequency within the R_1 inhomogeneous line they observed the usual FLN signals (the sharp lines in Fig. 9) immediately after the excitation pulse. As the narrow lines decreased in intensity with time a broad background, identical with the R_1 inhomogeneous lineshape, grew, as Fig. 9 shows. This *spectral diffusion* of excitation from the FLN lines to the full inhomogeneous line is a manifestation of excitation transfer from the Cr^{3+} ions excited by the laser to the main body of Cr^{3+} ions. This nonresonant transfer process was analyzed by HOLSTEIN, LYO, and ORBACH /36/. The question of resonant transfer in ruby will be discussed in a later section.

6. Exchange-coupled Pairs of Cr^{3+} Ions

As the chromium concentration is increased the probability also increases that two Cr^{3+} ions in ruby may be situated sufficiently close to each other for a strong exchange interaction to occur between them. Such an exchange-coupled

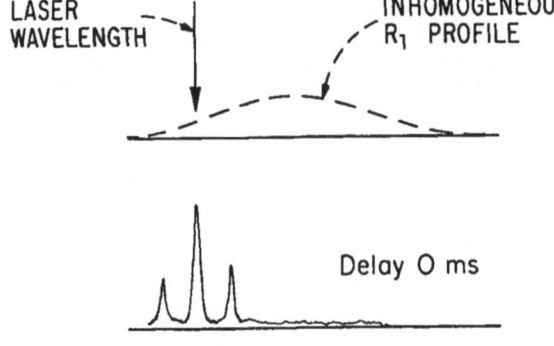

LASER WAVELENGTH

INHOMOGENEOUS R_1 PROFILE

Fig. 9. FLN signals from a 0.9 at.% ruby at 10 K. The laser excitation is on the high energy side of the line center. The delay indicates the time after the pulsed laser excitation (from Selzer et al./35/).

Delay 0 ms

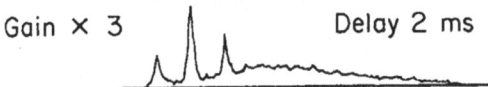

Gain × 3

Delay 2 ms

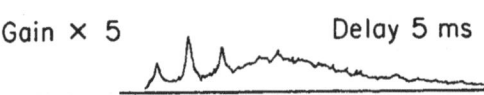

Gain × 5

Delay 5 ms

ion pair is a distinct spectroscopic center with distinct optical transitions. Some additional sharp luminescence lines are found in the vicinity of the R lines in ruby of medium to heavy doping, these were labelled N lines ("Nebenlinien") by Deutschbein /7/, and were correctly interpreted by SCHAWLOW, WOOD, and CLOGSTON /37/ as originating on exchange-coupled pairs of Cr^{3+} ions. In the Al_2O_3 lattice first-, second-, third-, and fourth-nearest Al^{3+} neighbors are sufficiently close for a strong exchange interaction to occur between Cr^{3+} ions occupying such near-neighbor sites. Thus there is a plethora of sharp lines due to exchange-coupled Cr^{3+} pairs in heavily-doped ruby. The process of identifying these lines with specific pair types was a formidable task which was addressed by a number of workers. The technique employed by MOLLENAUER and SCHAWLOW /38/ and by KAPLYANSKII and PRZHEVUSKII /39/ was to apply uniaxial stress along various crystallographic directions in ruby and observe the effects on the pair lines. The principle of these experiments is that when the stress is along the line joining the two ions of a specific pair the optical transitions on this pair are most strongly affected. In this way particular N lines can be ascribed to specific pair types.

When both Cr^{3+} ions of the pair are in the $^4A_2$ ground state the interaction between the ions can be written in the form

$$H_{EX} = -J\,\vec{S}_1 \cdot \vec{S}_2 + j(\vec{S}_1 \cdot \vec{S}_2)^2 . \tag{2}$$

The first term is the bilinear exchange term, where J is the exchange integral whose value is determined by experiment. The second term (biquadratic exchange) is small and arises from exchange striction effects. We will only concern ourselves with the values of J.

The eigenstates of this Hamiltonian can be classified according to the values of the total spin, $\vec{S} = \vec{S}_1 + \vec{S}_2$ of the two-ion system, and in the case of two-spin 3/2 ions S can have values 0, 1, 2, 3. If J is negative (positive) the exchange is antiferromagnetic (ferromagnetic). Fig. 10 shows the luminescence transitions on a fourth-nearest neighbor pair at 1.6 K. The pattern of lines corresponds to ferromagnetic coupling with J = 7.0 cm^{-1}.

Table 1 summarizes the findings for the first four types of exchange-coupled Cr^{3+} ion pairs in ruby. The values of dJ/dr for first- and second-nearest neighbors were obtained by HEBER and PLATZ /40/ from stress experiments. The data were used by them to compute values for J for the first- and second-nearest neighbors in the antiferromagnetic material Cr$_2$O$_3$. This has the same crystal structure as ruby, but the separations between ions are larger than in ruby. For the first-nearest neighbors the separation is 2.55 A in ruby and 2.65 A in Cr$_2$O$_3$. Taking the values of J and dJ/dr for ruby and using a linear extrapolation one estimates J = -78 cm^{-1} for the first-nearest neighbor pair in Cr$_2$O$_3$. The measured value is -86±6 cm^{-1}. One similarly calculates J = -30 cm^{-1} for the second-nearest neighbor pair in Cr$_2$O$_3$; the measured value is -38±6 cm^{-1}.

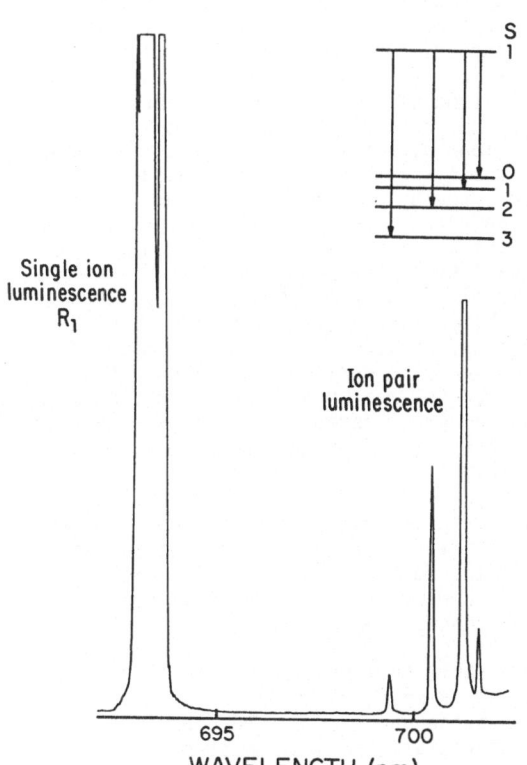

Fig. 10. Luminescence from a ruby of medium concentration at 2 K. The four weak lines originate on fourth-nearest neighbor ion pairs. The strongest pair line above is the N$_2$ line.

Table 1. Values of the exchange parameter for near-neighbor pairs in ruby

Neighbor type	Separation /A/	J /cm^{-1}/	dJ/dr /cm^{-1}A^{-1}/
1	2.55	-115	3.7×10^{-2}
2	2.78	-109	8.0×10^{-2}
3	3.18	-11.7	
4	3.50	+ 7.0	

7. More on Excitation Transfer

In their paper in which they identified the N lines as luminescence transitions on exchange-coupled pairs Varsanyi et al. /24/ pointed out that the intensity of the N lines was larger than one would expect from statistical considerations of the probability of the occurrence of pairs. They suggested that the pair lines might derive part of their intensity by energy transfer from single ions.   That such a transfer occurs is clear from a study of the decay patterns of R and N lines after broadband pulsed excitation (IMBUSCH /41/); the R line decay is single exponential while the N line pattern has an initial fast decay rate (the intrinsic decay rate of the pairs) followed by a slow decay indistinguishable from the R line decay. The slowly-decaying component was attributed to excitation transfer from the single ions.   A more direct demonstration of this transfer was provided by SELZER et al./42/ who used a pulse of narrowband laser excitation at the R_1 line frequency to excite single Cr^{3+} ions directly into the $\bar{E}(^2E)$ level in a ruby of 0.51 at.% chromium.   After the laser pulse was extinguished R_1 line and N line luminescence was observed; the N_2 luminescence pattern had an initial rise followed by a decay whose rate strongly resembled the R_1 decay rate, precisely what is expected if the pairs derive excitation only by transfer from the single Cr^{3+} ions excited by the laser pulse.   Figure 11, taken from a later study by SELZER et al./35/, shows these decay patterns.
   Because of the strong resemblance between the R_1 decay pattern and the slowly-decaying part of the N_2 line decay pattern, the view was advanced that in heavily-concentrated ruby samples the pairs derived excitation not only from nearby excited single ions but from the main body of excited single ions, and this came about through a strong *resonant* nonradiative transfer through the single ions. Not all research workers were agreed that such a strong resonant energy transfer occurs in ruby.

   That a rapid nonradiative resonant transfer might be occurring among the Cr^{3+} ions prompted intense theoretical interest in the resonant transfer process in ruby since it might permit the observation of an *Anderson localization* phenomenon.   This would show up when the chromium concentration was gradually decreased, as an *abrupt* change from a condition of excitation migration over a macroscopic portion of the ruby sample to a condition where the excitation was localized in microscopic sections of the sample. Indirect experiments to demonstrate the existence of an Anderson localization gave conflicting results and underlined the need for a more direct method of measuring the spatial extent of the nonradiative migration of chromium excitation.

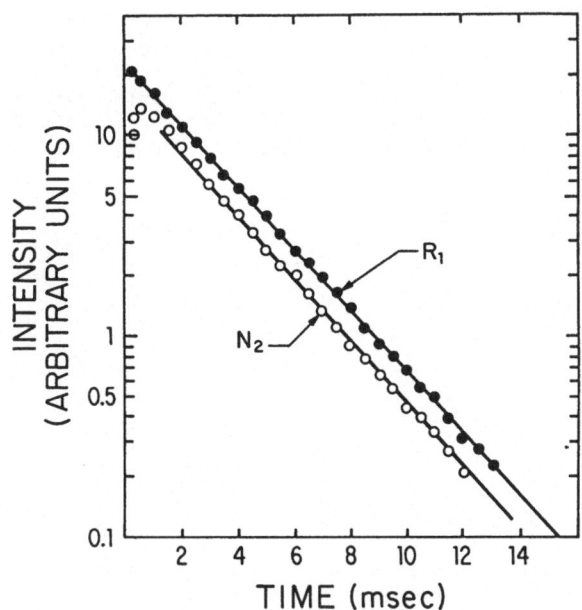

Fig. 11. Luminescence decay patterns of the R_1 and N_2 lines after pulsed laser excitation by a narrowband laser tuned to the R_1 transition (after Selzer et al./35/).

The first attempts to directly measure nonradiative spatial transfer employed the techniques of degenerate four-wave mixing (EICHLER et al./43/, LIAO et al./44/, HAMILTON et al./45/). In one such experiment two counter-propagating laser beams, $\vec{k}_1$ and $\vec{k}_2$, and a third beam, $\vec{k}_3$, making a small angle with $\vec{k}_1$, are directed at the ruby sample (Fig. 12). All the beams are derived from the same laser. The interference between beams $\vec{k}_1$ and $\vec{k}_3$ produces a spatial periodic variation in the density of excited Cr^{3+} ions which leads to a spatial periodic modulation in the refractive index in the ruby sample. This acts like a diffraction grating (Fig. 12) whose period is

$$d = \frac{\lambda}{2 \sin \frac{\theta}{2}} \cdot \tag{3}$$

where λ is the laser wavelength. Beam $\vec{k}_2$ is diffracted by this grating and by Bragg's law a diffracted beam ($\vec{k}_4$) will be observed along the $-\vec{k}_3$ direction. If beams $\vec{k}_1$ and $\vec{k}_3$ are turned off the gradual fading of the grating causes the intensity of the diffracted beam $\vec{k}_4$ to decrease in time as

$$I_4(t) = I_4(0) \exp(-t/\tau), \tag{4}$$

where

$$\frac{1}{\tau} = \frac{2}{\tau_r} + 2D \cdot (\frac{4\pi}{\lambda} \sin \frac{\theta}{2})^2, \tag{5}$$

τ_r is the radiative decay time of the Cr^{3+} ions, and D is the diffusion coefficient describing the transfer of excitation among the Cr^{3+} ions. All

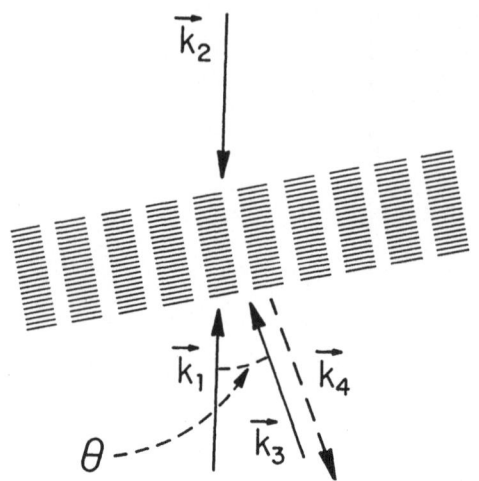

Fig. 12. Representation of the grating formed in a sample of ruby by laser beams $\vec{k}_1$ and $\vec{k}_3$. Laser beam $\vec{k}_2$ is diffracted by the grating, and a coherent diffracted beam is produced in the $\vec{k}_4$ direction.

the experiments concluded that there is no spatial energy diffusion within the limits of resolution; the upper limit for the diffusion was put at around 30 nm. This is still a large distance. What was needed was a much finer probe, and this was provided by the very ingenious experiment of CHU et al. /46/.

This experiment takes advantage of the inequivalence of the two types of Al^{3+} sites, labelled A and B in Fig. 2, of the $\alpha-Al_2O_3$ lattice. Because of the existence of the odd-parity $V_{trig}^{(u)}$ crystal field term there is an internal electric field, E_0, (pointing along the optic axis), acting at each Al^{3+} site, and this field acts in opposite directions on the Al^{3+} (or Cr^{3+}) ions in A and B sites. If an external electric field, E, is applied along the optic axis it adds to the internal field at one site but subtracts from the internal field at the other site. Because of the different resultant electric fields acting on them the Cr^{3+} ions in the A and B sites have their energies shifted relative to each other. The resultant linear *pseudo-Stark splitting*, which was originally discovered by KAISER, SUGANO, and WOOD /47/, is quite large. Thus, by applying an external electric field the R_1 lines from the A and B sites can be separated by about 1 cm^{-1}, enough to move the ions out of resonance with each other.

We can visualize the experiment of Chu et al. with the aid of Fig. 13. (a): With no external field the R_1 lineshapes of the A and B ions are identical. (b): With an external field applied the R_1 lines of the A and B ions are separated from each other. A fast pulse from a narrowband dye laser excites a subset of the A ions. An FLN signal is seen immediately after the pulse. This is indicated as a single line (shaded) in the figure, although in practice the FLN signal may consist of a number of sharp lines (because of ground state splitting). (c): The external field is switched off for a specified time; this brings the A and B ions back into resonance so that the excited A ions may transfer excitation to those B ions which are in resonance with them - if such a resonance transfer is possible. (d): The external field is switched on, again separating the A and B ions. One looks for a FLN signal from the B ions which is indicated by the second shaded component in the figure. The strength of this second component is a measure of the resonant transfer probability between the ions. The conclusion of Chu et al.

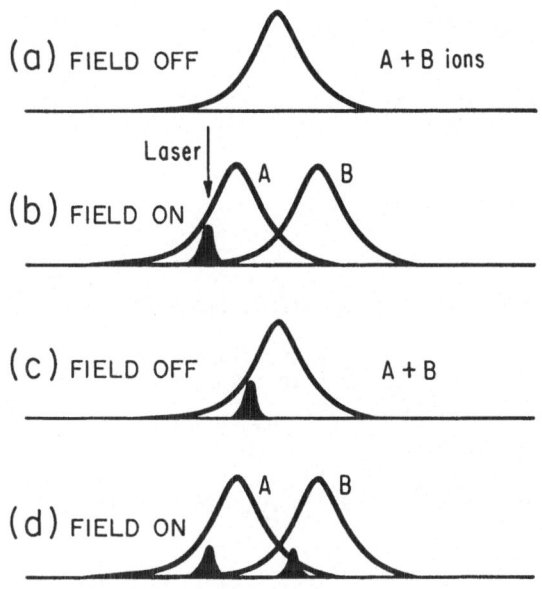

Fig. 13. Schematic representation of the separation of A and B ions by an electric field in the experiment of Chu et al. /46/

is that there is nonradiative resonant transfer but it is very slow, much slower than had been postulated. A similar study by JESSOP and SZABO /48/ gave the same result. Although these findings rule against the existence of a rapid resonance transfer between A and B sublattice ions do they also rule out rapid resonance transfer between Cr^{3+} ions within the A or B sublattices? MONTEIL and DUVAL /49/ claim that intrasublattice transfer can be rapid. However, the electric field results cited above, along with the theoretical analysis of GIBBS et al. /50/ are generally accepted as convincing evidence against a rapid resonance transfer among single Cr^{3+} ions in ruby.

Although in this instance ruby appears not to fulfil our expectations and exhibit rapid resonant excitation transfer and the much-sought Anderson transition, the theoretical and experimental investigations made in search of these elusive phenomena have greatly deepened our understanding of the general excitation transfer process. As a bonus, however, it was found in the course of the laser excitation and four-wave mixing experiments that when concentrated ruby was irradiated with an intense laser beam an internal electric field of high strength (MV/cm) was generated in ruby. This field manifests itself by causing a pseudo-Start splitting of the R lines /51,52/. Furthermore, this internal electric field persisted after the irradiation had ceased.

8. Conclusion

The impressive body of accumulated knowledge of the crystallographic and spectroscopic properties of ruby, the availability of large good-quality specimens, and the strong narrow easily excited luminescence lines have all contributed to the acceptance of ruby as an excellent material with which to test new spectroscopic techniques and to look for new spectroscopic phenomena. It has served us well in the past and continues to be an object of our investigations in new and interesting areas. For example, at the present time much attention is being focussed on the spectroscopic properties of amorphous and disordered materials. A recent interesting study in this area is the laser excitation experiments and detailed analysis reported by WASIELA, BLOCK, and MERLE D'AUBIGNE /53/ on α-Al$_{(2-x)}$Ga$_{2x}$O$_3$:Cr^{3+}, which can be regarded as disordered ruby in that some of the Al^{3+} sites are randomly occupied by Ga^{3+} ions.

References

1. L.F. Mollenauer: Microwave Laboratory Report No. 1325, Stanford University (1964)
2. E. Becquerel: La Lumière - ses causes et ses effets (Librairie de Firmin Didot Freres, Fils et Cie, Imprimeurs de L'Institut, Paris, 1867)
3. L. de Boisbaudran: Comptes Rendus 103, 1107 (1886)
4. H. du Bois and G.J. Elias: Annalen der Physik 27, 233 (1908)
5. J. Becquerel: Phys. Zeit. 8, 932 (1907)
6. C.E. Mendenhall and R.W. Wood: Phil. Mag. 30, 316 (1915)
7. O. Deutschbein: Annalen der Physik 14, 712 (1932); 20, 828 (1934); Zeitschrift für Physik 77, 489 (1934)
8. H. Bethe: Annalen der Physik 3, 133 (1929)
9. Y. Tanabe and S. Sugano: J. Phys. Soc. Japan 9, 753 (1954); J. Phys. Soc. Japan 11, 864 (1956); Y. Tanabe and H. Kamimura: J. Phys. Soc. Japan 13, 394 (1958); S. Sugano and Y. Tanabe: J. Phys. Soc. Japan 13, 880 (1958); also S. Sugano, Y. Tanabe, and H. Kamimura: in Multiplets of Transition-Metal Ions in Crystals (Academic Press, New York 1976)
10. S. Sugano and I. Tsujikawa: J. Phys. Soc. Japan 13, 899 (1858)
11. A.A. Manenkov and A.M. Prokhorov: Sov. Phys. - JETP 1, 611 (1955)
12. G. Makhov, C. Kikuchi, J. Lambe, R.W. Terhune: Phys. Rev. 109, 1399 (1958)
13. T.H. Maiman: Nature 187, 493 (1960)
14. A. Szabo: Phys. Rev. Lett. 25, 924 (1970)
15. T. Muramoto, S. Nakamishi, T. Hashi: Opt. Comm. 21, 139 (1977)
16. P.E. Jessop, T. Muramoto, A. Szabo: Phys. Rev. B21, 926 (1980)
17. A.L. Schawlow: In Advances in Quantum Electronics, ed. by J.R. Singer (Columbia University Press, New York and London 1961) p.50
18. D.E. McCumber and M.D. Sturge: J. Appl. Phys. 34, 1682 (1963)
19. K.S. Gibson: Phys. Rev. 8, 38 (1916)
20. A.L. Schawlow: J. Appl. Phys. Suppl. 33, 395 (1962)
21. P.E. Jessop and A. Szabo: Opt. Comm. 33, 301 (1980)
22. E. Hahn: Phys. Rev. 80, 580 (1950)
23. N.A. Kurnit, I.D. Abella and S.R. Hartmann, Phys. Rev. Lett. 13, 567 (1964).
24. F. Varsanyi, D.L. Wood, and A.L. Schawlow: Phys. Rev. Lett. 3, 544 (1959)
25. D.E. Nelson and M.D. Sturge: Phys. Rev. 137A, 1117 (1965)
26. R. Adde, S. Geschwind, L.R. Walker: in Proceedings of the Fifteenth Colloque Ampere, ed. by P. Averback (North Holland, Amsterdam 1969) p.460
27. K.F. Renk and J. Peckenzell: J. Phys. (Paris) 33, C4 (1972)
28. J.I. Dijkhuis, A. van der Pol, H.W. de Wijn: Phys. Rev. Lett. 37, 1554 (1976)
29. R.S. Meltzer and J.E. Rives: Phys. Rev. Lett. 38, 421 (1977)
30. A.A. Kaplyanskii, S.A. Basun, V.A. Rachin, R.A. Titov: JETP Lett. 21, 200 (1975)
31. E.O. Schulz-du Bois: Bell System Technical Journal 38, 271 (1959)
32. S. Geschwind, R.J. Collins, A.L. Schawlow: Phys. Rev. Lett. 3, 545 (1959)
33. S. Geschwind, G.E. Devlin, R.L. Cohen, S.R. Chinn: Phys. Rev. A137, 1087 (1965)
34. G.F. Imbusch and S. Geschwind: Phys. Rev. Lett. 17, 238 (1966)

35. P.M. Selzer, D.L. Huber, B.B. Barnett, W.M. Yen: Phys. Rev. B17, 4979 (1978)
36. T. Holstein, S.K. Lyo, R. Orbach: Phys. Rev. Lett. 36, 891 (1976)
37. A.L. Schawlow, D.L. Wood, A.M. Clogston: Phys. Rev. Lett. 3, 271 (1959)
38. L.F. Mollenauer and A.L. Schawlow: Phys. Rev. 168, 309 (1968)
39. A.A. Kaplyanskii and A.K. Przhevuskii: Sov. Phys. - Solid State 9, 190 (1967)
40. J. Heber and W. Platz: J. Luminescence 18/19, 170 (1979)
41. G.F. Imbusch: Phys. Rev. 153, 326 (1967)
42. P.L. Selzer, D.S. Hamilton, W.M. Yen: Phys. Rev. Lett. 38, 858 (1977)
43. H.J. Eichler, J. Eichler, J. Knof, C.H. Noak: Phys. Status. Solidi 52, 481 (1979)
44. P.F. Liao, L.M. Humphrey, D.M. Bloom, S. Geschwind: Phys. Rev. B20, 4145 (1979)
45. D.S. Hamilton, D. Heiman, J. Feinberg, R.W. Hellwarth: Opt. Lett. 4, 124 (1979)
46. S. Chu, H.M. Gibbs, S.L. McCall, A. Passner: Phys. Rev. Lett. 45, 1715 (1980)
47. W. Kaiser, S. Sugano, D.L. Wood: Phys. Rev. Lett. 6, 605 (1961)
48. P.E. Jessop and A. Szabo: Phys. Rev. Lett. 45, 1712 (1980)
49. A. Monteil and B. Duval: In Energy Transfer Processes in Condensed Matter, ed. by B. Di Bartolo (Plenum Press, New York and London 1984) p. 643
50. H.M. Gibbs, S. Chu, S.L. McCall, A. Passner: in Coherence and Energy Transfer in Glasses, ed. by P.A. Fleury and B. Golding (Plenum Press, New York and London 1984) p. 373.
51. P.F. Liao, A.M. Glass, L.M. Humphrey: Phys. Rev. B22, 2276 (1980)
52. S.A. Basum, A.A. Kaplyanskii, S.P. Feofilov: Z.E.T.F. 87, 2047 (1984)
53. A. Wasiela, Y. Merle d'Aubigne, D. Block: J. Luminescence 36, 11 (1986); A. Wasiela, D. Block, Y. Merle d'Aubigne: J. Luminescence 36, 24 (1986)

Four-Wave Mixing Spectroscopy
of Metastable Defect States in Solids

S.C. Rand

Hughes Research Laboratories, 3011 Malibu Canyon Road,
Malibu, CA 90265, USA

1. Introduction

Early four-wave mixing experiments[1] relied on the high power
available from pulsed lasers to produce third harmonic generation, field-induced second harmonic generation and a variety of
other weak processes which combined three optical waves to
yield a fourth. It was immediately recognized however that
electronic resonances of various kinds could enhance nonlinear
mixing of light waves. This was confirmed by a plasma
experiment[2] in which input beams of frequencies ω_1, ω_2 and ω_3
produced an enhanced intensity of the four-wave mixing output
wave at $\omega_4 = \omega_1 + \omega_2 - \omega_3$ when $\omega_2 - \omega_3$ and $k_2 - k_3$ were adjusted to match
the plasma resonance frequency and wave vector. Experiments
followed which showed that intermediate electronic states[3],
as well as Raman[4] resonances enhanced third order mixing.
Resonant enhancement soon permitted narrowband, continuous-wave
lasers to be used successfully for degenerate four-wave
mixing[5] ($\omega_4 = \omega_3 = \omega_2 = \omega_1$) and coherent anti-Stokes Raman (CARS)
generation in liquids[6] and gases[7], which demonstrated the
high spectral resolution capability and sensitivity of these
techniques. Also, important applications of four-wave mixing
which were non-spectroscopic in nature were found, including
phase conjugation and amplified reflection[8,9,10].

From the beginning, applications of four-wave mixing in
solids were equally varied. Numerous investigations were
carried out to obtain spectroscopic information on Raman[11],
libron[12], phonon[13], vibron[14], and polariton[15]
excitations. Continuous-wave degenerate four-wave mixing due
to saturated absorption was discovered in ruby[16]. In
dielectrics, transient grating techniques were developed to
measure the rate of spatial migration of electronic excitation
over distances comparable to the wavelength of light[17], as
well as mechanisms of energy transfer between impurities[18].
Other workers determined fast excited state relaxation[19] and
coherence dephasing[14] times for localized and delocalized[20]
excitations. In semiconductors, similar studies led to time-
resolved studies of picosecond carrier dynamics[21], carrier
concentration[22] and nonlinear spectroscopy of excitons[23].
Also, Landau levels in InSb were observed to produce Raman-like
resonances in four-wave mixing processes[24].

In gases, spectroscopic studies based on four-wave mixing
rapidly developed into high resolution, Doppler-free

techniques[25]. In condensed matter however, initial studies
were not able to exploit the high spectral resolution
capabilities of the new methods, even after continuous-wave,
backward-wave generation was observed[16]. The narrowest
features encountered in solids, for example the zero phonon
transitions of rare earth ions in crystals cooled to liquid
helium temperatures, were typically ~5 GHz wide. It therefore
seemed pointless to develop and apply methods in solids with
resolution in excess of ~1 GHz. However, when a method was
found to tune the relative frequency of input beams derived
from a single laser source while preserving the correlation of
their frequency fluctuations, extremely narrow resonances were
unexpectedly discovered which did justify high resolution
methodology[27]. Features as narrow as a few Hertz were
observed in a variety of crystals and were shown to provide
accurate measurements of very long decay times from metastable
states of impurities[28] and color centers[29] in solids.

   In this article the theoretical and experimental basis for
studying relaxation processes of defect metastable states in
solids by this nearly degenerate four-wave mixing (NDFWM)
technique are reviewed. Results are presented for defects
comprising transition metal ions, rare earth ions and color
centers in a variety of media. Although the method works
equally well in photorefractive media, results of experiments
we have performed in $LiNbO_3$ and $BaTiO_3$ are not discussed here.
The basic correspondence between theory and experiment is
verified for the simple three-level systems formed by dilute Cr
ions in $YAlO_3$ and Al_2O_3, and for F_2 color centers in LiF.
Linewidth and saturation measurements in $Nd:\beta''$-Na-Alumina
reveal a transition from closed to open quantum system behavior
as Nd density increases. For N3 color centers in diamond,
NDFWM spectroscopy provides the first evidence for an unknown
deep level of this structure. The results as a whole show that
this method provides information about metastable states which
can be quite different from that provided by direct
fluorescence or phosphorescence measurements. Temporal decay
experiments measure depopulation of metastable excited states
whereas high resolution NDFWM spectroscopy measures
repopulation kinetics of the ground state. NDFWM is therefore
sensitive to dark processes and system saturation. As shown in
this paper, an understanding of ground state behavior
complements the picture furnished by radiative decay methods,
giving either additional or completely new information on the
interaction of light with defects in condensed media.

2. Theory

Nearly degenerate four-wave mixing resonances arise from the
interaction of two pump beams at frequency ω with an
independently tunable probe beam at frequency $\omega+\delta$ through the
third order susceptibility $\chi^{(3)}$ of a medium. The geometry of
the light beams is shown in Fig. 1. The pump beams are chosen
to be counter-propagating so as to guarantee phase-matching,
independent of incident probe direction. They create a
standing wave excitation with which the forward-going probe
interacts to generate an output wave radiated in the backward
direction, as dictated by momentum conservation. By energy

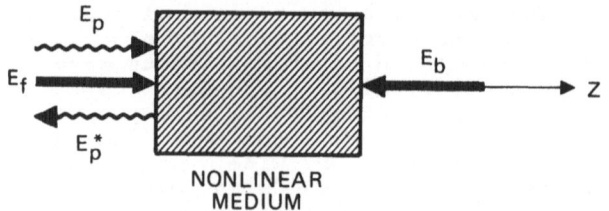

Fig.1. Geometry of the light beams assumed in nearly degenerate
four-wave mixing calculations. Forward and backward pump waves
E_f and E_b are counter-propagating along z so that a collinear
probe wave E_p gives rise to a collinear, conjugate signal E_p^*.

conservation the frequency of the fourth or signal wave is $\omega-\delta$
and it is phase conjugate to the probe.

Although there are 48 possible permutations of the light
fields, conjugate waves and collision events in the most
general formulation of four-wave mixing[30], we assume here
that only one or two terms dominate the mixing. Other
nonlinear processes in the sample are ignored. By virtue of
the geometry of the experiment and the near degeneracy of the
input frequencies, this simplification is well justified in the
study of many systems because only one or two terms are phase-
matched and therefore strong. Accidental occurrence of states
with energies near 2ω or 3ω above the ground state or the
generation of magnetic sub-level coherences can naturally
complicate the simple picture presented here.

In Sections 2.1 and 2.2, perturbation theory for NDFWM in 3-
and 4-level systems illuminated continuously by weak,
monochromatic beams is developed. The frequency of the pump
waves is assumed to be in resonance with an allowed one-photon
transition to an excited state which is connected by fast
relaxation processes to some metastable state of interest.
This state creates a population bottleneck, leading to
saturation of ground state absorption, generation of a large
third order susceptibility and efficient four-wave mixing at
zero pump-probe detuning. The NDFWM spectrum, obtained by
tuning the probe frequency relative to that of the pumps, is
shown to consist of a single Lorentzian-shaped resonance with a
width numerically related to the inverse lifetime of the
metastable state.

The proper interpretation of the NDFWM spectrum is, however,
that its width reflects the decay time of ground state spatial
hole-burning, since the resonance first appears in the second
order perturbation expression for the ground state density
matrix element, $\rho_{11}^{(2)}$. It does not appear in any other matrix
elements in second order. The NDFWM spectrum therefore
measures a repopulation rate, something quite different from
the customary radiative relaxation or population decay times of
excited states. Direct measurements of relaxation are
therefore possible on states which emit no light. The
technique should also be sensitive to the onset of non-

radiative processes which do not affect the population decay
time of radiating metastable states, a point discussed further
in Sections 3 and 4.

In Section 2.3, a non-perturbative approach to the theory is
presented which extends previous work[31] on saturation effects
due to pump waves of arbitrary intensity to 3-level systems.
Power-broadened NDFWM spectra are shown to furnish useful
information for the accurate determination of saturation
intensity in homogeneous or inhomogeneously broadened systems.
Results presented in Section 3 illustrate this application in
an inhomogeneously broadened material.

2.1 Perturbation Theory: 3-Level System

In Fig.2 we show the 3-level model to be considered in this
section. Absorption on the $|1\rangle$-$|2\rangle$ transition at frequency ω
is followed by decay either to the ground state or to
metastable state $|3\rangle$. The associated relaxation rates are
given in the figure and lead to the following starting
equations for the density matrix.

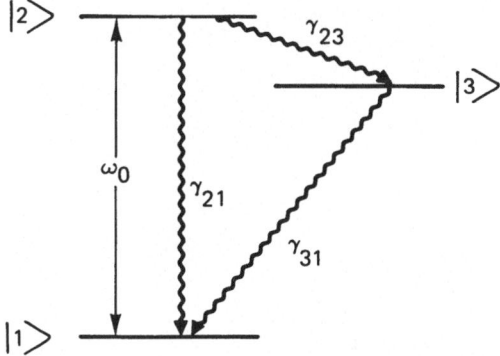

Fig.2. Schematic diagram of a 3-level system with an allowed
one-photon absorption at $\omega_\emptyset$. The γ_{ij} refer to relaxation rates
from state i to state j.

$$i\hbar\frac{d}{dt}\rho = [H,\rho] + [V,\rho] + i\hbar\frac{d}{dt}\rho\Big|_{decay} \tag{2.1.1}$$

$$i\hbar\frac{d}{dt}\rho_{11} = (V_{12}\rho_{21} - \rho_{12}V_{21}) + i\hbar\gamma_{31}\rho_{33} + i\hbar\gamma_{21}\rho_{22} \tag{2.1.2}$$

$$i\hbar\frac{d}{dt}\rho_{22} = (V_{12}\rho_{21} \; \rho_{12}V_{21}) - i\hbar(\gamma_{21}+\gamma_{23})\rho_{22} \tag{2.1.3}$$

$$i\hbar\frac{d}{dt}\rho_{33} = i\hbar\gamma_{23}\rho_{22} - i\hbar\gamma_{31}\rho_{33} \tag{2.1.4}$$

$$i\hbar\frac{d}{dt}\rho_{12} = -\hbar\omega_0\rho_{12} + (V_{12}\rho_{22} - \rho_{11}V_{12}) + i\hbar\Gamma_{21}\rho_{12} \tag{2.1.5}$$

$$\rho_{21} = \rho_{12}{}^* \qquad\qquad (2.1.6)$$

Here H is the atomic Hamiltonian in the absence of light, $\Gamma_{21} = (\gamma_{21}+\gamma_{23})/2 + \gamma_{ph}$ and V is the optical perturbation given by

$$(V_{12})_j = -\tfrac{1}{2}\,\mu_{12}(E_j e^{i(\omega_j t - k_j z)} + c.c.), \qquad\qquad (2.1.7)$$

where the subscript $j=f,b,p$ denotes the forward pump, backward pump and probe waves respectively. Let

$$\Omega_j = \mu_{12}E_j/2\hbar, \qquad\qquad (2.1.8)$$

$$\phi_j = \omega_j t - k_j z, \qquad\qquad (2.1.9)$$

and

$$\Delta_j = \omega_j - \omega_\emptyset. \qquad\qquad (2.1.10)$$

Because there are three distinct incident fields, we proceed to third order in the perturbation sequence. According to (2.1.2)-(2.1.5) a third-order off-diagonal matrix element $\rho_{12}{}^{(3)}$ can only be obtained from second-order elements $\rho_{11}{}^{(2)}$ and $\rho_{22}{}^{(2)}$ which are in turn derived from a first order $\rho_{12}{}^{(1)}$. From (2.1.5) we find directly that

$$(\rho_{12}^{(1)})_j = \frac{\Omega_j}{-\Delta_j + i\Gamma_{21}}\, e^{i\phi_j}\rho_{11}^{(0)}. \qquad\qquad (2.1.11)$$

Using (2.1.11) and (2.1.3) we obtain

$$i\hbar\frac{d}{dt}(\rho_{22}^{(2)})_{jk} = -\hbar \sum_{j,k} \Omega_j^* e^{-i\phi_j} e^{i\phi_k}(\frac{\Omega_k}{-\Delta_k + i\Gamma_{21}})\rho_{11}^{(0)}$$

$$+ \hbar \sum_{j,k} \Omega_j e^{-i\phi_k}e^{i\phi_j}(\frac{\Omega_k^*}{-\Delta_k - i\Gamma_{21}})\rho_{11}^{(0)} - i\hbar(\gamma_{21}+\gamma_{23})\rho_{22}^{(2)}. \qquad (2.1.12)$$

Setting $\gamma_2 = \gamma_{21}+\gamma_{23}$ and retaining only phase-matched terms which contain conjugate wave $\Omega_p{}^*$ we find

$$\rho_{22}^{(2)} = \frac{\rho_{11}^{(0)}}{\delta + i\gamma_2}\,[\Omega_p^*\Omega_f e^{-i(\phi_p - \phi_f)}(\frac{1}{-\Delta_p - i\Gamma_{21}} - \frac{1}{-\Delta_f + i\Gamma_{21}})$$

$$+ \Omega_p^*\Omega_b e^{-i(\phi_p - \phi_b)}(\frac{1}{-\Delta_p - i\Gamma_{21}} - \frac{1}{-\Delta_b + i\Gamma_{21}})]. \qquad (2.1.13)$$

Substituting (2.1.13) and the solution for $\rho_{33}{}^{(2)}$ in terms of $\rho_{22}{}^{(2)}$ from (2.1.4) into (2.1.2) we also find

$$\rho_{11}^{(2)} = \frac{\rho_{11}^{(0)}}{\gamma_2 - \gamma_{31}} \left(\frac{\gamma_{23}}{\delta + i\gamma_{31}} + \frac{\gamma_{21} - \gamma_{31}}{\delta + i\gamma_2} \right)$$

$$\cdot \left[-\Omega_p^* \Omega_f e^{-i(\phi_p - \phi_f)} \left(\frac{1}{-\Delta_p - i\Gamma_{21}} - \frac{1}{-\Delta_f + i\Gamma_{21}} \right) \right.$$

$$\left. -\Omega_p^* \Omega_b e^{-i(\phi_p - \phi_b)} \left(\frac{1}{-\Delta_p - i\Gamma_{21}} - \frac{1}{-\Delta_b + i\Gamma_{21}} \right) \right]. \qquad (2.1.14)$$

Notice that resonant denominator $(\delta + i\gamma_{13})^{-1}$ appears in the ground state element $\rho_{11}^{(2)}$, but not in $\rho_{22}^{(2)}$. Finally, we use (2.1.13) and (2.1.14) in (2.1.5) to obtain

$$\rho_{12}^{(3)} = \rho_{11}^{(0)} \left[\left(1 + \frac{\gamma_{21} - \gamma_{31}}{\gamma_2 - \gamma_{31}}\right) \left(\frac{1}{\delta + i\gamma_2}\right) + \left(\frac{1}{\gamma_2 - \gamma_{31}}\right) \left(\frac{1}{\delta + i\gamma_{31}}\right) \right]$$

$$\cdot \left[\frac{\Omega_p^* \Omega_f \Omega_b e^{i(\omega - \delta)t + ikz}}{(-\Delta_b + \delta + i\Gamma_{21})} \left(\frac{1}{-\Delta_p - i\Gamma_{21}} - \frac{1}{-\Delta_f + i\Gamma_{21}} \right) \right.$$

$$\left. + \frac{\Omega_p^* \Omega_b \Omega_f e^{i(\omega - \delta)t + ikz}}{(-\Delta_f + \delta + i\Gamma_{21})} \left(\frac{1}{-\Delta_p - i\Gamma_{21}} - \frac{1}{-\Delta_b + i\Gamma_{21}} \right) \right]. \qquad (2.1.15)$$

In this derivation we implicitly assume an optically thin sample, since field amplitudes throughout the sample are taken to be constant. From (2.1.15) it is clear that for excitation on resonance (Δ-0), a system with large homogeneous broadening ($\Gamma_{21} \gg \gamma_{31}, \delta$) and a slow decay rate from |3> to ground ($\gamma_{31} \ll \gamma_{21}, \gamma_{23}$) exhibits an NDFWM spectrum which is just a single Lorentzian peak with a full width at half maximum (FWHM) equal to γ_{31}.

2.2 Perturbation Theory: 4-Level System

If a fourth level is considered, as shown in Fig.3, and it is assumed that $\gamma_{31} = \gamma_{24} = 0$, we start with the equations below.

$$i\hbar \frac{d}{dt}\rho_{11} = (V_{12}\rho_{21} - \rho_{12}V_{21}) + i\hbar\gamma_{21}\rho_{22} + i\hbar\gamma_{41}\rho_{44} \qquad (2.2.1)$$

$$i\hbar \frac{d}{dt}\rho_{22} = -(V_{12}\rho_{21} - \rho_{12}V_{21}) - i\hbar(\gamma_{21} + \gamma_{23})\rho_{22} \qquad (2.2.2)$$

$$i\hbar \frac{d}{dt}\rho_{33} = i\hbar\gamma_{23}\rho_{22} - i\hbar\gamma_{34}\rho_{33} \qquad (2.2.3)$$

$$i\hbar \frac{d}{dt}\rho_{44} = i\hbar\gamma_{34}\rho_{33} - i\hbar\gamma_{41}\rho_{44} \qquad (2.2.4)$$

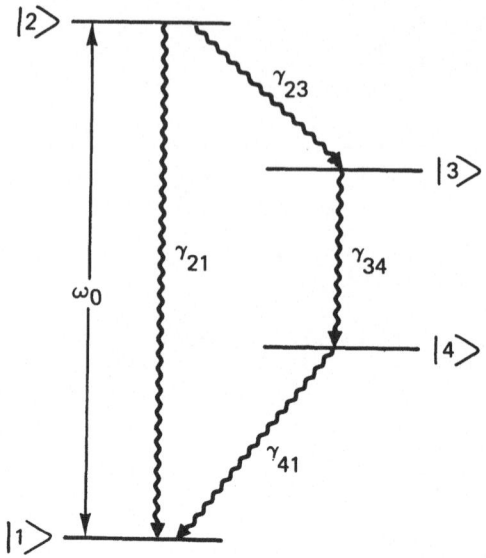

Fig.3. Schematic diagram of a 4-level system. The decay
pathways are restricted by the assumption that $\gamma_{31}=\gamma_{24}=0$.

$$i\hbar\frac{d}{dt}\rho_{12}=-\hbar\omega_0\rho_{12}+V_{12}(\rho_{22}-\rho_{11})-i\hbar\Gamma_{21}\rho_{12} \qquad (2.2.5)$$

(2.2.3) and (2.2.4) may be solved for ρ_{44} in terms of ρ_{22}
whereupon the solution for $\rho_{12}^{(3)}$ proceeds exactly as in the
previous section with the result that

$$\rho_{12}^{(3)}=\rho_{11}^{(0)}\{(\frac{1}{\delta+i\gamma_2})(1+[\gamma_{34}\gamma_{41}(\gamma_{34}-\gamma_{41})+(\gamma_2\gamma_{41}-\gamma_{41}\gamma_{23})(\gamma_{41}-\gamma_2)$$

$$+(\gamma_2\gamma_{34}-\gamma_{34}\gamma_{23})(\gamma_2-\gamma_{34})]/D)$$

$$+(\frac{1}{\delta+i\gamma_{34}})\gamma_{23}\gamma_{41}(\gamma_{41}-\gamma_2)/D$$

$$+(\frac{1}{\delta+i\gamma_{41}})\gamma_{23}\gamma_{34}(\gamma_2-\gamma_{34})/D\}$$

$$\cdot[\frac{\Omega_p^*\Omega_f\Omega_b e^{i(\omega-\delta)t+ikz}}{(-\Delta_b+\delta+i\Gamma_{21})}(\frac{1}{-\Delta_p-i\Gamma_{21}}-\frac{1}{-\Delta_f+i\Gamma_{21}})$$

$$+\frac{\Omega_p^*\Omega_b\Omega_f e^{i(\omega-\delta)t+ikz}}{(-\Delta_f+\delta+i\Gamma_{21})}(\frac{1}{-\Delta_p-i\Gamma_{21}}-\frac{1}{-\Delta_b+i\Gamma_{21}})], \qquad (2.2.6)$$

where $D=\gamma_{34}\gamma_{41}(\gamma_{34}-\gamma_{41})+\gamma_2\gamma_{41}(\gamma_{41}-\gamma_2)+\gamma_2\gamma_{34}(\gamma_2-\gamma_{34})$.

274

Notice that the presence of the fourth level introduces an additional resonance with a width γ_{41}, which corresponds to the decay rate from the new level. However, if only state $|3\rangle$ is metastable, a single resonance again dominates. That is, if $\gamma_{34} \ll \gamma_{21}$, γ_{23}, γ_{41} the NDFWM spectrum consists of a single Lorentzian with a width (FWHM) of γ_{34}. However, anticipating results for the N3 color center given in Section 3.2, we remark that if $|4\rangle$ is also metastable and $\gamma_{41} < \gamma_{34} \ll \gamma_{21}, \gamma_{23}$ the resonance due to $|4\rangle$ will dominate and the NDFWM linewidth will be narrower than γ_{34}.

2.3 NDFWM Theory Including Saturation Effects

In this section we extend earlier calculations of NDFWM[31] to show that for low intensities power-broadening in a 3-level system is linear in intensity for the homogeneous case but varies with the square root of incident intensity in inhomogeneous systems. This broadening occurs at intensities far below those necessary to cause AC Stark shifting of the levels and is therefore not accompanied by AC Stark sidebands, although the formalism incorporates the AC Stark effect. Even in the absence of AC Stark effects, when pump waves approach the 3-level saturation intensity, the perturbation theory presented in Sections 2.1 and 2.2 understandably fails. It is then necessary to take a different approach, one in which the pump waves are accounted for exactly in the first step of the calculation. The probe and signal waves are subsequently added in as perturbations in the second step.

The zero-order optical interaction Hamiltonian with equal amplitude pump waves is taken to be

$$V^{(0)} = -\frac{1}{2} \mu_{12} E_0 (e^{ikz} + e^{-ikz}) e^{-i\omega t} + c.c. \qquad (2.3.1)$$

while the first order term contains only probe and signal waves

$$V^{(1)} = -[\frac{1}{2} \mu_{12} E_1 (e^{ik_1 z - i\omega_1 t}) + \frac{1}{2} \mu_{12} E_2 (e^{-ik_2 z - i\omega_2 t})] + c.c. \qquad (2.3.2)$$

The probe wave is assumed to have frequency $\omega_1 = \omega + \delta$ as before, and the signal wave frequency is $\omega_2 = \omega - \delta$. To develop the starting equations for each order, we set

$$\rho = \rho^{(0)} + \lambda \rho^{(1)}, \qquad (2.3.3)$$

$$V = V^{(0)} + \lambda V^{(1)}, \qquad (2.3.4)$$

where λ is a perturbation series parameter. We use (2.3.3) and (2.3.4) in (2.1.1) to find

$$i\hbar \frac{d}{dt} \rho^{(0)} = [H, \rho^{(0)}] + [V^{(0)}, \rho^{(0)}] + i\hbar \frac{d}{dt} \rho^{(0)} \Big|_{decay}, \qquad (2.3.5)$$

$$i\hbar \frac{d}{dt} \rho^{(1)} = [H, \rho^{(1)}] + [V^{(1)}, \rho^{(0)}] + [V^{(0)}, \rho^{(1)}] + i\hbar \frac{d}{dt} \rho^{(1)} \Big|_{decay} \qquad (2.3.6)$$

275

with

$$V_{21}^{(0)} = - \hbar\Omega_{21}(e^{ikz}+e^{-ikz})e^{i\omega t} + \text{c.c.} , \qquad (2.3.7)$$

$$V_{21}^{(1)} = - \hbar\Omega_{21}(\epsilon_1 e^{i(k_1 z-\omega_1 t)}+\epsilon_2 e^{-i(k_2 z+\omega_2 t)}) + \text{c.c.}, \qquad (2.3.8)$$

where $\epsilon_1 = E_1/E_\emptyset$, $\epsilon_2 = E_2/E_\emptyset$ and $\Omega_{21}=\mu_{21}E_0/2\hbar$. Making use of the rotating-wave approximation (RWA), we drop complex conjugate terms in (2.3.7)-(2.3.8). The zero-order starting equations for a 3-level system are

$$i\hbar\frac{d}{dt}\rho_{11}^{(0)} = (V_{12}^{(0)}\rho_{21}^{(0)} - \rho_{12}^{(0)}V_{21}^{(0)}) + i\hbar\gamma_{21}\rho_{22}^{(0)} + i\hbar\gamma_{31}\rho_{33}^{(0)} , \qquad (2.3.9)$$

$$i\hbar\frac{d}{dt}\rho_{22}^{(0)} = -(V_{12}^{(0)}\rho_{21}^{(0)} - \rho_{12}^{(0)}V_{21}^{(0)}) - i\hbar(\gamma_{21}+\gamma_{23})\rho_{22}^{(0)} , \qquad (2.3.10)$$

$$i\hbar\frac{d}{dt}\rho_{33}^{(0)} = i\hbar\gamma_{23}\rho_{22}^{(0)} - i\hbar\gamma_{31}\rho_{33}^{(0)} , \qquad (2.3.11)$$

$$i\hbar\frac{d}{dt}\rho_{21}^{(0)} = \hbar\omega_0\rho_{21}^{(0)} + V_{21}^{(0)}(\rho_{11}^{(0)}-\rho_{22}^{(0)}) - i\hbar\Gamma_{12}\rho_{21}^{(0)} , \qquad (2.3.12)$$

$$\rho_{11}^{(0)} + \rho_{22}^{(0)} + \rho_{33}^{(0)} = N . \qquad (2.3.13)$$

Proceeding as in Section 2.1, we use (2.3.9)-(2.3.13) to obtain

$$\rho_{21}^{(0)} = \frac{-iN(V_{21}^{(0)}/\hbar)(+i\Delta+\Gamma_{12})}{\Delta^2+\Gamma_{12}^2[1+(1+[\gamma_{23}/2\gamma_{31}])|\Omega|_{21}^2)/\Gamma_{12}\gamma_2]} , \qquad (2.3.14)$$

$$\rho_{11}^{(0)} - \rho_{22}^{(0)} = \frac{N(\Delta^2 + \Gamma_{12}^2)}{\Delta^2+\Gamma_{12}^2 + |\Omega_{21}|^2(1+[\gamma_{23}/2\gamma_{31}])\Gamma_{12}/\gamma_2} . \qquad (2.3.15)$$

The first-order starting equations are

$$i\hbar\frac{d}{dt}\rho_{11}^{(1)} = (V_{12}^{(0)}\rho_{21}^{(1)} - \rho_{12}^{(1)}V_{21}^{(0)}) + (V_{12}^{(1)}\rho_{21}^{(0)} - \rho_{12}^{(0)}V_{21}^{(1)})$$

$$+ i\hbar\gamma_{21}\rho_{22}^{(1)} + i\hbar\gamma_{31}\rho_{33}^{(1)} , \qquad (2.3.16)$$

$$i\hbar\frac{d}{dt}\rho_{22}^{(1)} = -(V_{12}^{(0)}\rho_{21}^{(1)} - \rho_{12}^{(1)}V_{21}^{(0)}) - (V_{12}^{(1)}\rho_{21}^{(0)} - \rho_{12}^{(0)}V_{21}^{(1)})$$

$$- i\hbar\gamma_{21}\rho_{22}^{(1)} - i\hbar\gamma_{23}\rho_{22}^{(1)} , \qquad (2.3.17)$$

$$i\hbar\frac{d}{dt}\rho_{33}^{(1)} = i\hbar\gamma_{23}\rho_{22}^{(1)} - i\hbar\gamma_{31}\rho_{33}^{(1)} , \qquad (2.3.18)$$

$$i\hbar\frac{d}{dt}\rho_{21}^{(1)} = \hbar\omega_0\rho_{21}^{(1)} + V_{21}^{(0)}(\rho_{11}^{(1)} - \rho_{22}^{(1)}) + V_{21}^{(1)}(\rho_{11}^{(0)} - \rho_{22}^{(0)}) - i\hbar\Gamma_{12}\rho_{21}^{(1)}$$

$$(2.3.19)$$

$$\rho_{11}^{(1)} + \rho_{22}^{(1)} + \rho_{33}^{(1)} = 0. \qquad (2.3.20)$$

Because frequencies are no longer degenerate in first order, complex conjugates must be retained in expressions for purely real, diagonal elements of the density matrix.

$$\rho_{11}^{(1)} = \frac{1}{2}(\tilde{\rho}_{11}^{(1)}e^{i\delta t} + c.c.) \qquad (2.3.21)$$

$$\rho_{22}^{(1)} = \frac{1}{2}(\tilde{\rho}_{22}^{(1)}e^{i\delta t} + c.c.) \qquad (2.3.22)$$

$$\rho_{33}^{(1)} = \frac{1}{2}(\tilde{\rho}_{33}^{(1)}e^{i\delta t} + c.c.) \qquad (2.3.23)$$

The off-diagonal matrix element is

$$\rho_{21}^{(1)} = \frac{-iN\Omega_{21}|\Omega_{21}|^2}{1+I'}(e^{ikz}+e^{-ikz})^2\{L_2L_3e^{-i\omega_2 t}(2+\frac{\gamma_{23}}{\gamma_{31}-i\delta})$$

$$\bullet\ [(L+L_1^*)\epsilon_1^*e^{-ik_1 z}+c.c + (L+L_2)\epsilon_2 e^{-ik_2 z}+c.c.]+ L_1L_3^*e^{-i\omega_1 t}$$

$$\bullet\ (2+\frac{\gamma_{23}}{\gamma_{31}+i\delta})[(L^*+L_1)\epsilon_1 e^{ik_1 z}+c.c.+(L^*+L_2^*)\epsilon_2 e^{ik_2 z}+c.c.]\}$$

$$+ \frac{iN\Omega_{21}}{1+I'}\{\epsilon_1 e^{i(k_1 z-\omega_1 t)}L_1 + L_2\epsilon_2 e^{-i(k_2 z+\omega_2 t)}\}, \qquad (2.3.24)$$

where the Lorentzian resonant factors are

$$L = (\Gamma_{21}-i\Delta)^{-1}, \qquad (2.3.25)$$

$$L_1 = (\Gamma_{21}-i\Delta-i\delta)^{-1}, \qquad (2.3.26)$$

$$L_2 = (\Gamma_{21}-i\Delta+i\delta)^{-1}, \qquad (2.3.27)$$

$$L_3 = (\gamma_2+\frac{\gamma_2\gamma_{31}}{2\gamma_{31}+\gamma_{23}}(2+\frac{\gamma_{23}}{\gamma_{31}-i\delta})I'+i\delta)^{-1}, \qquad (2.3.28)$$

and $I' = \dfrac{2|\Omega_{21}|^2(2\gamma_{31}+\gamma_{23})}{\gamma_2\gamma_{31}}(1+\cos 2kz)(L+L_1^*) \doteq 2(1+\cos 2kz)I/I_{sat}.$

Dropping all but the phase-matched term, we get

$$\rho_{21}^{(1)} = \frac{-iN\Omega_{21}|\Omega_{21}|^2}{1+I'}(e^{ikz}+e^{-ikz})^2 L_2 L_3$$

$$\cdot \; (2+\frac{\gamma_{23}}{\gamma_{31}-i\delta})\epsilon_1^* e^{-i(k_1 z + \omega_2 t)}(L+L_1^*) \qquad (2.3.29)$$

The saturation behavior of the microscopic polarization is contained in L_3. Its magnitude squared varies as

$$|L_3|^2 = (\frac{1}{\gamma_{31}^2 + \delta^2}) \; \gamma_2^2 \; (\gamma_{31}^2 \; [1+I']^2 + \delta^2), \qquad (2.3.30)$$

where we have assumed $\gamma_2 \gg \gamma_{31}, \delta$.

For optically thin samples, we calculate signal intensity by using (2.3.29) as the nonlinear polarization source term in Maxwell's equations. This procedure[32] performs the appropriate spatial average of the sinusoidal distribution of pump light intensity in (2.3.30). This merely changes the intensity-dependent lineshape in (2.3.30) for the homogeneous case from $|\rho_{21}^{(1)}|^2 \; \alpha \; (\gamma_{31}^2[1+I/I_{sat}]^2+\delta^2)^{-1}$ to

$$I_{sig} \; \alpha \; (\gamma_{31}^2[1+4I/I_{sat}]^2+\delta^2)^{-1}. \qquad (2.3.31)$$

For inhomogeneous 3-level systems, an additional integration of $\rho_{21}^{(1)}$ over the frequency distribution of the absorption line is necessary. The result for a Gaussian distribution is that $|\rho_{21}^{(1)}|^2 \; \alpha \; (\gamma_{31}^2[1+I/I_{sat}]+\delta^2)^{-1}$. After spatial averaging this gives

$$I_{sig} \; \alpha \; (\gamma_{31}^2[1+4I/I_{sat}]+\delta^2)^{-1}, \qquad (2.3.32)$$

at low intensities. These results are illustrated in Fig. 4.

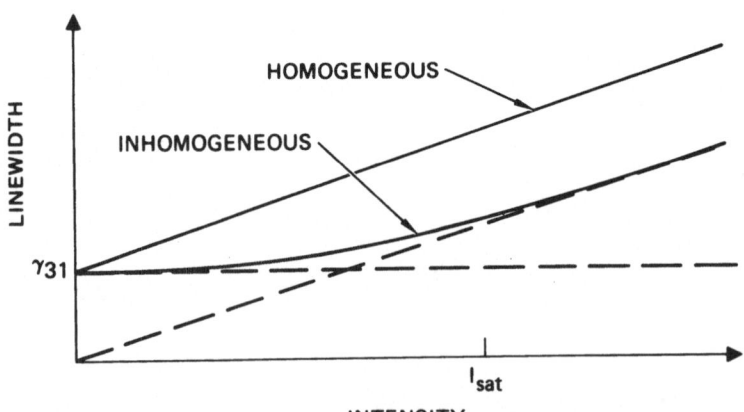

Fig.4. Broadening of the 3-level NDFWM Lorentzian linewidth as a function of incident intensity for homogeneous and inhomogeneous systems.

3. Experiments

3.1 Rare Earth and Transition Metal Dopants

Two sets of results are presented in this section. First, data on Cr ions in $YAlO_3$ and Al_2O_3 is used to verify the basic theory of Section 2.1, establishing the foundation of the high resolution NDFWM technique. Second, the interesting case of trivalent Nd ions in an unusual crystalline phase of alumina is discussed. β''-Na-Alumina accepts higher densities of rare earth ions than other materials without experiencing fluorescence quenching of the dopant. Curiously, the saturation intensity of $Nd:\beta''$-Na-Alumina is found to depend strongly on Nd density. NDFWM experiments reveal that this is due to the onset of non-radiative pair interactions between Nd neighbors which cause little change in upper state lifetime but increase the ground state decay time substantially.

The experimental configuration is depicted in Fig.5. Measurements on $Cr:YAlO_3$ were performed using a dye laser to irradiate a 3.9 mm-thick sample doped with approximately 0.05% Cr and exhibiting an optical density of 0.29 at 570 nm. The crystal was positioned with the c axis parallel to the counter-propagating pump beams. The probe beam was nearly collinear with the forward pump beam and intersected the pump beams in the sample at an angle of 0.2^{θ}. As shown in the figure, two acousto-optic modulators were used to synthesize the appropriate wavelengths for the pump and probe beams and to permit tuning through zero offset frequency. The two synthesizers were phase-locked together and stable to better than 1 Hz throughout the duration of each experiment.

Typically, ω_1 was held fixed at 40 MHz offset, and ω_2 was scanned over a range of a few hundred Hertz centered at 40 MHz offset frequency. The fundamental laser frequency was tuned near 570 nm for peak signal from the sample. The probe beam

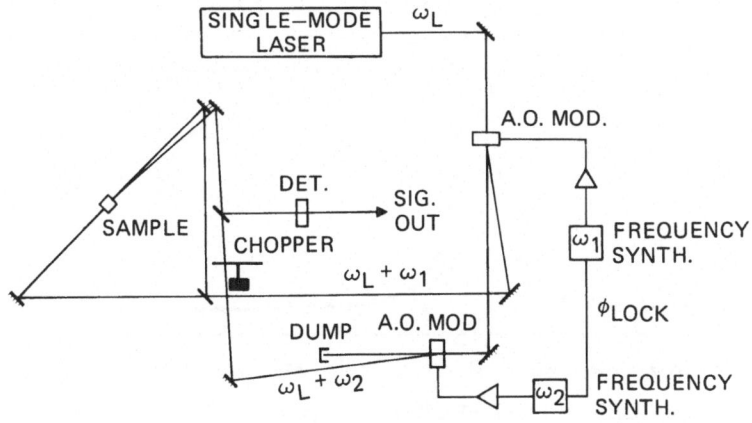

Fig.5. Experimental configuration for NDFWM spectroscopy using synthesized frequencies for the three input beams.

was chopped at a low frequency and the signal was detected phase-sensitively with a lock-in amplifier. The signal was recorded on a signal averager and then sent to a VAX 11-780 computer for analysis. The measurements were confirmed to be independent of intensity in the range 1-20 W/cm^2.

Figure 6 shows the NDFWM signal in dilute Cr:YAlO$_3$ versus pump-probe detuning δ together with the best fit theoretical curve and the ^{2}E fluorescence decay curve. When the incident lasers are tuned to the ^{4}A$_2$-^{4}T$_2$ absorption, Cr^{3+} forms a simple 3-level system with $|1\rangle=^4$A$_2$, $|2\rangle=^4$T$_2$ and $|3\rangle=^2$E. It is not surprising then to find that the observed Lorentzian linewidth is 9.8 Hz, in excellent agreement with the fluorescence decay data which yields $\gamma_{31}^{-1}= 33\pm1$ ms or a linewidth of 9.7 Hz. There is a single pathway open for relaxation from the metastable state to ground, so agreement with the theory of Section 2.1 is expected. Similarly, for Cr:Al$_2$O$_3$ in which the fluorescence decay is somewhat shorter, the observed NDFWM spectral width of 138 Hz agrees well with the fluorescence measurements (τ_{31}=2.6 ms or $\Delta\nu$=123 Hz).

Now consider the 4-level system formed by Nd:β''-Na-Alumina and shown in Fig.7. This material has the unusual property that Nd fluorescence is not quenched even at densities as high as 10^{21} Nd/cm^3. Because β''-Na-Alumina has significant potential for new applications in nonlinear optics and laser technology it is important to understand why the fluorescence lifetime is constant and what limitations might be imposed on uses of the material due to saturation effects.

NDFWM spectra were recorded as a function of incident intensity for five samples, together with fluorescence decay

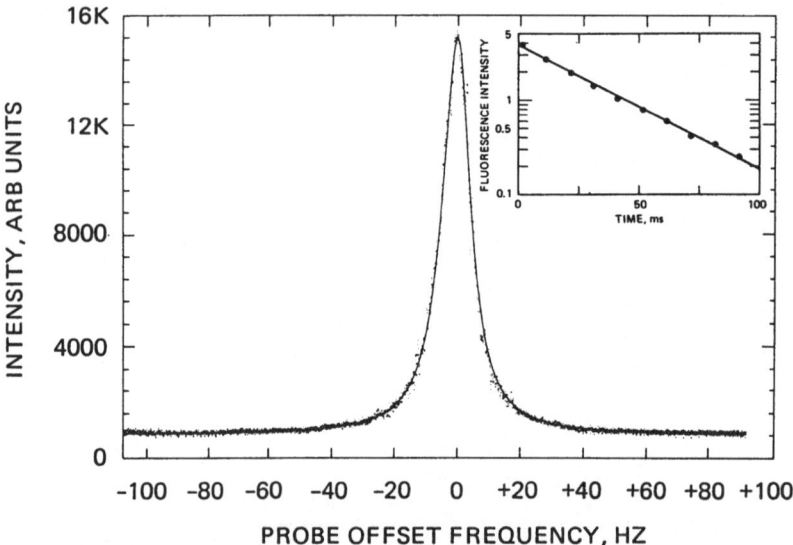

Fig.6. The NDFWM spectrum in Cr:YAlO$_3$ and (inset) the corresponding ^{2}E fluorescence decay curve.

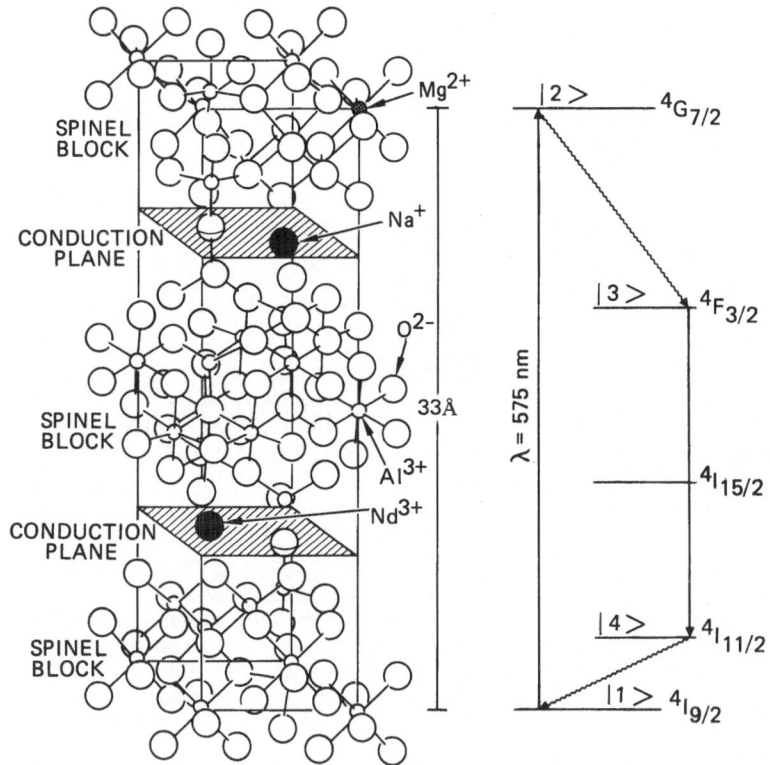

Fig.7. Crystal structure of β''-Na-Alumina ($Na_{1+x}Mg_xAl_{11-x}O_{17}$) with Nd^{3+} dopant ions exchanged for Na^+ via the conduction plane. Energy levels and the excitation/relaxation pathway of a single Nd^{3+} ion are shown on the right.

curves measured at a wavelength of 1.06 μm. In Fig.8a, several spectra recorded at different intensities are shown for a sample containing 6.4×10^{20} Nd/cm^3. Power broadening is clearly evident at the higher intensities and in Fig.8b the Lorentzian linewidth of these traces is seen to scale with the square root of the incident intensity, in excellent agreement with (2.3.32).

A lineshape of the form (2.3.32) was fitted to data similar to that in Fig.8 to determine the unsaturated linewidth and saturation intensity of each sample. The results are shown in Fig.9. The signal here originates from spatial hole-burning in the ground state and exhibits saturation behavior identical to that of the gain coefficient in an inhomogeneous system [33].

The NDFWM linewidth fluctuates below 5×10^{20} Nd/cm^3 in response to crystal field variations[28] and exhibits a small overall decline as density increases. The saturation intensities show a much larger decrease as a function of

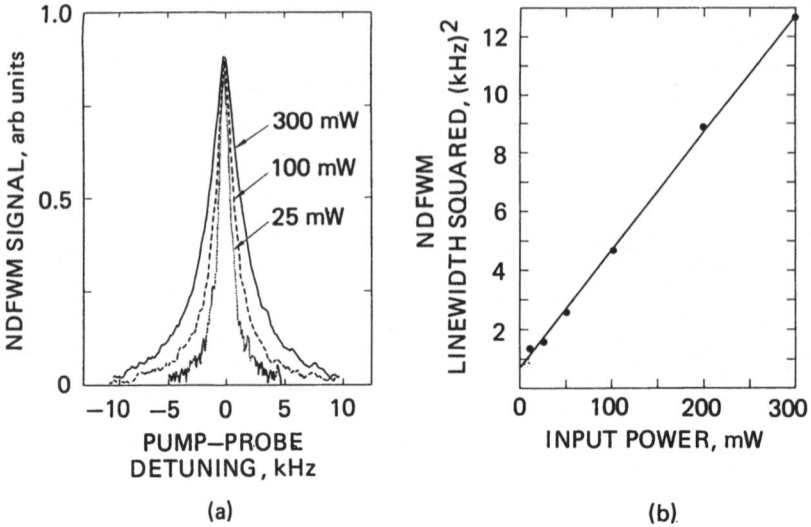

Fig.8. (a)NDFWM signal versus pump-probe detuning for total incident powers of 25, 100 and 300 mW, showing power-broadening. Nd^{3+} density=6.4×10^{20} cm^{-3}. (b)The square of the NDFWM linewidth (FWHM) versus total input power. The size of the data points is indicative of the experimental uncertainties and the solid curve is a least squares fit to the data.

density, diminishing to a value of 12 kW/cm^2 at $8.4 \times 10^{20} Nd/cm^3$, in excellent agreement with Boyd[34]. This corroborates the trend in the linewidth data and confirms a genuine <u>increase</u> in the ground state grating lifetime, albeit small, over a density range in which the excited state decay time actually <u>decreases</u> slightly[28,35]. In other words, the relaxation of the excited state is qualitatively different from that of the ground state above a density of $\sim 6 \times 10^{20} cm^{-3}$.

These NDFWM results are surprising. Ordinarily the onset of the process of cross-relaxation, familiar as the mechanism of fluorescence quenching in virtually all other highly doped Nd materials[36], would cause more rapid decay of the emitting state |3>. In a system following a closed relaxation path, the ground state grating would then wash out faster at higher densities. The NDFWM linewidth and saturation intensity would therefore increase with increasing Nd density. Instead, just the opposite is observed in $Nd:\beta"$-Na-Alumina. The reason for this is that as density increases, Nd-Nd pairs consisting of one optically excited $^4F_{3/2}$ ion and one $^4I_{9/2}$ neighbor interact more readily in a nearly resonant, mutual decay process yielding two $^4I_{15/2}$ ions which relax more <u>slowly</u> to the <u>ground</u> state than the $^4I_{11/2}$ ions usually produced by spontaneous emission. At high enough densities, the ground state grating which gives rise to four-wave mixing can therefore persist longer due to pair interactions. Processes other than this pair interaction would enhance the ground state relaxation

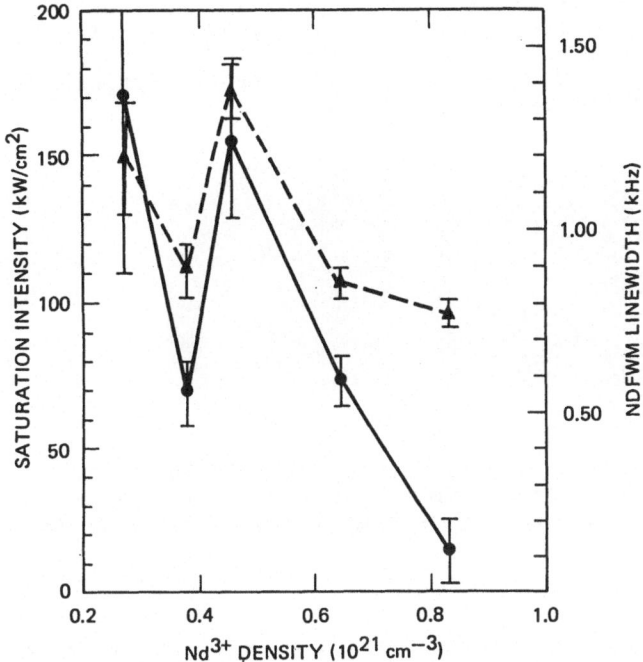

Fig.9. NDFWM saturation intensity (solid) and linewidth (dashed) measurements versus Nd^{3+} dopant density. The linewidth data give the best fit values of Lorentzian full widths at half maximum intensity.

state, in disagreement with our measurements. The NDFWM results provide clear evidence of a transition at high Nd density from a closed to an open quantum system, mediated by pair interactions which permit the ground and excited states to evolve separately.

3.2 Color Center Studies

Two sets of data are again used in this section to illustrate the simplicity and unique capabilities of NDFWM spectroscopy in solids. NDFWM techniques are applied to F_2 color centers to establish intersystem crossing as the mechanism for efficient, continuous-wave phase conjugation with these structures. Similarly, signals from N3 color centers in diamond arise due to intersystem crossing, but the NDFWM spectrum is narrower than expected from temporal decay measurements. This indicates the presence of a hidden deep level for the N3 defect.

When pure LiF receives a dose of 100 MRad ^{60}Co γ-rays, high densities of F-aggregate color centers are formed, particularly F_2 centers. These radiation products consist of vacancy pairs occupied by two electrons, and are stable for years in the presence of low intensity light, although high intensities typically ionize the centers, forming short-lived F_2^+ defects. Cw phase conjugation is relatively efficient at the first resonance energy of the F_2 defects[29] and corrects optical aberrations in the usual manner (Fig.10).

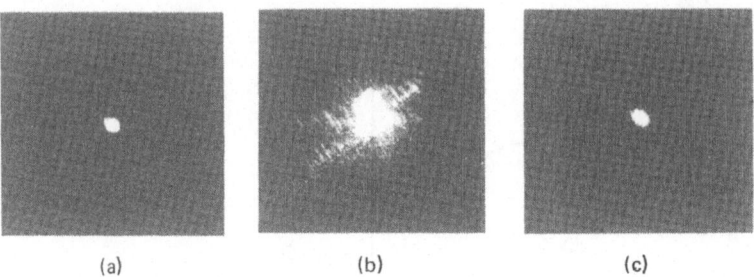

(a)   (b)   (c)

Fig.10. (a)Phase-conjugate signal from F_2:LiF without aberrator. (b)Aberrated probe beam with mirror replacing the sample. (c)Restored phase-conjugate signal with aberration as in (b).

The two electrons form singlet and triplet manifolds with the three lowest states constituting a 3-level system similar to Cr^{3+}[37]. In this case however, intersystem crossing can populate the metastable triplet level and phosphorescence accounts for relaxation to the ground state. Observations of the NDFWM spectrum[29] have shown that the linewidth agrees very well with the inverse of the phosphorescence decay time measured at the same temperature, confirming that this picture of saturated absorption due to intersystem crossing is correct. In Fig.11, the linewidth also narrows as temperature is decreased, in accord with previous studies of F_2 phosphorescence decay in KCl[37].

In diamond, color centers can be formed from aggregates of substitutional nitrogen atoms and are more stable to light and heat than the F-centers in alkali halides. For example the N3 center consists of three nitrogen atoms bonded to a common carbon or vacancy[38] with the absorption spectrum shown in Fig.12. The energy levels of this center are not fully known but it has been proposed[39] that the small features around 450-500 nm are due to partially allowed absorption to a state with the same symmetry as the ground state. Light absorption out of this state has been reported and the lifetime measured to be 0.73±.03 ms by a double resonance technique[40], although no emission from this state has been reported.

Figure 13a presents the energy level diagram determined by these earlier measurements. By analogy with the Cr and F_2 systems, it is apparent that given the one known metastable level (and perhaps others) below the main resonance, the ground state absorption of the N3 center should be easily saturable under continuous excitation. With a single metastable state, NDFWM spectroscopy should reveal a linewidth corresponding to

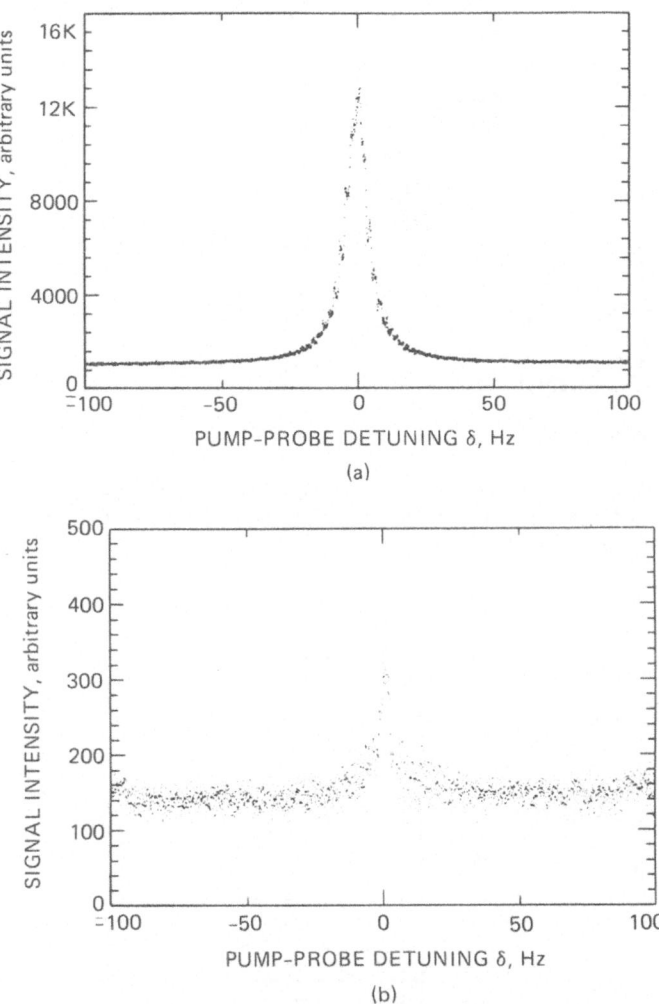

Fig.11. NDFWM spectra of γ-irradiated LiF. (a)T=20°C. The least squares fit Lorentzian linewidth is 4.70 Hz(FWHM). (b)T=-25°C. The spectral width has narrowed to 2.1 Hz (FWHM).

0.73 ms. With an additional deep level however, the linewidth could be narrower, as explained in Section 2.2.

   Indeed this is the case for the N3 center in diamond which is accessible to study using several lines of the Kr$^+$ laser. As indicated by the preliminary data in Fig.14, the observed width of 265 Hz is roughly a factor of two narrower than expected on the basis of the energy levels of Fig.13a and the earlier decay measurement[40]. It therefore provides the first evidence of additional metastable energy levels for the N3 center below the $^2A_2(^2A_1)$ excited state.

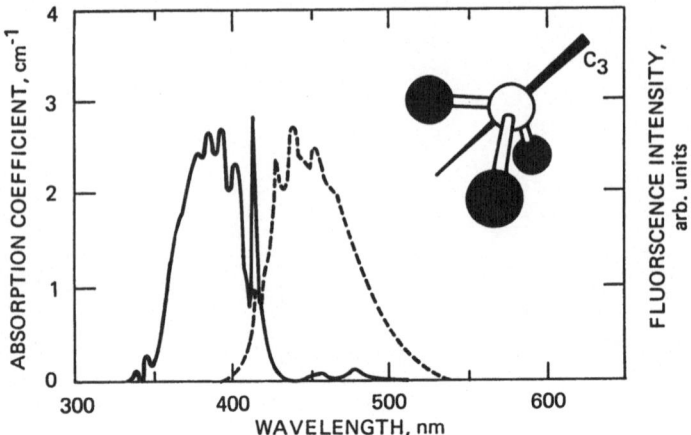

Fig.12. Absorption (solid) and emission (dashed) spectra of the N3 center with structural model inset. The solid circles are substitutional nitrogen impurities.

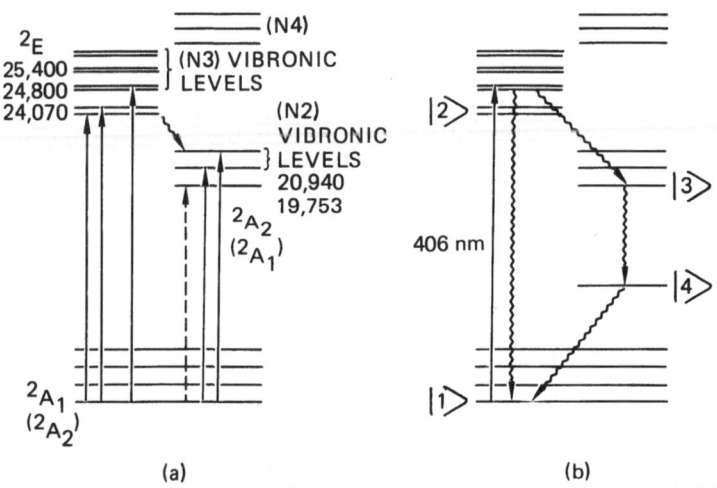

Fig.13. Energy levels of the N3 center in diamond, (a) as in Ref.[39] and (b) with an additional, metastable level as indicated by the present NDFWM results. Solid (dashed) lines indicate allowed (forbidden) transitions.

Although a complete interpretation must await a full analysis of the data, it can be concluded that the energy level diagram should be modified as shown in Fig.13b. On the basis of the NDFWM measurement, an additional deep level (or levels) with a decay rate of $(265\pi)^{-1}$ must be present and in keeping with the current ground state assignment[39] it should transform as a singlet representation of the $C_{3v}(3m)$ group.

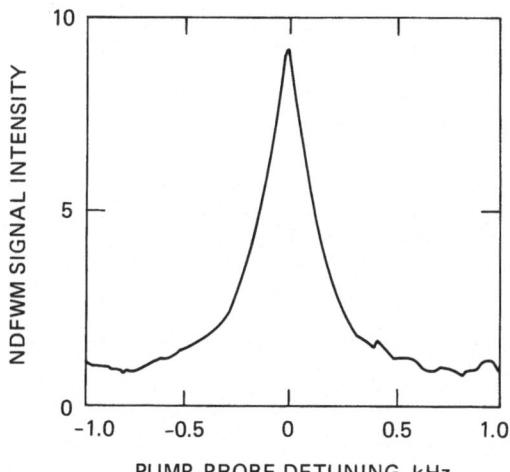

Fig.14. NDFWM signal versus pump-probe detuning for the N3 center. The sample was 1.75 mm thick and exhibited a peak optical density in the N3 absorption band of 0.23. Excitation wavelength was 406 nm.

4. Discussion

The excellent agreement between theory and the measured NDFWM spectrum of Cr establish the equivalence of fluorescence decay experiments and ground state relaxation measurements in simple "closed" systems. This correspondence can elucidate four-wave mixing mechanisms, as illustrated by the results on the 3-level system formed by F_2 color centers in LiF. However, in more complex systems with several metastable states which may emit no radiation at all, NDFWM spectroscopy provides unique information, inaccessible to conventional optical techniques. This is illustrated by detecting the onset of weak non-radiative pair interactions in Nd:β"-Na-Alumina over a dopant density range in which the fluorescence decay rate is constant. Also, the existence and relaxation rate of a hidden level of the N3 color center in diamond are established despite the fact that no emission from the metastable manifold has ever been reported.

5. Acknowledgements

The author wishes to thank D.G. Steel and J. Lam for many useful discussions regarding the theory of four-wave mixing. R.S. Turley and R.A. McFarlane contributed invaluable assistance with data analysis. R.A. McFarlane also gave a critical reading of the manuscript which was prepared for publication by J. McNulty. β"-Na-Alumina samples were grown by D. Dunn of UCLA and O.M. Stafsudd kindly furnished unpublished data on this material. Natural diamonds were loaned by V. Manson of the Gemological Institute of America. J. Brown and R. Cronkite provided superb technical assistance. The author also wishes to thank A. Au for the generous loan of a Kr ion laser. This research was funded under AFOSR contract F49620-85-C-0058.

References

1. P.D. Maker and R.W. Terhune, Phys.Rev. A137, 801(1965); R.L. Carman, R.Y. Chiao and P.L. Kelley, Phys.Rev.Lett. 17, 1281(1966); R.Y. Chiao, P.L. Kelley and E. Garmire, Phys. Rev.Lett. 17, 1158(1966).
2. B.L. Stansfield, R. Nodwell and J. Meyer, Phys.Rev.Lett. 26, 1219(1971).
3. G.C. Bjorklund, J.E. Bjorkholm, P.F. Liao, and R.H. Storz, Appl.Phys.Lett. 29, 729(1976).
4. F. DeMartini, F. Simoni, E. Santamato, Opt.Comm.9, 176(1973).
5. P.F. Liao, D.M. Bloom, and N.P. Economou, Appl.Phys Lett. 32, 813(1978).
6. S.A. Akhmanov, A.F. Bunkin, S.G. Ivanov, N.I. Koroteev, A.I. Kovrigin and I.L. Shumay, in Tunable Lasers and Applications, eds. A. Mooradian, T. Jaeger, P. Stoketh (Springer, Berlin, Heidelberg, New York, 1976), p.389, and references therein; S.A. Akhmanov, F.N. Gadjiev, N.I. Koroteev, R. Yu Orlov, I.L. Shumay, JETP Lett. 27, 243(1978).
7. J.J. Barrett, R.F. Begley, Appl.Phys.Lett. 27, 129(1975); M. Henesian, L. Kulevskii, R.L. Byer, J. Chem. Phys. 65, 5530(1976).
8. D.M. Bloom and G.C. Bjorklund, Appl.Phys.Lett. 31,592(1977.
9. D.M. Bloom, P.F. Liao, N.P. Economou, Opt.Lett.2, 58(1978).
10. A. Yariv and D.M. Pepper, Opt.Lett. 1, 16(1977).
11. M.D. Levenson and N. Bloembergen, Phys.Rev. B10,4470(1974).
12. K.Duppen, B.M.Hesp, D.A.Wiersma, Chem Phys.Lett.79, (1981).
13. M.D. Levenson, C.Flytzanis, N. Bloembergen, Phys.Rev. B6,3962(1972).
14. D.D. Dlott, C.S. Schosser and E.L. Chronister, Chem.Phys. Lett. 90, 386(1982).
15. J.J. Wynne, Phys.Rev.Lett.29, 650(1972); F. DeMartini, G.Giuliani, P. Mataloni, E. Palange, Y.R. Shen, Phys.Rev.Lett. 37, 440(1976).
16. P.F. Liao and D.M. Bloom, Opt.Lett. 3, 4(1978).
17. J.R. Salcedo, A.E. Siegman, D.D. Dlott and M.D. Fayer, Phys.Rev.Lett. 41, 131(1978); D.S. Hamilton, D. Heiman, J. Feinberg and R.W. Hellwarth, Opt.Lett 4, 124(1979).
18. J.K. Tyminski, R.C. Powell and W.K. Zwicker, Phys.Rev. B29, 6074(1984).
19. D. W. Phillion, D.J. Kuizenga, and A.E. Siegman, Appl. Phys. Lett. 27, 85(1975).
20. R.W. Olson, F.G. Patterson, H.W. Lee, and M.D. Fayer, Chem. Phys.Lett. 79, 403(1981).
21. R.K. Jain, Opt.Eng. 21, 199(1982); K. Jarasiunas and J. Vaitkus, Phys.Stat.Solidi a44, 793(1977).
22. P. Kupacek, M. Comte, and D.S. Chemla, Appl.Phys.Lett. 38, 44(1981).
23. A. Maruani and D.S. Chemla, Phys.Rev. B23, 841(1981).
24. E. Yablonovich, N. Bloembergen, and J.J. Wynne, Phys.Rev.B3, 2060(1971).

25. R.K. Raj, D. Bloch, J.J. Snyder, G. Camy and M. Ducloy, Phys. Rev. Lett. $\underline{44}$, 1251(1980).
26. P.F. Liao and D.M. Bloom, Opt.Lett. $\underline{3}$, 4(1978).
27. D.G. Steel and S.C. Rand, Phys.Rev.Lett. $\underline{55}$, 2285(1985).
28. S.C. Rand, J. Lam, R.S. Turley, R.A. McFarlane and O.M. Stafsudd, to be published.
29. S.C. Rand, Opt. Lett. $\underline{11}$, 135(1986) and to be published.
30. N. Bloembergen, in Quantum Electronics, Proc. 3rd Quantum Electronics Conference, Paris, 1963, ed. by N. Bloembergen,P. Grivet (Dunod, Paris 1964)pp.1501-1512; also Proc.IEEE $\underline{51}$, 124-131(1963).
31. R.L.Abrams and R.C. Lind, Opt.Lett. $\underline{2}$, 94(1978); D. Harter and R. W. Boyd, IEEE J.Q.E. $\underline{QE-16}$, 1126(1980).
32. Y.R. Shen, The Principles of Nonlinear Optics, J. Wiley (New York, 1984), p.48.
33. See for example A. Yariv, Quantum Electronics, 2nd Edition Wiley (New York, 1975), p.170.
34. R.W. Boyd, M.T. Gruneisen, P. Narum, D.J. Simkin, B. Dunn and D.L. Yang, Opt.Lett. $\underline{11}$, 162(1986).
35. M. Jansen, A. Alfrey, O.M. Stafsudd, B. Dunn, D.L. Yang and G.C. Farrington, Opt.Lett. $\underline{10}$, 119(1984).
36. A.A. Kaminskii, Laser Crystals, Springer-Verlag (Berlin,1981), p.327.
37. Y. Farge, J.M. Ortega and R.H. Silsbee, J.Chem.Phys $\underline{69}$, . 3972(1978) and references therein.
38. G. Davies, Diamond Res.15-24(1977);M.D. Crossfield, G.Davies, A.T. Collins and E.C. Lightowlers, J.Phys.C$\underline{7}$, 1909(1974); J.A.Van Wyk, J.Phys.C$\underline{15}$, L981(1982).
39. G. Davies, C.M. Welbourn and J.H.N. Loubser, Diamond Research (1977), pp. 23-30; M.F. Thomaz and G. Davies, Proc.Roy.Soc. London A$\underline{362}$, 405(1978).
40. S.C. Rand and L.G. DeShazer, Opt.Lett.$\underline{10}$,481(1985).

The Turvy Topsy Contest

In 1975, when Schawlow was President of the Optical Society of America, he proposed a contest for the best "Turvy Topsy." A "Turvy Topsy" is, of course, the inverse of a Topsy Turvy, whereas a Topsy Turvy is a picture which looks like some recognizable object when right side up or upside down. A "Turvy Topsy", therefore, must show an object which is always obviously upside down.

The purpose of the contest was to advance the art of slidesmanship, by demonstrating how to produce a slide that can never be presented the right way up. Four prizes were offered: First Prize, $10, Second Prize, $5, Third Prize, a copy of Schawlow's latest paper, and Fourth prize, copies of Schawlow's two most recent papers.

The First Prize was won by the figure below, submitted by Bruce D. Hansche of the University of Michigan, which shows a mathematical integral, that cannot be made to end in "dx".

$$xd^{8}\left[\frac{\delta H}{\Delta \phi}\right]\int\left[\frac{\phi \nabla}{H \mathcal{G}}\right]_{8}px$$

Linn F. Mollenauer with a large ruby rod used at Stanford in 1965

Part IV

Miscellaneous Ideas

A scientist typically does his best work on the back of an envelope. To enhance productivity, provide envelopes with two backs. Such envelopes were available from "Doublethink, Inc." a division of "Nocturnal Aviation", A.L. Schawlow, Proprietor

Using a Tokamak to Study Atomic Physics: Brightness Ratios of Transitions Within the $n=2$ Levels of Be I-like Ions (C III to Cr XXI)

H.W. Moos

Department of Physics and Astronomy, Johns Hopkins University, Baltimore, MD 21218, USA

Abstract

Although the primary motivation for constructing devices which produce magnetically confined high-temperature plasmas is controlled thermonuclear fusion, these devices can also be used to study the physics of highly ionized atoms. This paper reviews an example of such a study in which the n=2 to n=2 transition lines in the Be I isoelectronic sequence were measured with photometrically calibrated extreme ultraviolet instrumentation. The brightness ratios of the 2s2p ^{3}P-2p^2 ^{3}P (R) and the 2s^2 ^{1}S$_0$-2s2p ^{3}P$_1$ (R*) lines to the resonance 2s^2 ^{1}S$_0$-2s2p ^{1}P$_1$ line were determined and compared with theoretical predictions. At low nuclear charge, Z, the experimental values agree with those computed by the R-matrix method which includes the effect of resonances. At higher Z, the data agrees very well with calculations based on the distorted wave approximation. In addition to confirming the theoretical methods, this agreement lends confidence to the use of line ratios for diagnostic studies of high-temperature plasmas.

1. Introduction

A. L. Schawlow frequently has noted the mutually beneficial interaction between science and technology. Advances in scientific knowledge are used to produce improved and new technology. In turn, these technological improvements are utilized by the scientific community for further advances, etc. Similar relationships also exist between less and more applied branches of science. Although this observation was based on a different field of endeavor, it is also true of controlled thermonuclear fusion research. A reliable understanding of the excitation physics of atoms and ions in a high-temperature plasma is necessary both to minimize the radiative losses from a plasma and to utilize spectroscopic data for plasma diagnostics. In turn, the existence of these devices and improvements in the electron temperature make possible new experimental checks of the theoretical techniques used to compute the atomic parameters for the highly ionized species present in these plasmas. Indeed, it can be argued that measurements performed in plasmas (as against beam experiments) are extremely important in that they check both the accuracy of the atomic physics calculations and for previously unsuspected excitation mechanisms such as inner shell ionization [1].

Over the past decade, a new generation of spectroscopic instruments has come into use for diagnostic studies of the magnetically confined high temperature plasmas produced for fusion research by devices such as tokamaks. These spectrographs use microchannel plate image converters optically coupled to photo-diode or CCD arrays to provide simultaneous time resolved measurements of a large number of spectral emission lines. In addition, these instruments can be calibrated against absolute photometric standards such as synchrotron radiation. Although designed for diagnostics, these instruments have provided the opportunity to check the theoretical calculations of excitation rates. It is expected that the experimental checks on the theory will lead to a more confident application of the spectroscopic data for density and temperature measurements in both laboratory and astrophysical plasmas. In magnetic fusion plasmas, these diagnostics will be particularly important for complex magnetic geometries where electron density and temperature measurements are not always available at the point where the spectroscopic measurements are made.

Stimulated by the requirements of both astrophysics and fusion, a number of workers have computed relative level populations in ionized atoms as a function of both density and temperature. The relative population of any pair of levels can be determined by measuring the brightness ratio of a pair of emission lines. Thus, if the calculations are correct, it is possible to use these ratios to measure the electron density or temperature (depending on the line pair selected) in the plasma. This laboratory has used diagnostic instrumentation to measure these ratios in laboratory fusion plasmas where the electron temperature and density have been well determined by other means as a check on the theoretical calculations. Experiments of this type have been performed on the Princeton Large Torus at the Princeton Plasma Physics Laboratory [2,3], the Tandem Mirror Experiment at Lawrence Livermore National Laboratory [4,5] and the TEXT Tokamak at the University of Texas [4-8].

This article will review the work performed on the Be I-like ions ranging from C III to Cr XXI using the TEXT Tokamak [6,8]. This work has shown that although the more complex R-matrix calculations which include resonances are necessary at low nuclear charge, at higher Z very good agreement is found with the distorted wave approach.

Figure 1 shows a simplified Grotian diagram of the n=2 levels of Be I-like ions. It was possible to measure two line brightness ratios with some precision along the isoelectronic sequence:

$$R = B(2s2p\ ^3P_2-2p^2\ ^3P_2)/B(2s^2\ ^1S_0-2s2p\ ^1P_1) \tag{1}$$
and
$$R^* = B(2s^2\ ^1S_0-2s2p\ ^3P_1)/B(2s^2\ ^1S_0-2s2p\ ^1P_1). \tag{2}$$

(For low Z, all of the lines in the multiplet $^3P - ^3P$ were used to compute R.) These ratios are relatively insensitive to the electron temperature and those of low Z ions have been used to determine solar plasma densities [9]. At high Z, these ratios are not sensitive to changes in the density at the values near the center of a tokamak. Hence, the high Z ratios are not useful tokamak diagnostics, although the low Z ratios can be used at the plasma edge (a region where spectroscopic density diagnostics may be especially valuable). However, note that for both ratios the denominator is proportional to the spin-allowed excitation rate, whereas at tokamak densities the numerator for R* depends on the intercombination $^1S_0 - ^3P$ excitation rate and the numerator for R depends on both the $^1S_0 - ^3P$ and the $^3P - ^3P$ excitation rates. Consequently, the numerator and denominator brightnesses depend on very different types of electron excitations (spin allowed vs. intercombination). Thus, the experimentally determined values of the ratios over a wide range of Z serve as an important test of the theoretical calculations. The good agreement with theory shown by the experimental measurements lends confidence in using the theory both for edge diagnostics and to seek other line ratios more appropriate for tokamak density diagnostics near the center.

296

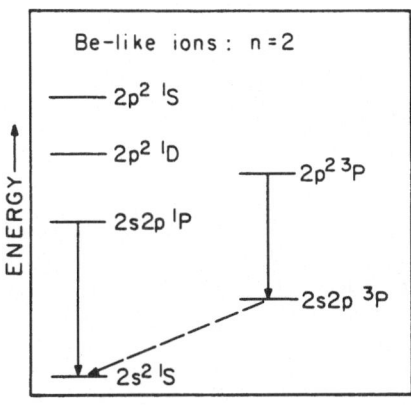

Fig. 1. Simplified Grotian diagram of the n=2 levels of Be I-like ions showing the transitions used.

2. Experimental Method

The plasma of the TEXT Tokamak [10] plasma is confined in a toroidal vacuum vessel. The major radius of the torus is 100 cm; the minor radius, is 27 cm. The toroidal magnetic field is 15-30 kG, and the plasma current is 150-400 kA (Fig. 2a). A TEXT discharge lasts ~ 400-600 ms, with a 200-400 ms steady state phase, during which the plasma current, the electron density, and the temperature are essentially constant (Fig. 2b). There is a slight rise in the electron density during injection of an impurity. In the standard TEXT discharge, the toroidal field is 28 kG, the plasma current is 300 kA, the average density along a chord through the center of the plasma is $3.5 \times 10^{13} \text{cm}^{-3}$ and the electron temperature at the toroidal axis is ~ 1 keV. The electron temperature and density across the minor radius of the torus are displayed in Figure 2c. In order to study the density dependence of the line ratios of intrinsic oxygen, plasmas were produced at three different chord-averaged densities, 1.5×10^{13}, 3.5×10^{13}, and $8 \times 13 \text{ cm}^{-3}$ (most of the data was obtained at the middle density), with central electron temperatures varying between 700 and 1100 eV. The electron density is measured by far infrared interferometry for six different chords across the plasma diameter and then inverted to obtain a radial profile. The electron temperature is obtained by Thomson scattering. Since some of the lines of interest were emitted near the plasma edge, electron temperatures and densities were measured in this region by Langmuir probes.

The spectra were recorded in the 50-2000 Å range by means of two time-resolving spectrometers: a grazing incidence instrument and a normal incidence instrument. Both are equipped with microchannel plate image converter detectors. The instruments, the detectors and their properties have been described in detail elsewhere; for the normal-incidence instrument see [11] and for the grazing-incidence instrument see [12]. The two instruments were located at the same toroidal position. Radial profiles of the ion emissivities were obtained by scanning the plasma on a shot by shot basis with the grazing incidence instrument and Abel inverting the brightness profiles of emission lines in this spectral region (Δn=1 transitions for low Z and Δn=0 for high Z).

Absolute brightness calibrations of the two instruments over their entire wavelength range were obtained by using the SURF synchrotron radiation source at the National Bureau of Standards [13,14]. The spectral resolution of the normal-incidence instrument varies from 0.7 to 4 Å

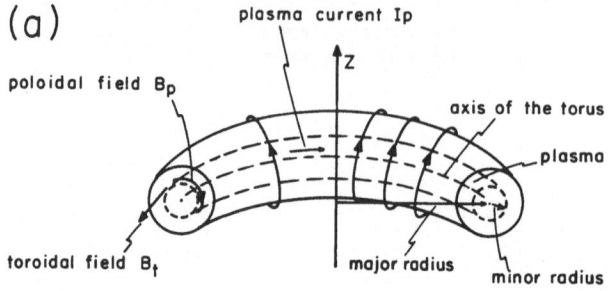

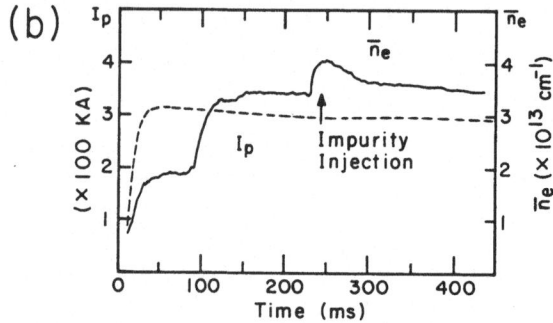

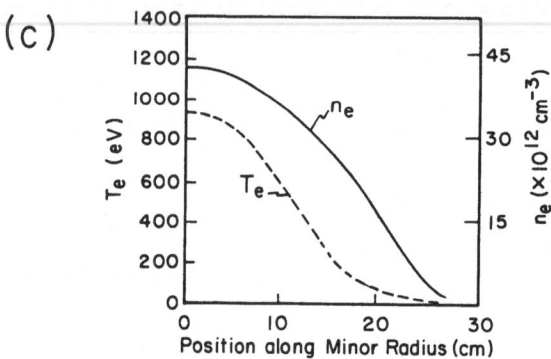

Fig. 2. TEXT Tokamak characteristics (a) Configuration of TEXT Tokamak;
(b) time histories of plasma current I_p (dashed line) and line average
density $\bar{n}_e$ (solid line) of a nominal discharge with impurity injection and
(c) radial profiles of electron temperature (dashed line) and electron
density (solid line) of a standard TEXT discharge. Reference [6].

depending on the breadth of the spectral range covered; several gratings
having the same radius of curvature but different ruling densities permit
the recording of either large spectral ranges with low resolution or smaller
spectral domains with higher resolution. The grazing-incidence spectrograph
has a resolution of 0.7 Å, a range of 15 to 360 Å and the detector
simultaneously covers ~70 Å. The time resolutions of the two instruments

are slightly different, 4.1 ms in the case of the normal-incidence instrument, 5.4 ms for the grazing-incidence one.

A number of intrinsic impurities such as carbon and oxygen were always present in the discharge, introduced by plasma-wall interactions. Elements which were not intrinsic could be introduced into the plasma either by a laser-ablation technique [15] or by a gas puff through a fast valve.

As the temperature decreases from the center to the edge, the ions are located in shells at a plasma radius with an electron temperature near that of ionization equilibrium. Thus the emission could be identified by wavelength, radius of the emission shell and time history. (Lower Z ions peak earlier if intrinsic and decay away more rapidly if laser injected.) Some of the spectral features were partially blended, but the linear properties of the detector permitted separation of the blends in many cases. In the case of injected impurities, the spectrum before the injection was subtracted from that after the injection to isolate the spectrum due to the injected elements. Fig. 3 presents an example for the case of titanium.

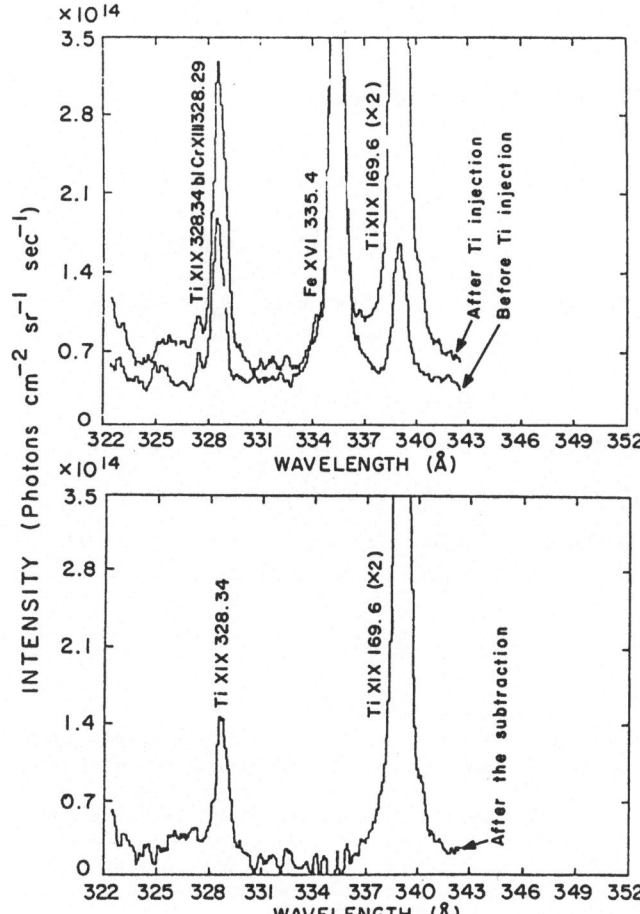

Fig. 3. Example of removal of blends by subtraction of a spectrum before the injection. The Ti XIX 328.34 Å line is blended with the Cr XIX 328.29 Å line. The signal due to the intrinsic Cr XIII 328.29 Å was eliminated after the subtraction as was that due to the intrinsic Fe XVI 335.4 Å. Reference [8].

3. Results

The wavelength range of the emissions divided the study into two parts; the low Z n=2 to n=2 emissions [6] were measured with the normal incidence instrument and the high Z emissions [8] with the grazing incidence instrument.

3.1. Low Z

Table 1 compares the measured values of R with R-matrix calculations which include the effect of resonances. The agreement between theory and experiment is quite good. It appears that the R-matrix calculations provide reliable predictions of the intensity ratios.

A widely used theoretical technique is the distorted wave method which is simpler to use than the R-matrix method but does not include the effects of resonances. This technique is expected to be unreliable at low Z but to improve with increasing charge of the Be I-like ion. As a comparison, a distorted wave calculation for C III (at T_e=5eV) gave a value of 1.2 for R, far outside the experimental uncertainty[6].

Table 1. Experimental and R-matrix Calculated Ratios

Ion	Wavelengths of transitions (Å)		$R = \dfrac{I(2s2p^3P-2p^2\ ^3P}{I(2s^2\ ^1S-2s2p^1P)}$			
			Experiment	Theory		
	$2s^2\ ^1S-2s2p^1P$	$2s2p^2P-2p^2\ ^3P$	Value at	Value	T_e(eV)	n_e(cm^{-3})
C III	977.0	1174.9-1176.4	0.56	0.54[a]	3	5×10^{11}
O V	629.7	758.7-762.0	0.82	0.81[a]	13	2×10^{12}
F VI	535.2	644.0-648.5	0.80	0.78[b]	22	4×10^{12}
Ne VII	465.2	558.6-564.5	0.80	0.75[c]	54	2×10^{13}

a) Reference [16]
b) Reference [17]
c) Reference [18]

Figure 4 displays the results for O V, F VI and Ne VII. Measurements of the O V emissions were obtained in discharges with three different densities, permitting a study of the density dependence. The agreement between the experimental values and the R-matrix calculations is quite good. Distorted wave calculations for O V [6] gave values (0.88 and 1.06 at an electron temperature of 15 eV and electron densities of 1.0×10^{12} and 1.0×10^{13} respectively) which were above those of the R-matrix method, but the differences were not as large as the case of C III.

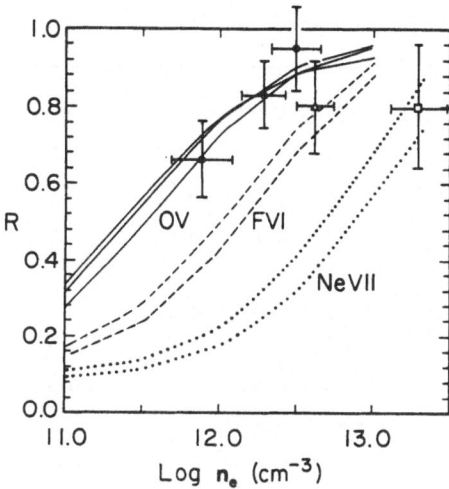

Fig. 4. Comparison of low Z experimental values with theoretical computations using the R-matrix method (1) O V - solid circles (experiment); solid curves (theory) for three electron temperatures, 14 eV, 22 eV, and 34 eV. (At the low-density limit the curves are ordered so that increasing R corresponds to decreasing temperature; at the high-density limit increasing R corresponds to 14 eV, 34 eV, and then 22 eV.) (2) F VI- empty triangle (experiment); dashed curves (theory) for two temperatures, 22 eV (upper curve) and 43 eV (lower curve). (3) Ne VII - empty square (experiment); dotted curves (theory) for two temperatures, 27 eV (upper curve) and 54 eV (lower curve). Reference [6].

3.2 High Z

The discrepancies between the R-matrix and the distorted wave methods are expected to decrease with increasing Z. A comparison of the relative brightnesses of fifteen lines in Ca XVII by [8] indicated agreement between R-matrix [19,20] and distorted wave [21] calculations of better than 25%. Of course, even though the predictions of two techniques agreed, this did not guarantee that they were correct, and it was extremely important to compare these predictions with experimental measurements. However, the convergence with Z indicated that a comparison with R-matrix calculations was necessary only at low Z and the appropriate comparison for high Z was with the more widely available distorted wave calculations.

For this range of Z, the line intensity ratios, R and R*, are expected to be insensitive to either the electron temperature or the electron density over the range of the tokamak plasma conditions [8]. A comparison of the experimental and theoretical values of R and R* for Sc XVII and Ti XIX at different electron densities showed that there was good agreement except for the case of R* for Sc XVIII, where two of the experimental values are significantly below the theoretical values. In no case was there evidence for a strong electron density dependence.

The measured and computed line ratios for Cl XIV to Cr XXI at an average line electron density near 3.5×10^{13} cm^{-3} are listed in Table 2 and presented as a function of Z in Fig. 5. Calculated line ratios were available for Ar XV, Ca XVII, Ti XIX and Cr XXI [21] and were extrapolated or interpolated for the other elements. Fig. 5 shows that the measured R line ratio points are scattered around a straight line connecting the points predicted by distorted wave calculations. The calculated R* ratios were scattered about a horizontal line at 0.034. The agreement between experiment and theory is quite good except for the case of R* for Sc XVIII. Table 2 lists the ratios of measured to calculated values, R_m/R_c and R^*_m/R^*_c. The two arithmetic means for the ratios of the experimental to calculated values are 1.01 ± 0.19 and 0.84 ± 0.17, respectively. Thus, it appears that the agreement between experiment and theory is quite good, well within the experimental uncertainties.

Table 2. Relative brightnesses of Be I-like ion lines at $\langle n_e \rangle$ = 3.5E13/cm³

	Cl XIV	Ar XV	K XVI	Ca XVII	ScXVIII	Ti XIX	V XX	Cr XXI	average
$R^*(\times 10^{-2})$:									
m					2.05	3.12	2.87	3.53	
c	3.21e	3.26	3.39i	3.57	3.33i	3.11	3.49i	3.81	
R^*_m/R^*_c:					0.62	1.00	0.82	0.93	0.84±.17
$R(\times 10^{-2})$:									
m	3.70		2.51	1.66	1.54	1.76	1.17	0.92	
c	2.78e	2.53	2.28i	2.02	1.77i	1.52	1.28i	1.05	
R_m/R_c:	1.33		1.10	0.82	0.87	1.16	0.91	0.88	1.01±.19

R and R* are defined in Eq. 1 and Eq. 2. m = measured, c = calculated
by reference [21], and interpolated between electron densities
i = interpolated along Z, e = extrapolated along Z

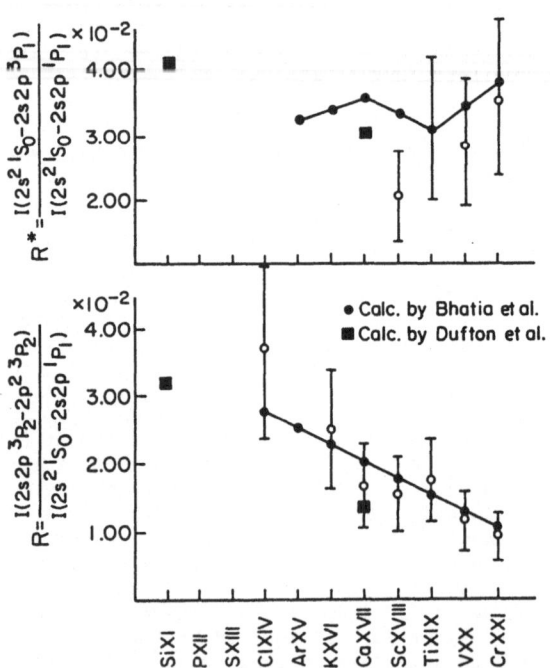

Fig. 5. Comparison of the measured line ratios, R and R*, at the average
electron density 3.5 × 10^{13}cm⁻³ with calculations based on the distorted
wave approximation [21] and the R-matrix method [17,19,20].   Reference [8].

R-matrix method calculations are available for only two cases in this range, Si XI [17] and Ca XVII [19,20]. The values are close to the distorted wave approximation results for Ca XVII and to an extrapolation of the distorted wave values for Si XI. A direct comparison with experimental data is available only in the case of Ca XVII for which the R line ratio was measured and is within 25% of both theoretical calculations.

Which relative collision strengths does an experiment such as this check and to what extent are the results of these experiments valid at other densities? Since one is normally interested in the percentage change in R for a percentage change in the collision strength, Ω_{ij}, the sensitivity of the ratio R to a given collision strength is best expressed by

$$S = abs[(\Omega_{ij}/R)(\partial R/\partial\Omega_{ij})] . \qquad (3)$$

A similar equation holds for R*. The quantity S was estimated for Ca XVII using the collision strengths of [22]. For R*, the values of S are about 1.0, 0.44 and 0.40 for the $2s^2$ $^1S_0$–$2s2p$ $^1P_1$, $2s^2$ $^1S_0$–$2s2p$ $^3P_1$ and $2s^2$ $^1S_0$–$2s2p$ $^3P_2$ transitions, almost independent of density. Other collision strengths contribute in negligible amounts. For R, the situation is slightly more complex. Fig. 6 shows the dependence of the most important

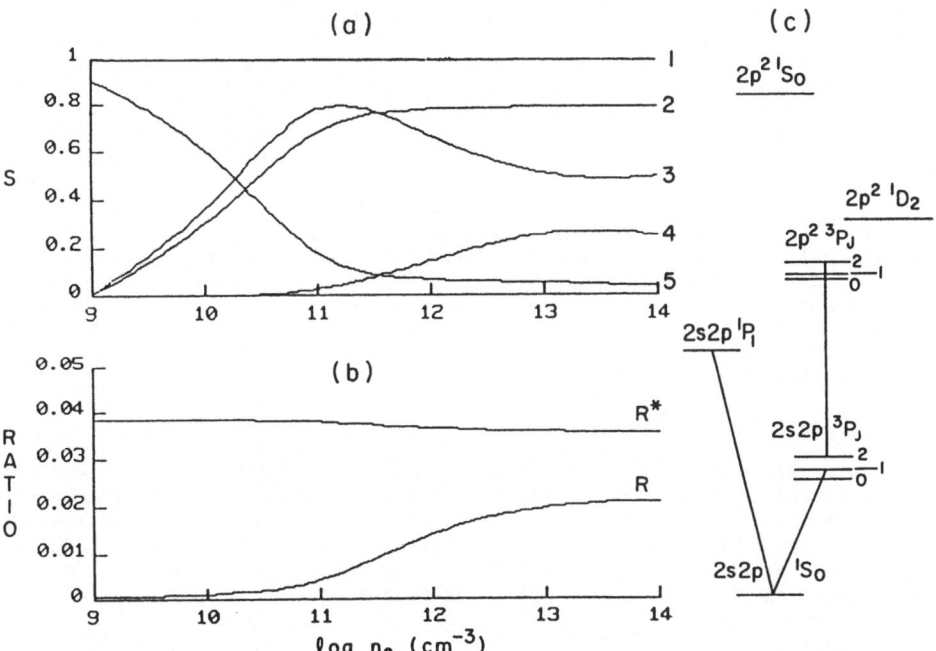

Fig. 6. S, the sensitivity of the value of R to uncertainties in the collision strengths. The values for the most important collision strengths in Ca XVII are plotted as functions of electron density. The curves are numbered with respect to the electron collisional transitions marked in part c. (b) The brightness ratios, R and R* defined in Eqs. 1 and 2, are plotted as functions of electron density for Ca XVII. (c) The Grotrian diagram of n=2 levels of Ca XVII. The important collisional transitions are numbered according to part a. Modified from reference [8].

sensitivities as a function of electron density along with the density dependence of R and R*. At a density of $3 \times 10^{13} \mathrm{cm}^{-3}$, where the tokamak measurements were made, the $2s^2$ $^1S_0$–$2s2p$ $^1P_1$, $2s^2$ $^1S_0$–$2s2p$ $^3P_2$ and $2s2p$ $^3P_2$–$2p^2$ $^3P_2$ transitions dominate followed by the $2s2p$ $^3P_2$–$2p^2$ $^3P_1$ transition. These transitions are important down to densities of about $3 \times 10^{10} \mathrm{cm}^{-3}$ where the $2s^2$ $^1S_0$–$2p^2$ $^3P_2$ transition becomes significant. The tokamak data provides no data on this collision strength, which involves a spin change and the promotion of two electrons from $2s^2$ to $2p^2$. Note, however, that at densities below this point, the value of R is near 10^{-3} and accurate spectroscopic measurements become more difficult. Thus, it appears that the tokamak experiments have checked the collision strengths for the most important transitions and it is likely that theoretical values of R will be accurate in the electron density range from 10^{14} down to $10^{11} \mathrm{cm}^{-3}$ for Ca XVII. For other ions, the lower limit will shift with Z.

4. Conclusions

Brightness ratios for emission lines ratios emitted between the n=2 levels of Be I-like ions ranging from C III to Cr XXI have been measured in a tokamak plasma in which the electron temperatures and densities were measured independently by non-spectroscopic methods. At low Z, the ratios agreed with those computed by the R-matrix method. These results at low Z confirm the importance of resonances by experimentally verifying the electron impact excitation rates computed by the R-matrix approach. For high Z (Cl XIV to Cr XXI), the experimentally determined line ratios, R and R*, agree extremely well with theoretical calculations based on the distorted wave approximation; the ratios of the measured to calculated values of R and R* were 1.01 ± 0.19 and 0.84 ± 0.17 respectively. For the one case where an R-matrix method value was available, Ca XVII, good agreement also exists.

Finally, note that this agreement also gives assurance that the theoretical calculations can be relied on for plasma diagnostics and predictions of radiative loss from plasmas. The relationship between science and technology discussed in the introduction has now come full circle. Indeed, the next step is to seek out additional ratios which are sensitive to electron density and temperature, and determine which are the most practical for high-temperature plasma diagnostics.

Acknowledgements

This article is a review of an extensive study with a large number of collaborators. (See [6] and [8].) I wish to acknowledge each of them and in particular the ideas and efforts of M. Finkenthal. The TEXT Tokamak research group produced and characterized the large number of similar discharges necessary for this study. This work was supported by the Department of Energy under grant DE-FG02-85ER53214 to the Johns Hopkins University.

References

1. M. Finkenthal, B. C. Stratton, H. W. Moos, A. Bar Shalom, and M. Klapisch: Physics Letters 108A, 71, (1985).
2. B. C. Stratton, H. W. Moos and M. Finkenthal: Ap.J. Letters 279, L31 (1984).
3. B. C. Stratton, H. W. Moos, S. Suckewer, U. Feldman, J. F. Seely and A. K. Bhatia: Phys. Rev. A 31, 2534 (1985).
4. T. L. Yu, M. Finkenthal and H. W. Moos: Ap.J. 305, 880 (1986).

5. M. Finkenthal, T. L. Yu, S. L. Allen, L. K. Huang, S. Lippmann, H. W. Moos, B. C. Stratton, P. L. Dufton and A. E. Kingston: Astron.Astrophys, submitted (1987).
6. M. Finkenthal, T. L. Yu, S. Lippmann, L. K. Huang, H. W. Moos, B. C. Stratton, A. K. Bhatia, R. D. Bengtson, W. L. Hodge, P. E. Phillips, J. L. Porter, T. R. Price, T. L. Rhodes, B. Richards, C. P. Ritz, and W. L. Rowan: Ap.J. 313, 920 (1987).
7. S. Lippmann, M. Finkenthal, L. K. Huang, H. W. Moos, B. C. Stratton, T. L. Yu, A. K. Bhatia and W. L. Hodge, Ap.J., in press 1987.
8. L. K. Huang, S. Lippmann, T. L. Yu, B. C. Stratton, H. W. Moos, M. Finkenthal, W. L. Hodge, W. L. Rowan, B. Richards, P. E. Phillips, A. K. Bhatia: Phys.Rev., in press, (1987).
9. P. L. Dufton and A. E. Kingston, Adv.Atom.Molec.Phys. 17, 355 (1981).
10. K. W. Gentle et al.: Plasma Physics and Contr.Fusion 26, 1407 (1984).
11. R. E. Bell, M. Finkenthal and H. W. Moos: Rev.Sci.Instrum. 52, 1806 (1981).
12. W. L. Hodge, B. C. Stratton and H. W. Moos: Rev.Sci.Instrum. 55, 16 (1984).
13. D. L. Ederer, E. B. Saloman, S. C. Ebner and R. P. Madden: Journal of Research of NBS-A.Physics and Chemistry 79A, 761 (1975).
14. D. L. Ederer and S. C. Ebner: A Users Guide to Surf, National Bureau of Standards, Washington, D.C. 20234 (1985).
15. D. R. Terry, W. Rowan, W. C. Connolly and W. K. Leung: Proc. 10th Symp. on Fusion Energy, 14, 959 (1983).
16. P. L. Dufton, K. A. Berrington, P. G. Burke and A. E. Kingston: Astron.Astrophys. 62, 111 (1978).
17. F. P. Keenan, D. A. Berrington, P. G. Burke, A. E. Kingston and P. L. Dufton: Mon.Not.R.Astron.Soc. 207, 459 (1984).
18. P. L. Dufton, J. G. Doyle and A. E. Kingston: Astron.Astrophys. 78, 318 (1979).
19. P. L. Dufton, A. E. Kingston, J. G. Doyle and K. G. Widing: Mon.Not.R.Astron.Soc. 205, 81 (1983).
20. P. L. Dufton, A. E. Kingston and N. S. Scott: J.Phys.B 16, 3053 (1983).
21. A. K. Bhatia, U. Feldman and J. F. Seely: Atomic Data and Nuclear Data Tables, in press.
22. A. K. Bhatia and H. E. Mason: Astron. Astrophys. Suppl. Ser. 52, 115 (1983).

How to Squeeze the Vacuum, Or, What to Do When Even No Quantum Is Half a Quantum Too Many

M.D. Levenson

IBM Research, Almaden Research Center, 650 Harry Road,
San Jose, CA 95120, USA

The vacuum is conventionally defined as the absence of matter and energy. One might imagine that such a vacuum would be simple and without interest to physicists. Quantum mechanics, however, provides a detailed and complex model of the vacuum, pregnant with possibilities and well worth careful study [1]. The quantum mechanical vacuum is defined as the ground state of all fields. The electromagnetic field is the field most familiar to spectroscopists, and is the model used to understand more complex forces. The electromagnetic field is conventionally modelled as an assembly of harmonic oscillator modes. It is well known that the ground state of a harmonic oscillator does not have zero energy; instead, it contains one half quantum. This vacuum energy allows an oscillator in the ground state to have a slightly fluctuating position and momentum and thus fulfil the uncertainty principle [2].

The analogous zero point motion of the electromagnetic field corresponds to fluctuations of the electric and magnetic fields around their zero average value [3]. The mean square of the fields are, however, nonzero. The uncertainty principle similarly allows vacuum fluctuations of all the other force and matter fields of quantum mechanics.

Vacuum fluctuations affect optical processes in a variety of ways. Students of laser physics are familiar with the explanation of spontaneous emission as stimulated emission due to the vacuum fluctuation input [4]. Spontaneous emission into a laser mode in turn causes fluctuations which broaden the laser linewidth according to the famous Schawlow-Townes formula [5]. One explanation of the Lamb shift is that it results from perturbations of an electron's motion due to fluctuating vacuum fields [6].

The modern theory of quantum optics also implies that the familiar "shot noise" from a high efficiency photon detector is a direct result of the zero point motion of the vacuum field [7,8]. This viewpoint is in direct contradiction to the conventional model where the "shot noise" appears in a light detector as the result of the quantization of the electron charge. That model would predict an irreducible noise floor for the detection of any optical field, even a perfectly constant classical wave. This noise floor is sometimes called the "standard quantum limit" or the "vacuum noise level." This vacuum noise limits the sensitivity of all optical measurements; in particular, absorption spectroscopy is especially sensitive to this noise and thus, has a much lower sensitivity than emission spectroscopy. The newer theory allows states of the electromagnetic field with less noise, a clear indication of nonclassical behavior [9]. Some of these low noise states are states where some of the vacuum fluctuations appear to be smaller than normal. Such states correspond to a "squeezed vacuum."

To understand why a squeezed vacuum is possible and and how it might be produced and detected, one must begin by explaining the modern theory of quantum noise. Light is an electromagnetic oscillation, and the electric field amplitude can be expressed as $E(t) = E_1 \cos \omega t + E_2 \sin \omega t$. The quantities E_1 and E_2 are termed "quadrature amplitudes" or, simply, "quadratures." In the quantum mechanical theory, these quadratures are conjugate operators like position and momentum. The Heisenberg uncertainty principle requires that the product of the uncertainty in the two quadrature amplitudes be greater than a dimensional quantity that is proportional to Planck's constant. The root mean square average of the zero point motion is the uncertainty in the quadrature amplitude.

The vacuum state must be independent of phase, and thus the uncertainties in the two quadrature amplitudes must be equal. The coherent states familiar from laser theory are constructed by adding a coherent amplitude to the fluctuating vacuum field [3]. If that coherent amplitude is in the cosine quadrature, $E(t) = (E_1 + \delta E_1) \cos \omega t + \delta E_2 \sin \omega t$. Where δE_1 and δE_2 are understood to represent the quantum mechanical fluctuations. On the average, the electric field is $<E(t)> = E_1 \cos \omega t$. The instantaneous intensity is $I = (c/8\pi)[(E_1 + \delta E_1)^2 + (\delta E_2)^2]$, but the average intensity is just $< I > = (c/8\pi)|E_1|^2$. The term containing δE_1 has zero average, and the terms quadratic in the vacuum fluctuations reflect only the vacuum energy. A quantum detector produces an instantaneous current proportional to the instantaneous power $P = IA$, where A is the detector area and the vacuum energy has been omitted. The proportionality constant for perfect quantum efficiency is $e/\hbar\omega$ where e is the electron charge. The number of electrons produced in time t is just $N = Pt/\hbar\omega$, the number of photons incident in time t.

The noise is best parameterized as the mean square deviation of the optical power (or current or number of detected electrons) from the corresponding average value. In terms of the field fluctuations, the mean square deviations are, respectively:

$$< \delta P^2 > = < P^2 > - < P >^2 = \frac{cA}{2\pi} < P > \; < \delta E_1^2 > ; \tag{1a}$$

$$< \delta i^2 > = \left(\frac{e}{\hbar\omega} \right)^2 < \delta P^2 > = \frac{ecA}{2\pi\hbar\omega} < i > \; < \delta E_1^2 > ; \tag{1b}$$

$$< \delta N^2 > = < N^2 > - < N >^2 = \frac{Act}{2\pi\hbar\omega} < N > \; < \delta E_1^2 > . \tag{1c}$$

These mean square deviations are proportional to the nonzero mean square vacuum fluctuations in the E_1 quadrature. For the vacuum state and coherent states, the mean square fluctuations in E_1 and E_2 are equal and constant [9]. In cgs units:

$$< \delta E_1^2 > = < \delta E_2^2 > = < \delta E_c^2 > = \frac{2\pi\hbar\omega}{Act} . \tag{2}$$

The mean square fluctuations depend only on the sum of the squares of the average quadrature amplitudes, that is, the average intensity.

$$< \delta P^2 > = \frac{\hbar\omega}{t} < P > ; \tag{3a}$$

$$< \delta i^2 > = \frac{e}{t} < i > ; \tag{3b}$$

$$< \delta N^2 > = N. \tag{3c}$$

The mean square fluctuation in the number of electrons produced in period t is equal to N, the average number produced. This is one signature of Poisson statistics, and is just the noise level predicted by the conventional "shot noise" model [9]. The time t is inversely proportional to the bandwidth of the measurement. That model correctly predicts the properties of coherent states and incoherent sums of coherent states. With a few exceptions, all experimental results up until 1985 could be explained in terms of such states.

A squeezed vacuum would have mean square fluctuations in one quadrature less than those characteristic of an unsqueezed vacuum. For example:

$$< \delta E_1^2 > = \ < \delta E_c^2 > e^{2r}; \tag{4a}$$

$$< \delta E_2^2 > = \ < \delta E_c^2 > e^{-2r}; \tag{4b}$$

$$< \delta E_1 \delta E_2 > = 0. \tag{4c}$$

The parameter r describes the degree of squeezing [8]. The product of the uncertainties remains equal or greater than the limit required by the uncertainty principle. If the average quadrature amplitudes are

$$< E_1 > = E_0 \cos \theta \quad \text{and} \quad < E_2 > = E_0 \sin \theta. \tag{5}$$

The quadrature fluctuations at phase angle θ ($\delta E_\theta = \delta E_1 \cos \theta + \delta E_2 \sin \theta$), are responsible for the mean square fluctuations in the power, current and number of detected electrons. For the quadrature fluctuations in (4), the relevant mean square quadrature fluctuation is $< \delta E_\theta^2 > = \ < \delta E_c^2 > \{e^{2r} \cos^2\theta + e^{-2r} \sin^2\theta\}$. The power, current and number fluctuations, then, depend on the phase angle θ as:

$$< \delta P_\theta^2 > = \ \frac{\hbar \omega}{t} < P > \{e^{2r} \cos^2\theta + e^{-2r} \sin^2\theta\}; \tag{6a}$$

$$< \delta i_\theta^2 > = \ \frac{e}{t} < i > \{e^{2r} \cos^2\theta + e^{-2r} \sin^2\theta\}; \tag{6b}$$

$$< N_\theta^2 > = N\{e^{2r} \cos^2\theta + e^{-2r} \sin^2\theta\}. \tag{6c}$$

For the correct choice of θ, the mean square fluctuation for a squeezed vacuum is reduced by a factor of exp(−2r) [8].

The squeezed vacuum has more energy than the vacuum state, but substantial reductions in the detected noise occur for minimal increase in field energy. The average value of the field remains zero. However, even one quantum radiated or scattered randomly into the detector can obliterate the squeezing effect. One of the major experimental challenges of squeezed state generation and detection is avoiding all additional noise sources.

Squeezed light is generated by nonlinear interactions in various kinds of materials. Strong pump waves are required for these interactions, and even very small scattering cross sections can scatter enough light to wipe out the noise reduction. Moreover, lasers are never perfect. Laser-produced noise above the quantum level must somehow be suppressed. Experimentally overcoming these difficulties can be just as challenging as squeezing the vacuum.

In attempts to avoid problems with unstable dye lasers or strongly scattering sodium vapor, we chose to squeeze the vacuum fluctuations using the nonlinear index of refraction of an optical fiber [10]. In an optical fiber, the index of refraction depends on the intensity of the light propagating through the core [11]. In terms of the electric field amplitude, the index is

$$n(E) = n_0 + n_2 |E|^2. \tag{7}$$

After propagating through a length ℓ of optical fiber, the phase of a light wave has shifted by $\phi = 2\pi n(E)\ell/\lambda$. Fluctuations in the amplitude of the pump wave cause correlated fluctuations in the phase of the wave:

$$\delta\phi(\ell) = \frac{4\pi n_0 n_2 \ell}{\lambda} <E_1> \delta E_1 + \delta\phi(0) \tag{8}$$

where $\delta\phi(0) = \delta E_2(0)/E_1$ is the phase fluctuation at the input to the fiber. This equation applies even for pump amplitude fluctuations caused by the vacuum fields [12].

After propagating through a length ℓ of optical fiber, part of the phase fluctuations of a light beam are correlated with the amplitude fluctuations. The amplitude fluctuations cannot be changed by a fluctuating index of refraction and are thus equal at input and output.

The result is illustrated in Fig. 1. At the input to the fiber the fluctuations around the average optical amplitude can be represented as a fuzzy circle, without phase dependence. After propagating through the fiber, correlated amplitude and phase fluctuations arise. These fluctuations can be represented by a tilted ellipse as in Fig. 1b. The minor axis of the ellipse is smaller than the radius of the initial circle. The fluctuations along that axis can be made to appear as intensity modulation of a wave by phase shifting the average amplitude to be parallel to the minor axis. The result is a light beam with lower noise than the noise level at the input to the fiber, even when the input noise is at the vacuum level [10].

This analysis can be made more formal by introducing operators for the creation and destruction of photons, Hamiltonians and similar paraphernalia. The physics of the interaction, however, is captured in the treatment here [13].

Propagating the light through a length ℓ of optical fiber causes fluctuations δE_1 and δE_2 to become correlated according to:

$$\delta E_2(\ell) = \frac{4\pi n_0 n_2 \ell}{\lambda} <E_1>^2 \delta E_1(\ell) + \delta E_2(0); \tag{9a}$$

$$\delta E_1(\ell) = \delta E_1(0). \tag{9b}$$

The fluctuations are detected at the output of the fiber by beating them against a phase-shifted local oscillator wave derived from the transmitted average amplitude. If the local oscillator is in phase with the transmitted amplitude, only δE_1 is detected. If it is phase shifted by $90°$, δE_2 appears. In general, the detected quadrature is a linear superposition of these two:

$$\delta E_\theta = \delta E_1 \cos\theta + \delta E_2 \sin\theta \tag{10}$$

where θ is the phase shift of the local oscillator wave. The mean square averages for the quadrature operators are equal for vacuum states, but not for the squeezed states produced by the fiber. The ratio of the mean square fluctuations $- <\delta E_\theta(\ell)^2>$ for the squeezed states and $<\delta E_c^2>$ for coherent and vacuum states is:

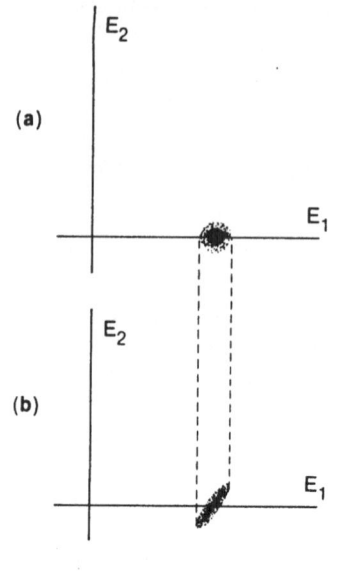

(a)

(b)

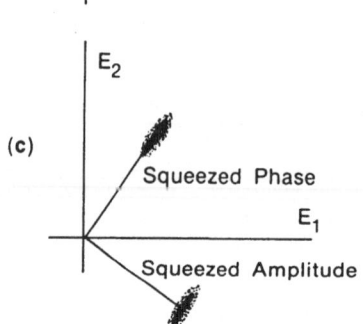

(c)

Squeezed Phase

E_1

Squeezed Amplitude

Figure 1. Quadrature amplitude plots for coherent and squeezed states. A coherent state with $< E_2 > = 0$ appears in Fig. 1a. The average amplitude is at the center of the fuzzy circle which represents the quantum fluctuations. The kind of squeezed state made by an optical fiber appears in Fig. 1b. The projection of the quantum fluctuations along the E_1 axis remains as in the coherent state in Fig. 1a, but fluctuations correlated to δE_1 are added to E_2. Figure 1c shows phase – shifted average amplitudes which produce squeezed states with phase or amplitude fluctuations less than a coherent state. Only the squeezed amplitude fluctuations reduce the noise as measured on a simple light detector.

$$\frac{< \delta E_\theta^2(\ell) >}{< \delta E_c^2 >} = V(\theta) = 1 + 2r \sin 2\theta + 4r^2 \sin^2\theta \qquad (11)$$

where $r = 2\pi n_0 n_2 \ell < E_1 >^2/\lambda$, is analogous to the squeeze parameter introduced in (4).

The mean square fluctuations in the optical power, detector current and photon number all vary with phase in this way. At $\theta = 0$, the mean square fluctuations for the fiber-produced squeezed state are equal to those for the coherent states in (3). The projection of the ellipse in Fig. 1 on a line at angle θ is $\sqrt{V}$.

Unfortunately, all of this can be disrupted by inelastic light scattering, fluorescence, absorption and loss. All of these processes add fluctuations. In optical fibers, the light scattering processes are most troublesome. It happens that the forward light scattering at low frequencies consists of correlated Stokes and anti-Stokes emission which together add to δE_2, but not δE_1 [14].

Figure 2 shows a spectrum of this light scattering for a bare optical fiber. The narrow peaks are resonances corresponding to vibrational modes of the cylindrical optical fiber.

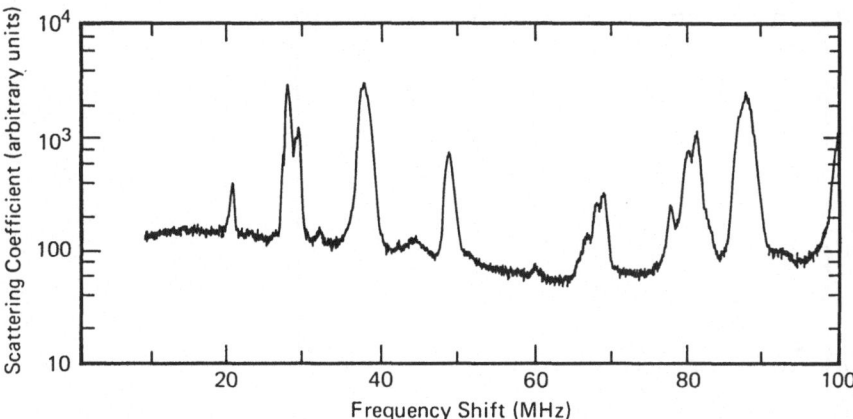

Figure 2. Spectrum of light scattering from a bare optical fiber at 60K. The narrow peaks correspond to Brillouin scattering by phonons guided by the cylindrical fiber structure (so-called Guided Acoustic Wave Brillouin Scattering or GAWBS). The noise floor between the peaks is due to another light scattering mechanism, probably related to structural relaxation.

Between those peaks is a background of light scattering due to structural relaxation. These two processes have different dependences on temperature. The phonon scattering varies linearly with temperature as shown in Fig. 3a; the background scattering is very different, as shown in Fig. 3b. At present, the details of this background light scattering are not well understood.

In the context of squeezing the vacuum, light scattering is a source of noise to be eliminated. Even at 2K, the noise from light scattering almost eliminates the noise reduction. At such a low temperature, the threshold for stimulated Brillouin oscillation is very low. Strong phase modulation is necessary to suppress the oscillation and allow a strong enough pump field [10].

The experimental noise level normalized to the vacuum noise level is plotted as a function of local oscillator phase in Fig. 4. For a narrow range of angles around $-22°$, the measured level is 12% below that of the vacuum. Other experiments with different technology have shown noise levels a factor of 3 below the vacuum [15].

A remarkable feature of these squeezed vacuum states is that attenuation increases the measured noise level. This is illustrated by Fig. 5 where the normalized noise level for a squeezed vacuum state and incandescent light (which is a sum of coherent states) are plotted as a function of the transmission of a neutral density filter [13]. When the transmission is 100%, the squeezed vacuum gives a noise level 10% below the vacuum level. As the transmission decreases, the normalized noise level for the squeezed vacuum rises. The normalized noise of the incandescent light remains constant at the vacuum level. At optical frequencies and ambient temperatures, black body radiation and vacuum fluctuations are synonymous. The partly absorbing filter radiates vacuum fluctuations ("darkness waves") and thus enforces the standard quantum limit.

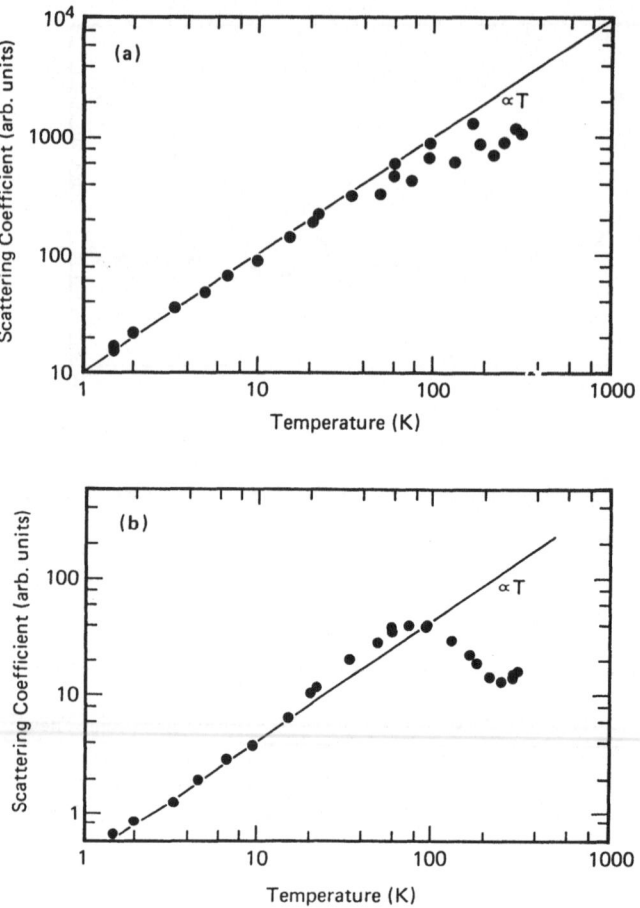

Figure 3. Temperature dependence of the light scattering in an optical fiber. The strong (GAWBS) phonon peak at 49.6 MHz in Fig. 2 shows linear temperature dependence as in Fig. 3a. The baseline noise level between the peaks (measured at 55 MHz) varies as shown in Fig. 3b.

The nonlinear interactions in an optical fiber can produce correlations among more than the two quadratures of (9a) and (9b). When there are two average amplitudes (E_x and E_y) at two frequencies in an optical fiber, the phase fluctuations analogous to those in (8) correlate with the amplitude fluctuations of both waves:

$$\delta\phi_x(\ell) = \delta\phi_x(0) + \frac{4\pi n_0 \ell}{\lambda} \left[n_2(\omega_x, \omega_x) < E_{1x} > \delta E_{1x} + n_2(\omega_x, \omega_y) < E_{1y} > \delta E_{1y} \right]. \quad (12)$$

In this equation, $n_2(\omega_x, \omega_y) = (24\pi/n_0)\chi^{(3)}(-\omega_x, \omega_x, \omega_y, -\omega_y)$ is a complex and spectroscopically interesting quantity which can be enhanced by Raman resonances that do not contribute to $n_2(\omega_x, \omega_x)$. If $|\omega_x - \omega_y|$ approaches a Raman mode frequency, the Raman resonance dominates all other effects. Figure 6 is a plot of the variation of

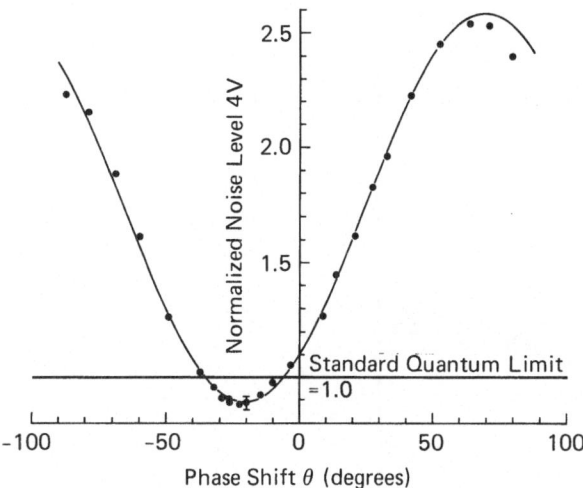

Figure 4. The noise-to-current ratio for the fiber produced squeezed state compared to the noise from an incandescent source (Standard Quantum Limit). The phase dependence of the noise with a net noise level below the standard quantum limit shows that this is a squeezed state. The radius of the solid circles is three times the estimated uncertainty of all of the measurements used at a particular phase while the error bracket shows the uncertainty of a single measurement.

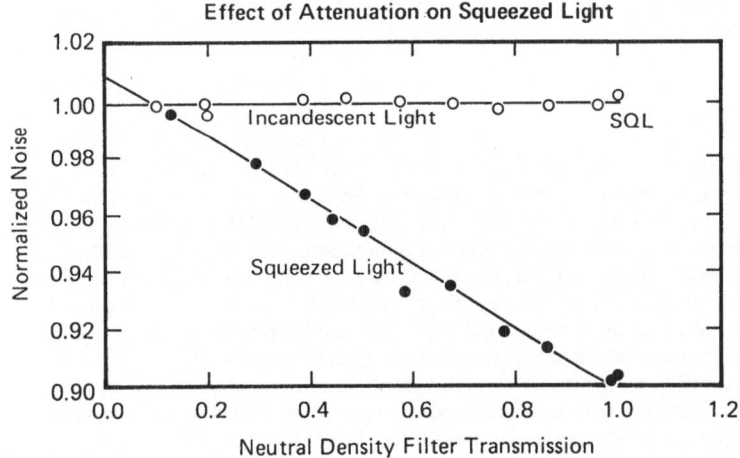

Figure 5. The effect of attenuation on squeezed light. The noise-to-current ratio is plotted here as a function of the transmission of a neutral density filter. The ratio is independent of attenuation for light at the standard quantum limit as produced by our incandescent source. However, the noise-to-current ratio of squeezed light rises toward the standard quantum limit as the transmission of the filter is decreased

313

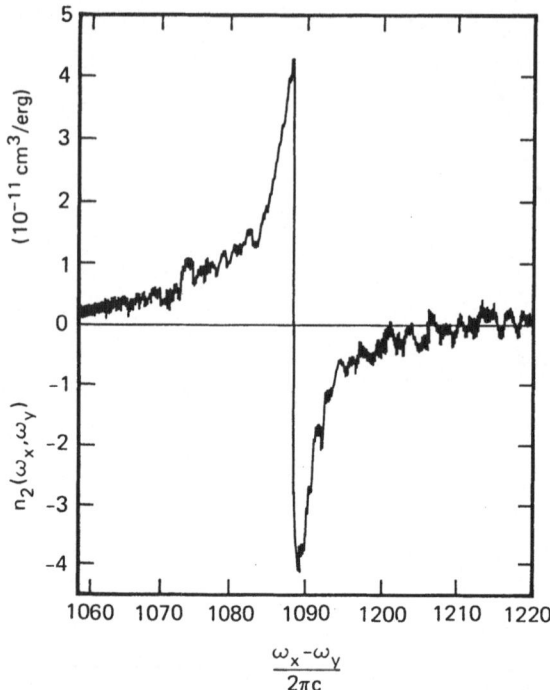

Figure 6. A plot of the Raman enhanced Kerr nonlinearity of calcite. The nonlinearity n_2 responsible for the ordinary squeezing interaction would be 4×10^{-13} esu. The resonance can be used to enhance the sensitivity of quantum nondemolition detection when the interacting waves are shifted by 1087 cm^{-1} from one another. For fused silica, $n_2 = 1.3 \times 10^{-13}$ esu.

$n_2(\omega_x, \omega_y)$ for a material with a strong Raman resonance. There is nearly a 100 fold enhancement of the nonlinearity over the value obtained when $\omega_x = \omega_y$. In many such materials, even $n_2(\omega_x, \omega_x)$ is larger than fused silica. The potential of this Raman enhanced quantum correlation has yet to be exploited, but the interactions observed in fused silica optical fibers show what might be expected. The fluctuations in $\delta\phi_x$ are correlated with the amplitude fluctuations of the other wave (*i.e.*, δE_{1y}). This effect allows one to measure the vacuum fluctuations - without perturbing them [16]. This is a form of back-action evading measurement [17] or quantum nondemolition detection (QND). Subsequent measurements of the amplitude fluctuation would result ideally in identical values. The uncertainty necessarily added by a quantum measurement appears in the phase of the measured wave, not the amplitude. Thus, not only can the vacuum be squeezed, it can be measured – repeatedly.

Squeezing and measuring the vacuum can enhance the sensitivity and precision of the techniques of laser spectroscopy. While remarkable, the techniques required are not unduly complex. The difficult thing is believing that an apparently intrinsic quantum limit can be transcended. The experimental proof is now at hand, the future will disclose the applications.

One spectroscopic application deserves to be mentioned in any volume honoring Arthur Schawlow: it is possible to build a dual beam spectrometer with this technology which makes absorption spectroscopy as sensitive as fluorescence spectroscopy. The basic idea is shown in Fig. 7. Light waves with two different frequencies are coupled in an optical fiber. The two wavelengths are separated at the output, and the phase fluctuations of one are measured as in squeezed state experiments. That phase fluctuation signal measures the amplitude fluctuations at the other wavelength. The second beam passes through a weakly absorbing sample. The absorption may be modulated by dithering the laser frequency, *etc.*,

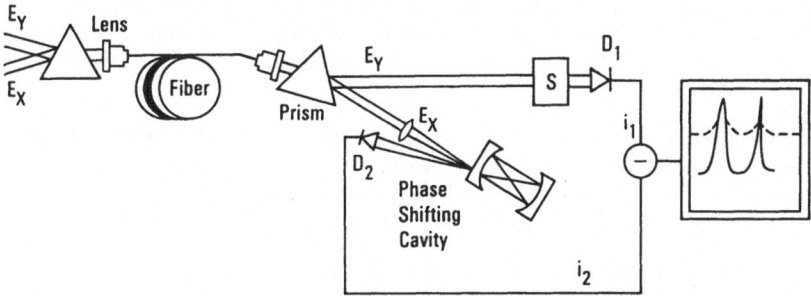

Figure 7. A dual beam "spectrometer" employing QND in an optical fiber. The phase fluctuations of the beam E_x become correlated with the amplitude fluctuations of the beam E_y in the optical fiber. The current produced by the detector D_2 is a QND measurement of the amplitude fluctuations. A weakly absorbing sample at S modifies the amplitude of the wave reaching detector D_1. The quantum fluctuations (which had been previously measured at D_2) are subtracted off allowing the detection of small changes caused by the sample.

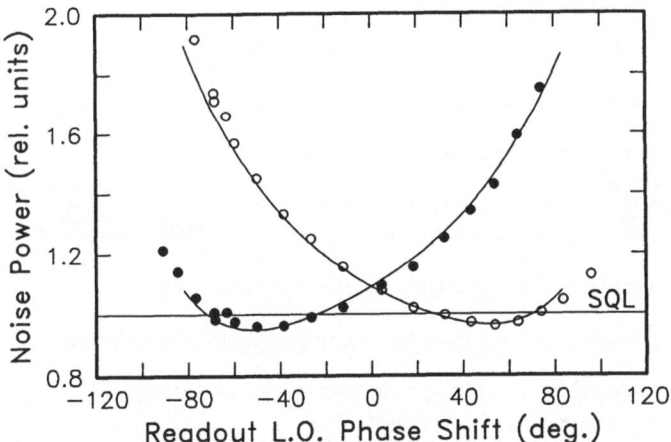

Figure 8. Noise level for the "QND Spectrometer" shown in Fig. 7 normalized to the standard quantum limit noise level on detector D_1, plotted as a function of the cavity phase shift. The noise level is reduced by 5% below the standard quantum limit for the appropriate choice of optical phase. The open circles correspond to subtracting the electrical currents from the two detectors as shown in Fig. 7, while the solid circles correspond to the sum of the currents.

315

and the transmitted intensity is detected. The signals from the detectors which monitor the transmitted intensity and the phase of the other beam are then subtracted. Since the quantum noise for the beam passing through the sample causes phase fluctuations on the other beam which can be accurately measured, the quantum noise of the beam passing through the sample can be subtracted from the detector signal. All that remains are the effects of the sample on the light that passed through.

Fluctuations from the light source have long been subtracted from spectroscopy signals using dual beam techniques. The two beams are conventionally split optically using a partially reflecting mirror. Such a technique can eliminate classical fluctuations, but the quantum noise would still constitute a noise floor – the standard quantum limit. The back-action evading spectrometer proposed here would get below that noise floor, enhancing sensitivity even further.

Figure 8 shows a very early result of this sort, the noise level in a system such as shown in Fig. 7. The ordinary quantum noise level is shown as a horizontal line labelled SQL. The curves and data points show the noise level in a back-action evading system. Even in this early case, the noise level is 5% below the ordinary limit [12].

Lasers have already revolutionized spectroscopy by providing sources of strong coherent light. The next great breakthrough might be light with much less noise than even fully coherent light. The future remains incredibly bright.

REFERENCES

1. E. G. Harris: In A Pedestrian Approach to Quantum Field Theory (Wiley-Interscience, 1972), Chapter 4.
2. L. I. Schiff: In Quantum Mechanics (McGraw-Hill, 1955), p.60.
3. B. R. Mollow: Phys. Rev. 188, 1969 (1969).
4. C. Cohen-Tannoudji, B. Diu, F. Laloe: In Quantum Mechanics (Wiley-Interscience/Hermann, 1977), p. 618.
5. A. L. Schawlow, C. H. Townes: Phys. Rev. 112, 1940 (1958).
6. T. A. Welton: Phys. Rev. 125, 804 (1962).
7. R. Bondurant, J. H. Shapiro: Phys. Rev. D30, 2548 (1984).
8. C. M. Caves: Phys. Rev. D23, 1693 (1981).
9. D. F. Walls: In Nature (London) 306, 141 (1983).
10. R. M. Shelby, M. D. Levenson, S. H. Perlmutter, R. G. DeVoe, D. F. Walls: Phys. Rev. Lett. 57, 691 (1986).
11. Y. R. Shen: In The Principles of Nonlinear Optics (John-Wiley and Sons, 1984), 303 ff.
12. M. D. Levenson, R. M. Shelby: In Four Mode Squeezing and Applications, Optica Acta (to be published).
13. G. J. Milburn, M. D. Levenson, R. M. Shelby, D. F. Wall, S. H. Perlmutter, R. G. DeVoe: J. Opt. Soc. Am. B (to be published).
14. R. M. Shelby, M. D. Levenson, P. W. Bayer: Phys. Rev. 31, 5244 (1985).
15. L. A. Wu, H. J. Kimble, H. Wu, J. L. Hall: Phys. Rev. Lett. 57, 2520 (1986).
16. M. D. Levenson, R. M. Shelby, M. Reid, D. F. Walls: Phys. Rev. Lett. 57, 2473 (1986).
17. C. M. Caves: In Quantum Optics, Experimental Gravitation and Measurement Theory, ed. by P. Meystre, M. O. Scully (Plenum Press, New York, 1983), p. 567; C. M. Caves, K. S. Thorne, R. W. P. Drever, V. D. Sandberg, M. Zimmerman: Rev. Mod. Phys. 57, 341 (1980).

Raman Spectroscopy of Biomolecules

T.J. O'Leary

Department of Cellular Pathology, Armed Forces Institute of Pathology,
Washington, DC 20306, USA

1. INTRODUCTION

Raman spectroscopy is now so highly dependent on excitation by laser light sources that it hardly seems possible that C.V. Raman could have first observed this faint scattering phenomenon using filtered sunlight [1-4]. First predicted in 1923 by Adolf Smekel [5], the Raman effect is the scattering of an incident photon shifted in frequency by an allowed vibrational frequency of the scattering molecule. Thus, the Raman spectrum, like the infrared spectrum, yields information about the vibrational frequencies of molecules. Unlike infrared absorption, which results from an interaction of the incident radiation with the electric dipole of the molecule, Raman scattering results from an interaction of this radiation with the polarizability tensor [6]. The practical result is that the Raman spectrum does not yield exactly the same vibrational information as does the infrared spectrum, since the selection rules are different. A second result is that the Raman effect results in a very weak scattering spectrum. In classical Raman spectroscopy, only about one photon in 10^8 which reaches a sample is scattered (hence the need for sources with high spectral brightness). In spite of this severe limitation, investigations of the Raman effect and its use in chemical spectroscopy proceeded relatively rapidly even prior to the availability of laser sources, and hundreds, if not thousands, of papers on Raman spectroscopy had been published by the end of the 1950's. After the laser replaced the Toronto arc as the standard light source [7] an even wider range of investigations became possible.

In this short review I summarize a small subset of the published work on Raman spectroscopy applied to polypeptides, proteins, and lipids which relates to my own work on the structure of biological membranes. I attempt to identify some of "firsts" in this field, to delineate some types of biological problems which can now be tackled using Raman spectroscopy, and to point out current developments in theory and instrumentation which may well bring about even greater utility. I will ignore a great many important contributions, particularly if they have not had a fairly direct impact on my own research. Readers who wish a more extensive exploration of these studies are referred to the many excellent review articles and monographs which have appeared in the last few years [8-12].

2. THEORY

When a collection of molecules is illuminated by monochromatic light with frequency ν_0, the exciting radiation induces oscillation of the electrons in the sample. The induced polarizability P is equal to the product of the molecular polarizability tensor α and the electric field E arising from the incident light:

$$P = \alpha E. \tag{1}$$

The intensities I_{nm} of the vibrational Raman bands are proportional to the square of the induced transition moment matrix elements P_{nm}

$$I_{nm} = [64\pi^2/(3c^2)](\upsilon_o + \upsilon_{nm})^4 P_{nm}^2, \tag{2}$$

where υ_{nm} is the frequency of the transition between state n and state m, and c is the speed of light. The P_{nm} in turn are given by

$$P_{nm} = (1/h) \sum_r \{[M_{nr}M_{rm}/(\upsilon_{rn} - \upsilon_o)] + [M_{nr}M_{rm}/(\upsilon_{rm} + \upsilon_o)]\}E, \tag{3}$$

where the various M are transition moments, and the subscript r refers to an infinite number of virtual states. The matrix elements P_{nm} may also be obtained by taking the appropriate quantum mechanical integral

$$P_{nm} = (\int \phi_m{}^* \alpha \phi_n dQ)E, \tag{4}$$

where υ_m and υ_n are the vibrational wave functions of the final and initial states respectively of the scattering molecule, and α is a function only of the nuclear normal vibrational coordinates Q. If the force constants and geometry of the scattering molecule are known accurately, then the vibrational frequencies may be calculated relatively easily using the Wilson GF matrix method [13]; from a knowledge of the molecular symmetry one can also usually determine which vibrations will be observed in the infrared spectrum and which will be seen in the Raman; the intensities are much more difficult to estimate on theoretical grounds.

If the scattering molecule has an absorption band that overlaps or is near the frequency of the exciting light, then the efficiency of the Raman scattering process may be increased by many orders of magnitude, as predicted by (3). The theory behind this process, which is known as resonance Raman scattering, will not be given here because it is both complex and incomplete. I have included several references for anyone who wishes to delve further [14-18]

3. INSTRUMENTATION

The basic Raman spectrometer system is illustrated in Fig. 1. The light source is almost always a laser; most commonly either the 488 or 514 nm argon ion laser line is employed. Nearby plasma lines are filtered out using a grating or a Claassen filter. The sample is placed in a capillary tube, and scattered light is typically analyzed using a double or triple grating spectrograph; since shortly after the inauguration of laser excitation a photomultiplier tube has been used to record the scattered photons.

Several difficulties plague the Raman experiment in comparison with the infrared absorption measurement. The extreme inefficiency of the scattering process often necessitates the use of relatively wide spectrometer slits to record a spectrum in a reasonable amount of time, with a necessary reduction of spectral resolution. In addition, even small amounts of fluorescence, common in biological materials, can swamp the small Raman signal. The use of quenching agents, such as iodine, is only sometimes successful in eliminating this interference, and the use of infrared excitation sources has been hampered by the lower sensitivity of detectors for this wavelength region and the much lower scattering efficiency which results from the fourth power dependence of scattering on the frequency of the incident light, as seen in (2).

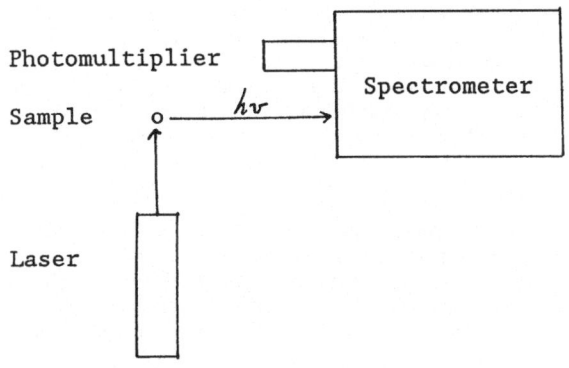

Fig. 1. Schematic illustration of typical Raman spectrometer system

4. AMINO ACIDS, POLYPEPTIDES AND PROTEINS

Edsall [19] credits Wright and Lee [20] with first investigating the Raman scattering properties of the amino acids glycine, alanine, tyrosine, and cystine. Nevertheless, the first systematic Raman spectroscopic investigations of these compounds really started with Edsall himself, who carried this work on for a number of years. His early studies [19,21,22] were concerned primarily with the effects of amino acid ionization on the Raman spectrum, but during his later studies he also analyzed the first spectra of polypeptides and proteins [23]. In the course of these studies, Edsall correctly deduced that bands centered at approximately 1650, 1526, 1400, 1250 and 915 cm^{-1} represented vibrational modes of the Amide bonds. Subsequent work by a number of people [24-30] demonstrated the sensitivity of Amide I (1630 - 1690 cm^{-1}) and Amide III (1220-1300 cm^{-1}) band positions and shapes to the secondary structure of the proteins (Table I). Lippert [31] developed a method for utilizing the spectra of proteins dissolved both in H_2O and D_2O to quantitate the components of the secondary structure. Recently, Williams[2] has developed "composite" spectra typical of the various types of protein secondary structure (α-helix, β-sheet, β-turn and disordered structures) and has demonstrated that they may be used to accurately determine the fractional contributions of these secondary structures to the spectra of proteins and polypeptides [32-35]. Empirical use of the Raman spectrum to deduce secondary structure has become an important tool for studying membrane systems, since circular dichroism spectroscopy, the most commonly used method for protein secondary structure determination, cannot easily be applied to turbid or absorbing solutions and dispersions.

Table 1. Typical Frequencies for Amide I and III Bands of Various Protein Structures

| | Frequency (cm^{-1}) | |
	Amide I	Amide III
α-helix	1645-1657	1260-1300
β-sheet	1660-1680	1225-1245
β-turn	1660-1690	1230-1290
Random coil	1660-1670	1240-1250

Figure 2 illustrates the use of the Amide I region Raman spectrum in studying systems which are not amenable to analysis by circular dichroism spectroscopy. The spectrum of the crystalline polypeptide immunosuppressant Cyclosporine A (spectrum A), for which a detailed molecular structure is known from X-ray diffraction experiments is very similar to that for the membrane associated polypeptide (spectrum B), demonstrating that no significant conformational changes occur when the polypeptide is incorporated into the membrane [36].

Although normal coordinate calculations of amino acids and polypeptides contributed greatly to our understanding and interpretation of amide region spectra [37-39], they have not until recently provided information which is particularly useful in analysing the secondary structure of particular molecules. This is changing, largely because of the extensive and careful

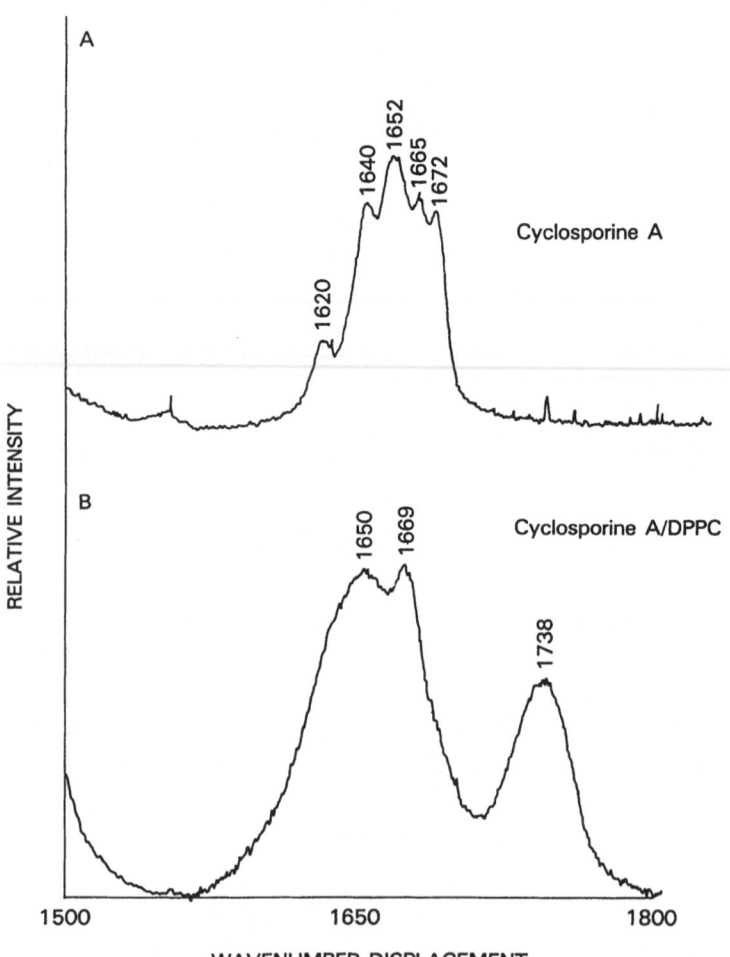

Fig. 2. Raman spectra of (A) polycrystalline cyclosporine A and (B) cyclosporine A/DPPC dispersion in the Amide I spectral region. The peak at 1738 cm^{-1} is due to a DPPC C=O stretching mode

studies of Krimm and his coworkers [40-43], who have recently applied the calculations to two very similar putative structures for a gramicidin A - ion complex, and have been able to unambiguously differentiate between the two [44]. This work strongly suggests that much more detailed structure determinations will be possible as the force fields are refined and measurements improved.

5. LIPIDS AND LIPID MEMBRANES

Lippert and Peticolas [45] first published the Raman spectrum of a lipid multilayer assembly, and demonstrated that the spectroscopic features of the 1000 - 1200 cm^{-1} C-C stretching mode region are characteristic of the state (gel or liquid crystalline) of the lipid assembly and can be used to follow the thermally induced phase transitions, such as the gel to liquid crystalline phase transition, and the perturbation of these transitions by non-phospholipid membrane constituents, such as cholesterol. Although Raman spectroscopy is most often used to study the structure of the hydrocarbon chains of lipid molecules, distinct spectral features are available which reflect the structure of all three regions of a lipid membrane - headgroup, interface, and acyl chain (Fig 3, Table 2).

Figure 5 illustrates schematically the major changes which occur during the phospholipid pretransition and gel to liquid crystalline (main) transition. The pretransition is characterized by a reduction of acyl chain tilt with respect to the plane of the bilayer, the appearance of "ripples", and the introduction of a few *gauche* isomers into the chain termini. The gel to liquid crystalline phase transition results in a loss of these ripples,

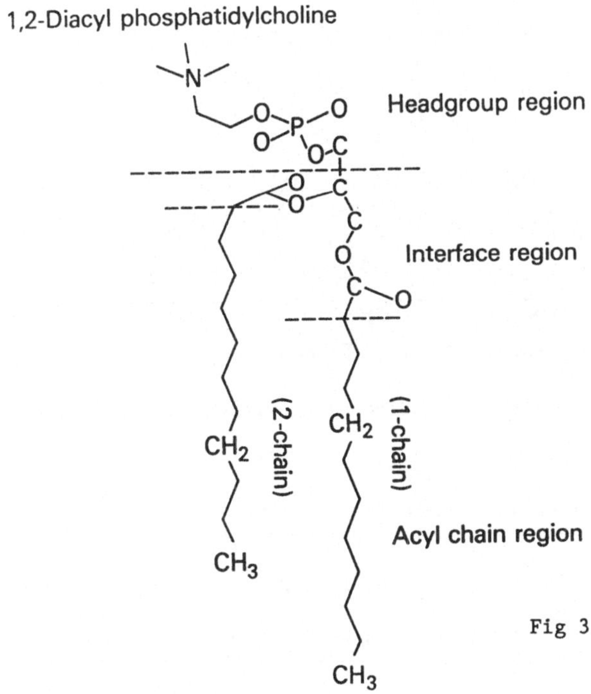

1,2-Diacyl phosphatidylcholine

Fig 3. Chemical structure of diacylphosphatidylcholine molecule

Table 2. Spectral regions particularly useful for the investigation of membrane structure [from 46]

Frequency range (cm^{-1})	Chemical Moiety	Assignment	
710- 720	Headgroup	C-N	symmetric stretch
750- 760	Headgroup	O-P-O	symmetric stretch
1700-1800	Interface	C=O	stretch
1000-1200	Acyl chain	C-C	stretch
1400-1500	Acyl chain	CH$_2$	deformation
2000-2300	Acyl chain	C-D	stretch
2800-3100	Acyl chain	C-H	stretch

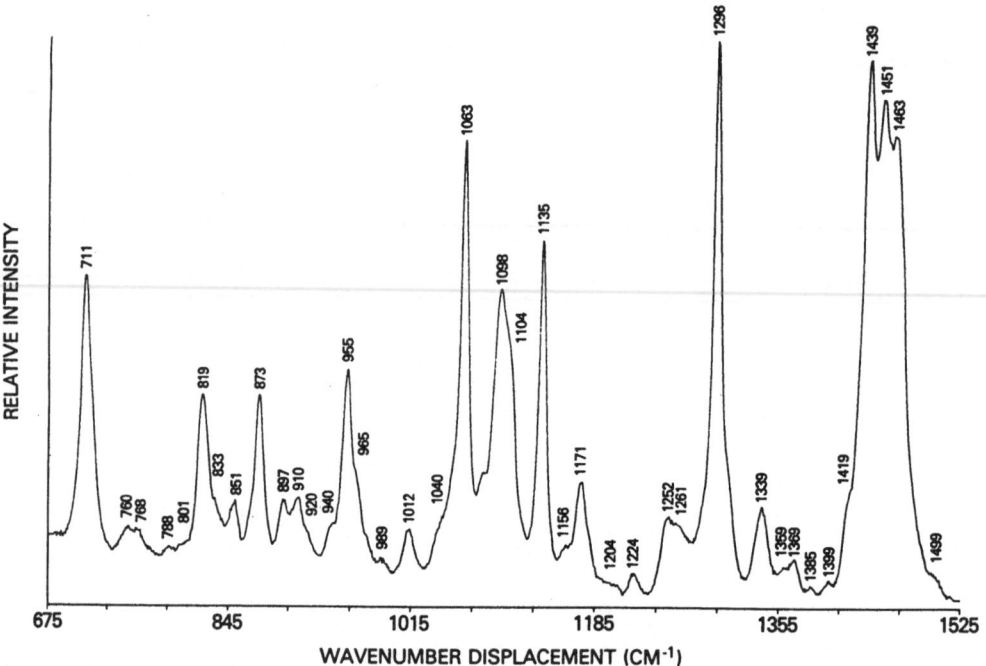

Fig 4. Raman spectrum of polycrystalline diisopalmitoylphosphatidylcholine

lateral expansion of the bilayer, and a much larger amount of *trans-gauche* isomerization. The CH stretching mode region has proven to be especially valuable for following the phase transitions because the dramatic change (Fig 6) in the Raman spectrum, as quantitated by the I_{2935}/I_{2880} cm^{-1} intensity ratio is linearly related to the entropy of transition [47], suggesting that this ratio is a direct probe of the number of gauche conformers in the lipid acyl chain, as well as providing confirmation of calorimetric measurements on lipid phase transitions.

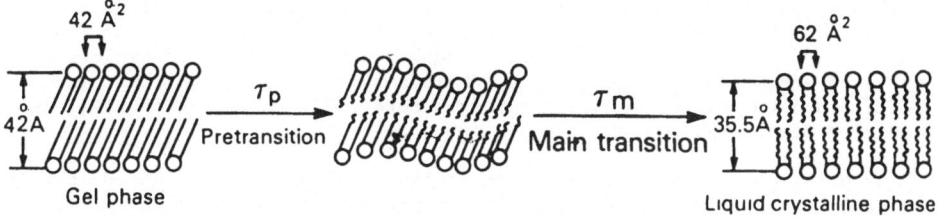

Fig 5. Schematic illustration of changes which occur at lipid bilayer phase
transitions

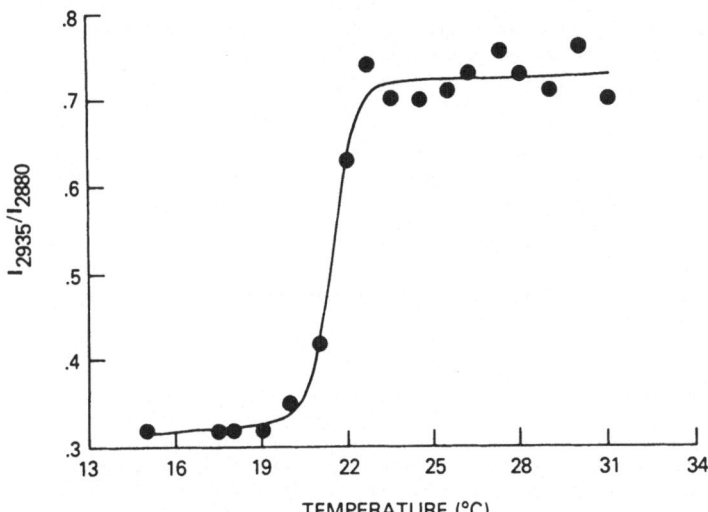

Fig. 6. Temperature profile of diisopalmitoylphosphatidylcholine CH stretching
mode region band intensity ratios, showing dramatic change at the gel to liquid
crystalline phase transition temperature

My own work on lipid spectroscopy has focused on understanding the role of
acyl chain and headgroup chemical structure in determining the physical state
and phase transitions of membrane assemblies, and in understanding how
incorporation of nonlipid membrane constituents influences the details of
membrane structure and phase transitions. For example, in work with John
Silvius of McGill, we were able to show that phosphatidylcholines with acyl
chains which are branched at preterminal methylene group have low temperature
gel phases in which the acyl chains are packed in an orthorhombic or
monoclinic subcell, rather than the quasihexagonal subcell typically formed by
nonbranched phosphatidylcholines under similar conditions. This conclusion can
be drawn by the appearance of a single spectroscopic feature, a shoulder at
approximately 1418 cm^{-1}, in spectra of the isobranched species which is not
present in spectra of the straight chain phosphatidylcholines [48]. By
determining a gel phase I_{2935}/I_{2880} ratio which is substantially lower for gel
phase diisopalmitoylphosphatidylcholine than for dipalmitoylphosphatidylcholine,
we could show that fewer *gauche* isomers were present in the acyl chains of the
former than of the latter. When used together with scanning calorimetry, to

more precisely determine phase transition temperatures, Raman spectroscopy allows a rather complete mapping of the phase diagram of lipids and lipid mixtures. For this reason, we have used Raman measurements to explore not only the effects of changes in acyl chain structure, as above, but also effects due to changes in headgroup structure [49], solvation [50-52], and addition of general anesthetics [53,54].

By using molecules in which isotopic labels have been introduced into lipid molecules at specific lipid sites, it is possible to increase the precision of the structural information obtained by Raman spectroscopy. For example, Bansil and coworkers [55] have shown that selective deuteration of lipid acyl chains at specific methylene sites can be used to probe the structure of the acyl chain region with great detail. Together with Ira Levin I was able to use these lipids to study the differences in how cholesterol alters the structure of phosphatidylcholine acyl chains near the headgroup and those near the terminal methyl group [56]. In earlier studies, Mendelsohn found that he could independently study the structure of two different types of lipid molecules within a lipid bilayer by completely deuterating the acyl chains of one while leaving unaltered those of the other species [57].

A second type of specific probe is provided by the availability of "reporter molecules" which demonstrate resonance Raman scattering. The carotenoids and similar molecules may, under some circumstances, be used to monitor the phase of lipid membranes. Mendelsohn [58] has shown that either the absolute intensity or the frequency of the 1525 cm^{-1} zeaxanthin feature may be used to monitor the physical state of phosphatidylcholine membranes. Similarly, the amphotericin B resonance Raman spectrum reflects the state of phosphatidylcholine-cholesterol membranes [59].

6. VIRUSES, CELLS AND TISSUES

Although there has not been as much work done on the Raman spectroscopy of intact organisms as there has on biomolecules, that which has been done provides encouragement that further studies may be worthwhile. Thomas [12] has shown that Raman spectroscopy is an extremely effective tool for studying the structure of both viral proteins and nucleic acids, both isolated and in intact virions. Raman spectra have been obtained on intact erythrocyte membrane [60], mitochondria [61], bacteria [62], and lymphocytes [63,64]. Although these studies have not yet demonstrated great utility for Raman spectroscopy in the study of living systems more complicated than viruses, they have clearly demonstrated that interpretable spectra may be obtained. It's a lot of fun to see the Raman spectrum of a bunch of lymphocytes come off the spectrometer! As reporter molecules are developed which reflect specific structural or metabolic characteristics of these organisms and their constituent molecules, it seems likely that Raman spectroscopy will provide another useful tool for cell biologists.

7. SUMMARY

Raman spectroscopy is a very useful tool for studying the structure of a number of biological structures, particularly proteins and membranes. Advances in the understanding and interpretation of these spectra make Raman spectroscopy second only to X-Ray crystallography in elucidating the structures of these compounds. A number of advances in instrumentation and technique are underway which may significantly enhance the usefulness of Raman spectroscopy in real biological problems.

Use of photodiode array detectors has greatly increased the speed with which spectra may be acquired, making possible ever more detailed studies of protein and membrane dynamics. By using resonance Raman active "reporter molecules" or probes, which are sensitive to the structural state of a biological molecule, or to such things as membrane potential and ion concentration, it is likely that Raman spectroscopy will find increasing use in studying kinetics of not only isolated biomolecules, but also of intact cells and tissues. In addition, resonance Raman studies of peptide bonds performed using ultraviolet excitation may improve both the speed and accuracy of protein secondary structure determination, making possible sensitivity superior to that currently afforded by circular dichroism techniques.

Although use of photodiode array detectors is limited at present to low resolution studies, the use of Fourier transform spectrometers to acquire Raman spectra promises to increase the obtainable resolution of Raman systems by several orders of magnitude, and, because of the "multiplex advantage", make possible the increasing use of infrared radiation for acquisition of Raman spectra. This will allow ready acquisition of conventional Raman spectra in preparations which fluoresce when excited with visible radiation, thus providing an alternative to coherent anti-Stokes Raman scattering (CARS), which has often been the only practical alternative for measuring Raman spectra on such compounds.

When I was a graduate student, Dr. Schawlow reminded me that bromine, a molecule on which I was working, was "a molecule with one atom too many." The molecules that I've been studying lately have a few more atoms than one atom too many. Through the laser, Dr. Schawlow has made it possible to learn a bit more even about messy molecules like these, and even perhaps, about how people are put together. I suspect that is pretty far beyond what he and Townes were thinking about when they first proposed "optical masers."

8. REFERENCES

1. C.V. Raman: Ind. J. Physics 2, 387 (1928)
2. C.V. Raman, K.S. Krishnan: Ind. J. Physics 2, 399 (1928)
3. C.V. Raman, K.S. Krishnan: Nature 121, 501 (1928)
4. C.V. Raman, K.S Krishnan: Proc. Royal Soc. (London) 122A, 23 (1928)
5. A. Smekal: Naturwiss. 11, 873 (1923)
6. G. Placzek: In Raleigh and Raman Scattering, UCRL Trans. 526L from Handbuch der Radiology, Vol 2, ed. by E. Marx,(Akademische Verlagsgesellshaft, Leipzig 1934), p 209
7. S.P.S. Porto, D.L. Wood: J. Opt. Soc. Amer. 52, 251 (1962)
8. A.T. Tu: Raman Spectroscopy in Biology: Principles and Applications (John Wiley and Sons, New York 1982)
9. F.S. Parker: Applications of Infrared, Raman, and Resonance Raman Spectroscopy in Biochemistry, (Plenum Press, New York 1983)
10. P.R. Carey: Biochemical Applications of Raman and Resonance Raman Spectroscopies (Academic Press, New York, 1982)
11. T.G. Spiro, B.P. Gaber: Annu. Rev. Biochem. 46, 553 (1977)
12. G.B. Thomas: In Infrared and Raman Spectroscopy, Ed. by E.G. Brahme and J.G. Grasselli, Vol 1 (Dekker, New York 1977), p717
13. E.B. Wilson, J.C. Decius, P.C. Cross: Molecular Vibrations, The Theory of Infrared and Raman Vibrational Spectra (McGraw-Hill, New York 1955)
14. A.C. Albrecht: J. Chem. Phys. 34, 1476 (1956)
15. A.C. Albrecht, M.L. Hutley: J. Chem. Phys. 55, 4438 (1971)
16. J. Tang and A.C. Albrecht: In Raman Spectroscopy, Ed. by H.A. Szymanski, Vol 2 (Plenum Press, New York 1970) p 33

17. L.A. Nafie, P. Stein and W. Peticolas, Chem. Phys. Lett. 12, 131 (1971)
18. W. Peticolas, L. Nafie, P. Stein and B. Fanconi, J. Chem. Phys. 52, 1576 (1970)
19. J.T. Edsall: J. Chem. Phys. 4, 1 (1936)
20. N. Wright, W.C. Lee: Nature 136, 300 (1935)
21. J.T. Edsall: J. Chem. Phys. 5, 225 (1937)
22. J.T. Edsall: J. Chem. Phys. 5, 508 (1937)
23. D. Garfinkel, J.T. Edsall: J. Amer. Chem. Soc. 80, 3318 (1958)
24. N.T. Yu, C.S. Liu: J. Amer. Chem. Soc. 94, 5127 (1972)
25. G.S. Bailey, J. Lee, A.T. Tu: J. Biol. Chem. 254, 8922 (1979)
26. R.L. Lord, N.T. Yu: J. Mol. Biol. 50, 509 (1970)
27. H. Brunner, M. Holz: Biochim. Biophys. Acta 379, 408 (1975)
28. W.S. Craig, B.P. Gaber: J. Amer. Chem. Soc. 99, 4130 (1977)
29. V.C. Lin, J.L. Koenig: Biopolymers 15, 203 (1978)
30. M. Pezolet, M. Pigeon-Gosselin, L. Coulombe: Biochim. Biophys. Acta 453, 502 (1976)
31. J.L. Lippert, D. Tyminski, P.J. Desmeules: J. Amer. Chem. Soc. 98, 7075 (1976)
32. R.W. Williams: J. Mol. Biol. 166, 581 (1983)
33. R.W. Williams, A.K. Dunker: J. Mol. Biol. 152, 783 (1981)
34. R.W. Williams, M.M. Teeter: Biochem. 23, 6796 (1984)
35. R.W. Williams: J. Biol. Chem. 260, 3937 (1985).
36. T.J. O'Leary, P.D. Ross, M.L. Lieber, I.W. Levin: Biophys. J. 49, 795 (1986)
37. S. Susuki, T. Shimanouchi, M. Tsuboi: Spectrochim. Acta 19, 1195 (1963)
38. K. Itoh, T. Shimanouchi: Biopolymers 9, 383 (1970)
39. K. Itoh, T. Shimanouchi: Biopolymers 10, 1419 (1971)
40. W.H. Moore, S. Krimm: Biopolymers 15, 2439 (1976)
41. W.H. Moore, S. Krimm: Biopolymers 15, 2465 (1976)
42. J. Bandekar, S. Krimm: Proc. Natl. Acad. Sci. USA 76, 774 (1979)
43. J.F. Rabolt, W.H. Moore, S. Krimm: Macromolecules 10, 1065 (1977)
44. V.M. Naik, S. Krimm, Biophys. J. 49, 1147 (1986)
45. J. L. Lippert, W.L. Peticolas: Proc. Natl. Acad. Sci. USA 68, 1572 (1971)
46. I. W. Levin: Adv. Infrared. Raman. Spect. 11, 1 (1984)
47. C.H. Huang, J.R. Lapides, I.W. Levin: J. Amer. Chem. Soc. 104, 5926 (1982)
48. J. R. Silvius, M. Lyons, P.L. Yeagle, T.J. O'Leary: Biochem. 24, 5388 (1985)
49. J.R. Silvius, P.M. Brown, T.J. O'Leary: Biochem. 25, 4249 (1986)
50. T.J. O'Leary, I.W. Levin: J. Phys. Chem. 88, 1790 (1984)
51. T.J. O'Leary, I.W. Levin: J. Phys. Chem. 88, 4074 (1984)
52. T.J. O'Leary, I.W. Levin: Biochim. Biophys. Acta 776, 185 (1984)
53. T.J. O'Leary, P.D. Ross, I.W. Levin: Biochem. 23, 4636 (1984)
54. T.J. O'Leary, P.D. Ross, I.W. Levin: Biophys. J. 50, 1053 (1986)
55. R. Bansil, J. Day, M. Meadows, D. Rice, E. Oldfield: Biochem. 19, 1938 (1980)
56. T.J. O'Leary, I.W. Levin: Biochim. Biophys. Acta 854, 321, (1986)
57. R. Mendelsohn, J. Maisano: Biochim. Biophys. Acta 506, 192 (1978)
58. R. Mendelsohn, R.W. van Holten: Biophys. J. 27, 221 (1979).
59. M.R. Bunow, I.W. Levin: Biochem. Biophys. Acta 464, 202 (1977)
60. J.L. Lippert, L.E. Gorczyca, G. Meiklejohn: Biochim. Biophys. Acta 382, 51 (1975)
61. F. Adar, M. Erecinska: Biochem. 17, 5484 (1978)
62. W.F. Howard, W.H. Nerslon, J.F. Sperry: Appl. Spectrosc. 34, 72 (1980)
63. L.V. DelPrior, A. Lewis, K.A. Shat: Membrane Biochem. 5, 97 (1984)
64. T.J. O'Leary, unpublished data
65. The opinions or assertions contained herein are the private views of the author, and are not to be construed as official or as reflecting the views of the Department of the Army or the Department of Defense

For Arthur Schawlow on His Sixty-fifth Birthday

1) There once was a man named Johannes Balmer,
   He lived in eighteen eighty-five.
   At the time he looked at the hydrogen spectrum
   None of us were alive.

2) Now we have a friend named Arthur Schawlow,
   This year he is sixty=five.
   When he last looked at the hydrogen spectrum
   All of us were alive.

3) When they invented the optical maser,
   And sent off a paper to the Physical Review,
   Then they went back to the lab and redid the calculation --
   They forgot a factor of 2.‡

4) Now we stimulate the atoms with coherent radiation
   M, A, S E R.
   It's a Money Acquisition Scheme for Expensive Research...
   But we won't use it for war.

5) Now back in the old days with the graduate students
   The Boss kept us busy and on the run.
   In those days the life was pretty exciting,
   We had some fun.*

6) We took a crystal of magnesium oxide,
   And doped it up with chromium.
   Lo and behold: there's a beautiful R line at
   Sixty nine eighty one.

7) Now when the Boss comes down the hall in the morning,
   He always has a story to tell;
   But with such a big laugh when he gets to the punch line,
   We miss the point of it all.

---

‡ Forgive the "poetic license". Of course they did <u>not</u> forget a factor of 2.

* For some of the verses the last line is a beat or so shorter, and the alternate music can be used. Also, of course, feel free to change the note values or the syncopation so as to fit the words in each of the verses.

8) Now our friend Arthur went up to the city,
    Got a double balloon from the S.F. zoo:
    Mickey Mouse on the inside and clear on the outside,
    Mickey Mouse is blue.*

9) So then he pulls the trigger on his Buck Rogers Laser,
    Flashes ruby crystal mounted inside.
    While the beam passes through the outer balloon,
    Mickey Mouse is fried.*

10) Now the Boss had the idea of the laser eraser,
    Helpin' out typists who make mistakes.
    Pull the trigger on the laser that is pointed at the character:
    It evaporates.*

11) We can measure the wavelength of light with a ruler
    In the optics class, we can demonstrate:
    From the pattern of diffraction we can calculate the wavelength:
    Sixty three twenty eight.

12) Now just last night I went down to the Safeway,
    And there were the lasers at the check-out stands.
    Reading the prices -- hopefully correctly,
    From the Zebra bands.*

13) Yes we have a friend in Arthur Schawlow,
    This year he is sixty-five,
    When he last looked at the hydrogen spectrum
    We were all alive.*

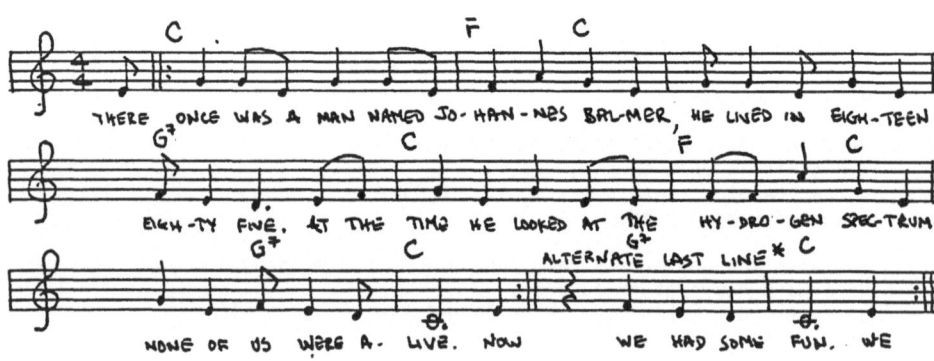

Peter Scott
U. C. Santa Cruz
June 1986

The Schawlow-Hänsch group when the Nobel Prize was announced, October 9, 1981.
Left to Right: Frans Alkamade, Gerard Morgan, Steven Rand, Philipp
Dabkiewics, Fred-a Jurian, Louis Bloomfield, Arthur Schawlow, Theodor
Hänsch, Edward Hildum, Antoinette Taylor, Kenneth Sherwin, Harald Gehrhardt,
Kevin Jones, and Li-Shing Lee.

The Authors

Valery M. Baev was born in Kamchatka in the USSR in 1948. He received his diploma in physics from the Moscow Physico-Technical Institute in 1972, and Ph.D. from the Lebedev Physical Institute in 1980. Since 1972, he has been a research fellow of the Lebedev Institute in Moscow. The main subject of his research has been intracavity laser spectroscopy, a field in which Arthur Schawlow published one of the pioneering papers in 1972, with Peter Toschek and Theodor Hansch as co-authors. Dr. Baev has visited Prof. Toschek's laboratory to collaborate on intra-cavity laser spectroscopy experiments, both at Heidelberg in 1979-1981 and at Hamburg in 1985-1987, most recently as an Alexander von Humbolt fellow.

Allister I. Ferguson graduated with a B.Sc degree in Physics from the University of St. Andrews, Scotland,in 1974. After spending a year at Imperial College, London, where he first met Art Schawlow, he completed the degree of Ph.D.in 1977 at the University of St. Andrews. His thesis work involved the development of a frequency doubled dye laser for spectroscopy. In 1977 he was awarded a Lindemann Fellowship and moved to Stanford University where he was a visiting scholar until February 1979. While at Stanford he worked on coherent multiple pulse spectroscopy using mode-locked dye lasers and spectroscopy of atomic hydrogen. He collaborated with Art Schawlow and others on Doppler-free optogalvanic spectroscopy. Since that time he has held posts in St. Andrews, Oxford and has lectured at Southampton University since 1984. His current interests include hydrogen spectroscopy, FM dye lasers, fibre lasers, diode pumped Nd:YAG lasers.

Theodor W. Hänsch is Professor of Physics at the University of Munich and Director at the Max-Planck-Institute for Quantum Optics in Garching. He returned to his native Germany in 1986, after spending 16 years at Stanford University. Joining the laboratory of Arthur L. Schawlow as a post-doc in 1970, he was appointed an associate professor at the Department of Physics in 1972 and a full professor in 1975. Working at times in close association with Arthur Schawlow, he has pursued research in laser physics, quantum electronics, nonlinear spectroscopy, and atomic physics. Among his principal research interests have been methods of high resolution Doppler-free laser spectroscopy and their applications for the measurement of fundamental constants and for the test of basic laws of physics. In 1973, he was named "California Scientist of the Year", jointly with Arthur Schawlow. He earned his doctorate from the University of Heidelberg, Germany, in 1969.

Serge Haroche, born in 1944, was a student at Ecole Normale Supérieure in Paris. He got his Ph.D. in Physics under the supervision of Claude Cohen-Tannoudji in 1971. His thesis was a theoretical and experimental study of optically pumped atoms interacting with strong radiofrequency fields. The concept of "dressed atom" was first introduced during this work, before being generalized in quantum optics to atoms interacting with intense laser fields. In 1972, Serge Haroche came as a postdoc to Stanford where, with A. Schawlow and J. Paisner, he performed the first quantum beat spectroscopy experiment using dye lasers.

At Ecole Normale since 1973, Serge Haroche and his co-workers have performed a large number of experiments in the general area of spectroscopy and quantum optics of Rydberg atoms. Recently, they have been interested in the field of cavity quantum electrodynamics, dealing with such topics as the enhancement of the spontaneous emission rate of excited atoms in microwave cavities, superradiance of small samples of Rydberg atoms, microscopic Rydberg masers, etc...

Serge Haroche was a visiting Professor at Harvard in 1981. Since 1984, he has spent part of his time at Yale where he has engaged in experiments studying the properties of atoms confined in small metallic structures (Rydberg atom-surface Van der Waals interactions, inhibition of optical spontaneous emission in micronsized cavities, etc...).

Wendell T. Hill, III is a National Science Foundation Presidential Young Investigator and an Assistant Professor in the Institute for Physical Science and Technology at the University of Maryland. As a Schawlow student from 1976 to 1980, he gained experience with several laser spectroscopic techniques such as two-photon and polarization spectroscopies. His thesis project, "Intracavity Absorption Spectroscopy", involved constructing one of the most sensitive apparatuses at the time for detecting very weak absorption and characterizing the peculiar lineshapes associated with this technique; the apparatus was used to study the visible absorption bands of molecular oxygen. After completing graduate school, he was a National Research Council postdoc at the National Bureau of Standards where he studied many-electron correlation effects in the vacuum ultraviolet spectra of atomic ions. His current research includes investigating the details of the atomic and molecular dynamics associated with photofragmentation of small molecules and developing techniques for quantitative measurements of population densities.

John F. Holzrichter is Deputy Associate Director for Advanced Lasers at the Lawrence Livermore National Laboratory. His early responsibilities included the design and construction of the SHIVA and NOVA lasers. From 1980 to 1984 he was responsible for the Inertial Fusion Program. Since 1984 he has been responsible for the Advanced Laser Development Program. Dr. Holzrichter received his Ph.D from Stanford in 1971 having worked under Professor Arthur Schawlow. During his thesis research period he collaborated closely with Dr. Roger Macfarlane (IBM) who was a research associate in the Schawlow group at that time. He built the first dye laser at Stanford (flashlamp pumped) to photo-induce magnetic signals in the anti-ferromagnetic $MnFe$. Other activities at Stanford included flashlamp studies and materials studies. In an interesting aside, he removed tatoos from a young lady and from a young pig (both living). He joined the Naval Research Laboratory in 1971 and LLNL in 1972. He is the author of over 30 papers and has given numerous invited talks on high power lasers and inertial fusion.

George Francis Imbusch was born in County Limerick, Ireland on October 7, 1935. He obtained his baccalaureate at the Galway campus of the National University of Ireland. He then obtained his Ph.D. degree from Stanford University on 1964. He attended Stanford under the aegis of a Joseph P. Kennedy, Jr. Fellowship (1959-61). He and Dr. Linn Mollenauer were the first two graduate students to join Professor Art Schawlow's group shortly after Art's move from Bell Telephone Laboratories in 1961. Dr. Imbusch is on record as holding the first Ph.D. granted by Professor Schawlow. After a short postdoctoral appointment with Dr.

S. Geschwind at Bell Labs, Dr. Imbusch was promoted to the technical staff and remained there until 1967 when he accepted the Chair of Experimental Physics at his Alma Mater in Galway. He is presently serving as the Chairman of Physics Department of University College, Galway, and he continues to maintain an active research program in the optical properties of solids. Dr. Imbusch has also established many collaborative programs with groups in the United States including those at Stanford, Wisconsin, Oklahoma State and Georgia. He was named to a Brittingham Professorship at the University of Wisconsin-Madison in 1977-78 and has served on numerous Scientific Committees and Boards.

James E. Lawler is currently a faculty member at the University of Wisconsin supervising research programs on optogalvanic spectroscopy and laser techniques for determining atomic transition probabilities. He received his B.S. degree from the University of Missouri at Rolla in 1973 and his M.S. and Ph.D. degrees from the University of Wisconsin in 1978. Between 1978 and 1980 he was a Research Associate in the Schawlow group at Stanford where he developed new techniques for high resolution spectroscopy of plasmas.

Marc D. Levenson is currently "head manager" in Optical Storage Technology at the IBM Almaden Research Center. He received his B.S. degree in physics from MIT in 1967 and his Ph.D. degree from Stanford in 1972. His Ph.D. thesis was conducted under the supervision of Prof. Schawlow and concerned saturated absorption spectroscopy of iodine. He was later a post-doctoral fellow at Harvard University and an Associate Professor of Physics at the University of Southern California.

Roger Macfarlane was born in New Zealand and received his formal education there, graduating with a Ph.D. in Physics from the University of Canterbury in 1964. From 1965-68 he was a Research Associate with Art Schawlow in the Physics Department at Stanford University. During that time he collaborated with John Holzrichter and Art Schawlow on one of the early applications of dye lasers to spectroscopy, measuring photomagnetism in MnF_2. This kindled an enthusiasm for resonant pumping experiments and high-resolution laser spectroscopy of solids which were carried out subsequently at the IBM Research Laboratory in San Jose. The application of frequency selective photochemistry to frequency domain information storage has been his major interest for the past several years. Much of the work at IBM was done in collaboration with Robert Shelby.

Warren Moos joined the fledgling laser spectroscopy group at Stanford as a Research Associate late in 1961, after completing his Ph.D. work at the University of Michigan. At present, he is Professor in the Department of Physics and Astronomy at the Johns Hopkins University. His current research interests include high-temperature plasmas, planetary atmospheres and magnetospheres, and space astronomy. He looks back on his stay at Stanford as a time of unusual excitement and creativity." We were always looking for new things to do with the laser. A person would come in the morning with an idea and within a day or two we would decide if it was within the experimental state of the art. A. L. Schawlow, of course, was most often the source for these ideas. Over the years, it has been fascinating to see how many of the ideas which had to be rejected because the technology was not available then have come to fruition with improvements in laser technology."

Munir Nayfeh is Associate Professor of Physics at the University of Illinois at Urbana-Champaign. He received his BSc from the American University of Beirut, and his PhD from Stanford University. He has been a research physicist at Oak

Ridge National Laboratory and a lecturer and research associate at Yale University. His early work at Stanford with the Schawlow-Hansch group involved a precision measurement of one of the fundamental constants of nature, the Rydberg constant. His research interests include experimental and theoretical studies of atomic collisions utilizing atomic scattering in the presence of laser fields, detection of low levels of atoms--including rare events--and the coherent interaction of light beams with atoms and molecules. More recently, he has been conducting experimental and theoretical work on the structure of highly excited atoms, including hydrogen, in strong external electromagnetic fields (dc electric, magnetic, microwave, and optical) and their interactions with other atoms and surfaces.

Timothy J. O'Leary received a B. S. in Chemistry from Purdue in 1972. Following a summer spent with Lowell Wood at the Lawrence Livermore Laboratory, during which time he became familiar with laser isotope separation, especially the Schawlow-Tiffany work, he came to Stanford as a Fannie and John Hertz Foundation Fellow. Although officially enrolled as a student in Chemistry, he persuaded John Brauman and Dr. Schawlow to "co-direct" his thesis work. During his stay he worked on developing some wideband chelate lasers (successful), hydrogen isotope enrichment by selective photodecomposition in HBr (successful), and enrichment of bromine by selective photofragmentation from a molecular beam (unsuccessful, and scooped by Bell Labs). After receiving his Ph.D. in 1976, he became a medical student at the University of Michigan (where he also served as a research associate in physiology). Since that time, he has served as a resident, fellow, attending pathologist and Chief of the Autopsy Service at the National Institutes of Health, medical officer in the Food and Drug Administration and, currently, as Chairman of the Department of Cellular Pathology, Armed Forces Institute of Pathology.

S. C. Rand received his doctorate in physics from the University of Toronto in 1978 for work with Boris Stoicheff on Brillouin light scattering in rare gas solids and molecular crystals. This was followed by optical coherent transient studies of "magic-angle" spin decoupling as a World Trade Fellow at IBM San Jose with Richard Brewer (1978-80). Later he was a Research Associate with Arthur Schawlow and Theodor Hansch at Stanford, where he performed high resolution spectroscopy of helium and investigated cooperative absorption and emission processes in rare earth solids (1980-82). Then in 1982 he joined Hughes Research Lab and developed the LiF color center Q-switch, the diamond laser, ultrahigh resolution techniques to study metastable defects in solids and also predicted the electric magic angle effect.

Stanley E. Stokowski , a native of Lewiston, Maine, received a Bachelor of Science degree from the Massachusetts Institute of Technology in 1963. While at M.I.T. he did an undergraduate thesis on laser-induced breakdown of air,under the direction of Professor C. H. Townes. In 1964 he received his masters degree from Stanford University after which he was accepted as a Ph.D. candidate in Arthur L. Schawlow's research group. After receiving his Ph.D. from Stanford in 1968, he did post-doctoral work at the National Bureau of Standards from 1968 to 1970. From 1970 to 1972 he was a member of the technical staff at Bell Telephone Laboratories in Murray Hill, N.J. In 1972 he joined the Martin Marietta Research Laboratories near Baltimore, Maryland. After coming to Lawrence Livermore National Laboratory in Livermore, California, in 1977, he has been actively involved in developing new materials for the large laser drivers for inertial confinement fusion. He is the author of about 50 published papers and two articles on Nd-doped laser glass and filter glasses in the CRC Handbook on Laser Science and Technology.

Satoru Sugano is currently Professor of Solid State Physics at the Institute of Solid State Physics, University of Tokyo, Japan. He received his Bachelor and Doctor of Science Degrees in Physics from the University of Tokyo in 1952 and 1959, respectively. From 1959 until 1961, he was a Member of the Technical Staff at Bell Telephone Laboratories, Murray Hill, N.J. where he met and collaborated with Dr. Schawlow. He has been a visiting Professor at the University of Colorado and at the Université Paris VII.

Peter E. Toschek was born in Hindenberg,Germany,and received his Ph.D. from the University of Bonn in 1961. Subsequently, he was research assistant and professor of physics at the applied physics laboratory of Heidelberg University. In 1981 he was appointed to the "I. Institut für Experimentalphysik" of Hamburg University. His topics of research include quantum optics, atomic physics, and laser spectroscopy. He first encountered Arthur Schawlow during a brief visit to Stanford in 1968. More extended visits in 1970 and 1972 led to more extensive interactions with Schawlow and with Toschek's former student, Ted Hänsch, as well as with Tony Siegman.

Zugeng Wong is presently Professor of Physics at East China Normal University, Shanghai, Peoples Republic of China. He was a visiting scholor at the Schawlow group at Stanford 1982-1983.

Hui-Ron Xia is Lecturer of Physics at the East China Normal University, Shanghai, Peoples Republic of China. She also was a visiting scholar at Stanford during 1982-83.

William M. Yen was born in Nanking, China on April 5, 1935. He obtained his B.S. degree from the University of Redlands in 1956 and his Ph.D. in Physics from Washington University (St. Louis) in January 1962. He joined Professor Schawlow's group at Stanford in the Summer of 1962 as a Research Associate, increasing the size of the team to four. He accepted a position as Assistant Professor at the University of Wisconsin, Madison, in the Fall of 1965 and was subsequently promoted to Associate and to Full Professor there. He has been the recipient of a Guggenheim Fellowship (1979) and of a Humboldt Senior US Scientist Award. He has held visiting positions at the University of Tokyo, the University of Paris, the Australian National University, the Goethe University of Frankfurt, Harvard and UCSB. In 1986, he was named to the Graham Perdue Chair of Physics at the University of Georgia, Athens. He has served as Chairman of a number of international scientific conferences including the International Conference on Luminescence held in Madison in 1984. His research interests have centered on the optical properties of ordered and disordered material.

Pei-Lin Zhang was born in Cheffoo, Shandung, China, on June 13, 1933. He received undergraduate degree in Radioelectonics and graduate degree in Physics from Tsinghua University, Beijing, China, in 1953 and 1956 respectively. He is the recipient of National Invention Award in 1981 by National Scientific and Technological Commission of China. From November 1982 to December 1983 he was with Department of Physics, Stanford University, as a visiting scholar, where he did parametric wave-mixing processes research with Prof, A. L. Schawlow. Since 1956, he has been with Tsinghua University, Beijing, People's Republic of China, where he is currently Professor of Physics. His research interests are in the area of lasers, spectroscopy and nonlinear optics.

Shuo-Yan Zhao received undergraduate degree in Engineering Physics from Tsinghua University, Beijing, People's Republic of China, in 1958. Presently she is an Associate Professor of Physics at Tsinghua University. Her research interests are in the area of laser spectroscopy and its applications.

Index of Contributors

Springer Series in Optical Sciences

Editorial Board: J.M. Enoch D.L. MacAdam A.L. Schawlow K. Shimoda T. Tamir